Proceedings of the Fourth International Colloquium on Differential Equations

Proceedings of the Fourth International Colloquium on Differential Equations

Plovdiv, Bulgaria, 18 - 22 August 1993

Editors: D. Bainov and V. Covachev

///VSP///
Utrecht, The Netherlands, 1994

VSP BV
P.O. Box 346
3700 AH Zeist
The Netherlands

First published in 1994

ISBN 90-6764-169-3

Printed in The Netherlands by Koninklijke Wöhrmann, Zutphen

CONTENTS

Preface

The Fourth International Colloquium on Differential Equations was organized by UNESCO and the Mathematical Society of Japan, with the cooperation of the Association Suisse d'Informatique, the Canadian Mathematical Society, Hamburg University, Kyushu University, the London Mathematical Society, Plovdiv University "Paissii Hilendarski", Technical University - Berlin, Technical University - Plovdiv, Union of the Bulgarian Mathematicians -Plovdiv Section, the University of Texas -Pan American, and sponsored by the firm "Eureka". It was held on August 18-23, 1993, in Plovdiv, Bulgaria. The subsequent colloquia will take place each year from 18-23 August in Plovdiv, Bulgaria.

The address of the organizing committee is:

Stoyan Zlatev, Mathematical Faculty of the Plovdiv University,
Tsar Assen Str.24, Plovdiv 4000, Bulgaria.

The Editors

4th Int. Coll. on Differential Equations, pp. 3-9
D. Bainov and V. Covachev (Eds)

Periodic solutions for multi-phase Stefan problems with dynamic boundary conditions

TOYOHIKO AIKI
Department of Mathematics Faculty of Education, Gifu University
1-1, Yanagido, Gifu, 501-11, Japan

ABSTRACT. We consider periodic solutions to multi-phase Stefan problems for a class of nonlinear parabolic equations with dynamic boundary conditions, that is, with boundary conditions containing time derivatives on the fixed boundary. In this paper, we investigate existence, order property and asymptotic stability of periodic solutions.

Key words: periodic solutions, dynamic boundary conditions, asymptotic behavior

Introduction

In this paper, we consider periodic stabilities for the solutions to Stefan problems with periodic condition in time in the enthalpy formulation with dynamic boundary conditions: Find functions $u = u(t,x)$ on $R \times \Omega$ and $V = V(t,x)$ on $R \times \Gamma$ satisfying that

$$u_t - \Delta\beta(u) = 0 \quad \text{in } R \times \Omega, \tag{0.1}$$

$$\begin{cases} -\dfrac{\partial\beta(u)}{\partial\nu} = g(t,x,V) + V_t \\ \beta(u) = V \end{cases} \quad \text{on } R \times \Gamma, \tag{0.2}$$

where Ω is a bounded domain in $R^N (N \geq 2)$ with smooth boundary $\Gamma = \partial\Omega$; $\beta : R \to R$ is a given nondecreasing function; $g = g(t,x,\xi) : R \times \Gamma \times R \to R$ is a given function which is nondecreasing in $r \in R$ for a.e. $(t,x) \in R \times \Gamma$; $(\partial/\partial\nu)$ denotes the outward normal derivative on Γ. We denote by $SP = SP(\beta, g)$ the above system (0.1) $\sim$ (0.2).

By many authors initial-boundary value problems for (0.1) with usual boundary conditions have been studied. In case the flux condition is of the form $-\partial\beta(u)/\partial\nu = g(t,x,\beta(u))$, the problem was uniquely solved in the variational sense by Visintin [1], Niezgodka, Pawlow & Visintin [2] and Niezgodka & Pawlow [3]. Also, some interesting results dealing with the boundary condition (0.2) are found in Cannon [4], Hintermann [5] and Grobbelaar & Dalsen [6]. Recently, boundary conditions similar to (0.2) were discussed by Primicerio & Rodrigues [7] and in Aiki [8] for one-dimensional Stefan problems with dynamic boundary conditions the local in time existence and uniqueness of classical solutions were shown. In particular, the existence and uniqueness for the solution to the initial-boundary value problems for $SP(\beta, g)$ are studied in Aiki [9]. The purpose of this paper is to establish the existence, the order property and the asymptotic stability of periodic solutions under the periodic condition in time, that is, $g(t,x,\xi) = g(t+T,x,\xi)$ a.e. on $R \times \Gamma \times R$ for some positive constant T. In papers by Haraux and Kenmochi [10], Aiki, Kenmochi and Shinoda [11, 12] and Aiki [13] similar questions to those above were discussed.

In this paper, we omit proofs of main results and shall show those in the author's forthcoming paper [14].

1. Statements of main results

Throughout this paper we assume that the function $\beta : R \to R$ satisfies the following conditions (β1) and (β2):

(β1) β is non-decreasing and Lipschitz continuous on R with $\beta(0) = 0$.

(β2) There are some positive constants L_β, l_β such that

$$|\beta(r)| \geq L_\beta |r| - l_\beta \quad \text{for all } r \in R.$$

Also, let $g = g(t,x,\xi) : R \times \Gamma \times R \to R$ be a function and suppose that:

(g1) $g(t,x,\xi)$ is non-decreasing in $\xi \in R$ for a.e. $(t,x) \in R \times \Gamma$;

(g2) for any $\xi \in R$, $g(\cdot,\cdot,\xi) \in L^2_{loc}(R; L^2(\Gamma))$;

(g3) $g(t,x,\xi)$ is locally Lipschitz continuous in ξ uniformly with respect to $(t,x) \in R \times \Gamma$, that is, for each $M > 0$ there is a positive constant $C_g(M)$ such that

$$|g(t,x,\xi) - g(t,x,\xi')| \leq C_g(M)|\xi - \xi'|$$

for all ξ, ξ' with $|\xi| \leq M, |\xi'| \leq M$ and a.e. $(t,x) \in R \times \Gamma$ and there are constants m_1, m_2 with $m_1 \leq m_2$ such that $g(t,x,\beta(m_1)) \leq 0, g(t,x,\beta(m_2)) \geq 0$ for a.e. $(t,x) \in \Sigma$.

For the sake of simplicity of notations we put

$$X = H^1(\Omega), \quad H = L^2(\Omega) \times L^2(\Gamma),$$

$$A(u,v) = \int_\Omega \nabla u \cdot \nabla v dx \text{ for } u, v \in X,$$

$$(u,v)_X = A(u,v) + (\int_\Omega u dx + \int_\Gamma u d\Gamma)(\int_\Omega v dx + \int_\Gamma v d\Gamma) \text{ for } u, v \in X,$$

$$(\vec{u}, \vec{v})_H = \int_\Omega uv\, dx + \int_\Gamma u_\Gamma v_\Gamma\, d\Gamma \quad \text{for } \vec{u} = (u, u_\Gamma), \vec{v} = (v, v_\Gamma) \in H,$$

and C_Ω is a positive constant satisfying that

$$|v|_{L^2(\Omega)} \leq C_\Omega |v|_X, \quad |v|_{L^2(\Gamma)} \leq C_\Omega |v|_X \quad \text{for any } v \in X.$$

Also, we define an operator $E : X \to H$ by putting

$$Ev = (v, v|_\Gamma) \quad \text{for } v \in X.$$

Clearly, X and H are Hilbert spaces with inner product $(\cdot,\cdot)_X$ and $(\cdot,\cdot)_H$, respectively, and the range of E, $R(E)$, is a dense subspace of H and E is linear and compact. We identify H with its dual H^* and denote by X^* the dual space of X. Therefore, denoting by E^* the dual operator of E we have

$$\langle E^*(v, v_\Gamma), \eta\rangle_X = \int_\Omega v\eta \, dx + \int_\Gamma v_\Gamma \eta \, d\Gamma \quad \text{for any } (v, v_\Gamma) \in H \text{ and } \eta \in X,$$

where $\langle\cdot,\cdot\rangle_X$ is a duality pairing between X^* and X.

We now formulate problem SP in variational sense.

Definition 1.1. Let $J = [t_0, t_1]$ be a compact interval, $Q = (t_0, t_1) \times \Omega$, $\Sigma = (t_0, t_1) \times \Gamma$. Then a couple $\{u, V\}$ of functions $u : J \times \Omega \to R$ and $V : J \times \Gamma \to R$ is called a solution of SP on J if $u \in C_w(J; L^2(\Omega)) \cap L^\infty(Q)$, $\beta(u) \in L^2(J; X)$, $V \in C_w(J; L^2(\Gamma))$, $\beta(u) = V$ a.e. on Σ and the following variational equation is satisfied:

$$-\int_Q u\eta_t dx dt - \int_\Sigma V\eta_t d\Gamma dt + \int_Q \nabla\beta(u) \cdot \nabla\eta dx dt + \int_\Sigma g(t, x, V)\eta d\Gamma dt = 0$$
$$\text{for any } \eta \in Z,$$

where $Z = \{\eta \in C^1(J; X); \eta(t_0) = \eta(t_1) = 0\}$.

Remark 1.1. From the above definition of solution to SP on $J = [t_0, t_1]$ it follows that a couple of functions $\{u, V\}$ is a solution of SP on J if and only if $E^*(u, V) \in W^{1,2}(J; X^*)$, $u \in L^\infty(J \times \Omega)$, $\beta(u) \in L^2(J; Y)$, $V \in L^\infty(J \times \Gamma)$, $\beta(u) = V$ a.e. on $J \times \Gamma$ and for a.e. $t \in J$

$$\langle \frac{d}{dt} E^*(u(t), V(t)), \eta\rangle_X + A(\beta(u(t)), \eta) + \int_\Gamma g(t, \cdot, V(t, \cdot))\eta d\Gamma = 0 \quad \text{for any } \eta \in X.$$

Definition 1.2. Let J' be any interval in R. Then a couple $\{u, V\}$ of functions $u : J' \times \Omega \to R$ and $V : J' \times \Gamma \to R$ is called a solution of SP on J' if for every compact subinterval $J = [t_0, t_1]$ of J' the couple $\{u, V\}$ is a solution of SP on J in the sense of Definition 1.1.

Next, we formulate the Cauchy problem and the problem with the periodic condition in time for SP.

Definition 1.3. (i) Let $J' = [t_0, t_1]$ or $[t_0, t_1)$, and let $u_0 \in L^2(\Omega)$, $V_0 \in L^2(\Gamma)$. Then a couple $\{u, V\}$ of functions $u : J' \to L^2(\Omega)$ and $V : J' \to L^2(\Gamma)$ is a solution of the Cauchy problem with initial condition $u(t_0) = u_0$, $V(t_0) = V_0$, denoted by $CSP(u_0, V_0)$ on J', for problem SP on J', if $\{u, V\}$ is a solution of SP on J' with $u(t_0) = u_0$, $V(t_0) = V_0$.

(ii) Let T be a positive number, and let $\{u, V\} : J' \to V$ is a solution of SP on R such that $u(t + T) = u(t)$, $V(t + T) = V(t)$ for all $t \in R$. Then $\{u, V\}$ is called a T-periodic solution of SP on R.

We now recall an existence-uniqueness result for $CSP(u_0, V_0)$.

Theorem 1.1. *(cf. [9, Theorems 1.1 and 1.2]) We suppose that ($\beta 1$), ($\beta 2$) and (g1) $\sim$ (g3) hold. Let t_0 be any number in R. Then, for any functions $u_0 \in L^\infty(\Omega)$ and $V_0 \in L^\infty(\Gamma)$, there exists one and only one solution $\{u, V\}$ of $CSP(u_0, V_0)$ on $J' = [t_0, \infty)$.*

The main results of this paper are stated in the following theorems. The theorem is concerned with the existence of T-periodic solutions.

Theorem 1.2. *Let T be a positive number, and suppose*

$$(g4) \qquad g(t+T,x,\xi) = g(t,x,\xi) \quad \text{for all } \xi \in R \text{ and a.e. } (t,x) \in R \times \Gamma.$$

Then there exists a T-periodic solution $\{u, V\}$ of SP on R.

We denote by $\mathcal{P}_T$ the set of all T-periodic solutions of SP on R; in general $\mathcal{P}_T$ is not a singleton (cf. [11, Example 1.1]). The following two theorems are concerned with the structure of $\mathcal{P}_T$.

Theorem 1.3. *Under the same assumptions as in theorem 1.2 we have*

$$g(\cdot,\cdot,V_1) = g(\cdot,\cdot,V_2) \quad \text{a.e. on } R \times \Gamma \text{ for any } \{u_1,V_1\},\{u_2,V_2\} \in \mathcal{P}_T,$$

and

$$\int_0^T \int_\Gamma g(t,x,V)d\Gamma dt = 0 \quad \text{for any } \{u,V\} \in \mathcal{P}_T.$$

Theorem 1.4. *Under the same assumptions of theorem 1.2 we have the following statements (a) $\sim$ (c):*

(a) If $\{u_1,V_1\}$, $\{u_2,V_2\} \in \mathcal{P}_T$ and

$$\int_\Omega u_1(0,x)dx + \int_\Gamma V_1(0,x)d\Gamma = \int_\Omega u_2(0,x)dx + \int_\Gamma V_2(0,x)d\Gamma,$$

then

$$\beta(u_1) = \beta(u_2) \quad \text{a.e. on } R \times \Omega \quad \text{and} \quad V_1 = V_2 \quad \text{a.e. on } R \times \Gamma.$$

(b) If $\{u_1,V_1\}$, $\{u_2,V_2\} \in \mathcal{P}_T$ and

$$\int_\Omega u_1(0,x)dx + \int_\Gamma V_1(0,x)d\Gamma > \int_\Omega u_2(0,x)dx + \int_\Gamma V_2(0,x)d\Gamma,$$

then

$$\beta(u_1) \geq \beta(u_2) \quad \text{a.e. on } R \times \Omega \quad \text{and} \quad V_1 \geq V_2 \quad \text{a.e. on } R \times \Gamma.$$

(c) If $\{u_1,V_1\}$, $\{u_2,V_2\} \in \mathcal{P}_T$ and

$$\int_\Omega u_1(0,x)dx + \int_\Gamma V_1(0,x)d\Gamma > \int_\Omega u_2(0,x)dx + \int_\Gamma V_2(0,x)d\Gamma,$$

then for any $a_0 \in R$ with

$$\int_\Omega u_1(0,x)dx + \int_\Gamma V_1(0,x)d\Gamma > a_0 > \int_\Omega u_2(0,x)dx + \int_\Gamma V_2(0,x)d\Gamma$$

there exists a T-periodic solution $\{u,V\}$ of SP on R such that

$$a_0 = \int_\Omega u(0,x)dx + \int_\Gamma V(0,x)d\Gamma.$$

Finally, as to the asymptotic stability of T-periodic solutions we prove the following.

Theorem 1.5. *Suppose that all conditions of theorem 1.2 are satisfied. Let t_0 be any* *number in R and let $\{u, V\}$ be any solution of SP on $[t_0, \infty)$. Then there is $\{\hat{u}, \hat{V}\} \in \mathcal{P}_T$ such that*

$$\beta(u(nT + \cdot)) \to \beta(\hat{u}) \quad \in L^2(0,T; W^{1,2}(\Omega)) \text{ as } n \to \infty.$$

2. Known results

In this section we show some known results which will use in the proof of theorems 1.2 ~ 1.5

First, we recall a comparison result for $CSP(u_0, V_0)$.

Theorem 2.1. *(cf. [9, Theorem 1.3]) Suppose that the same assumptions as in Theorem 1.1 hold and a pair $\{\overline{u}_0, \overline{V}_0\} \in L^\infty(\Omega) \times L^\infty(\Gamma)$. Let t_0 be any number in R and $\{u, V\}$(resp. $\{\overline{u}, \overline{V}\}$) be a solution of $CSP(u_0, V_0)$ (resp. $CSP(\overline{u}_0, \overline{V}_0)$) on $[t_0, \infty)$. Then for any $t \in [t_0, \infty)$,*

$$\begin{aligned} & |[u(t) - \overline{u}(t)]^+|_{L^1(\Omega)} + |[V(t) - \overline{V}(t)]^+|_{L^1(\Gamma)} \\ \leq\ & |[u(t_0) - \overline{u}(t_0)]^+|_{L^1(\Omega)} + |[V(t_0) - \overline{V}(t_0)]^+|_{L^1(\Gamma)}. \end{aligned} \tag{2.1}$$

In particular, if $u_1(t_0, \cdot) \leq u_2(t_0, \cdot)$ a.e. on Ω and $V_1(t_0, \cdot) \leq (t_0, \cdot)$ a.e. on Γ, then

$$u_1 \leq u_2 \quad \text{a.e. on } R \times \Omega \quad \text{and} \quad V_1 \leq V_2 \quad \text{a.e. on } R \times \Gamma.$$

Next, we consider the degenerate parabolic equation with linear dynamic boundary condition:

$$\begin{cases} u_t - \Delta\beta(u) = 0 & \text{in } (t_0, \infty) \times \Omega, \\ \dfrac{\partial\beta(u)}{\partial\nu} + \dfrac{\partial V}{\partial t} + h = 0 & \text{on } (t_0, \infty) \times \Gamma, \\ \beta(u) = V & \text{on } (t_0, \infty) \times \Gamma, \end{cases} \tag{2.2}$$

where h is a function given in $L^2_{loc}(R; L^2(\Gamma))$ and $-\infty \leq t_0 < \infty$; $\beta : R \to R$ is a function satisfying ($\beta 1$) and ($\beta 2$).

Following [13, section2], we say that for any compact interval $J = [t_0, t_1]$, a couple of functions $u : J \to L^2(\Omega)$ and $V : J \to L^2(\Gamma)$) is a solution of (2.2) on J, if $E^*(u, V) \in W^{1,2}(J; X^*)$, $u \in L^\infty(J; L^2(\Omega))$, $\beta(u) \in L^2(J; X)$, $V \in L^\infty(J; L^2(\Gamma))$, $\beta(u) = V$ a.e. on $J \times \Gamma$ and

$$\langle \frac{d}{dt} E^*(u(t), V(t)), \eta\rangle_X + A(\beta(u(t)), \eta) + \int_\Gamma h(t, \cdot)\eta d\Gamma = 0 \quad \text{for any } \eta \in X \text{ and a.e. } t \in J.$$

For a general interval $J' \subset R$, solutions of (2.2) on J' are defined in a manner similar to definition 1.2. Also, solutions of Cauchy problems and the problems with T-periodic condition are defined just as definition 1.3.

Next, we mention some results on T-periodic solutions to (2.2) on R.

Theorem 2.2. *We suppose that $h \in L^2_{loc}(R; L^2(\Gamma))$. Let T be a positive number, and assume that*

$$h(t + T, \cdot) = h(t, \cdot) \text{ a.e. on } \Gamma \text{ for any } t \in R,$$

and

$$\int_0^T \int_\Gamma h(t, x) d\Gamma dt = 0.$$

Then the following statements (i) ~ (iv) holds.

(i) For each $a_0 \in R$ there exists a T-periodic solution $\{u, V\}$ of (2.2) on R such that

$$\int_\Omega u(0,x)dx + \int_\Gamma V(0,x)d\Gamma = a_0.$$

(ii) Let $\{u, V\}$ be a solution of (2.2) on R. Then $\{u, V\}$ is T-periodic solution of (2.2) if and only if $u \in L^\infty(R; L^2(\Omega))$ and $V \in L^\infty(R; L^2(\Gamma))$.

(iii) Let $\{u_1, V_1\}$, $\{u_2, V_2\}$ be T-periodic solutions of (2.2) on R such that

$$\int_\Omega u_1(0,x)dx + \int_\Gamma V_1(0,x)d\Gamma = \int_\Omega u_2(0,x)dx + \int_\Gamma V_2(0,x)d\Gamma.$$

Then

$$\beta(u_1) = \beta(u_2) \quad \text{a.e. on } R \times \Omega,$$

and there exist functions $w \in L^2(\Omega)$, $w_\Gamma \in L^2(\Gamma)$ with $\int_\Omega w dx + \int_\Gamma w_\Gamma d\Gamma = 0$ such that

$$\left.\begin{array}{ll} u_1(t,\cdot) - u_2(t,\cdot) = w(\cdot) & \text{a.e. on } \Omega \\ V_1(t,\cdot) - V_2(t,\cdot) = w_\Gamma(\cdot) & \text{a.e. on } \Gamma \end{array}\right\} \text{ for any } t \in R.$$

(iv) Let $\{u_1, V_1\}$, $\{u_2, V_2\}$ be two T-periodic solutions of (2.2) on R such that

$$\int_\Omega u_1(0,x)dx + \int_\Gamma V_1(0,x)d\Gamma > \int_\Omega u_2(0,x)dx + \int_\Gamma V_2(0,x)d\Gamma.$$

Then

$$\beta(u_1) \geq \beta(u_2) \quad \text{a.e. on } R \times \Omega.$$

Finally, as to the asymptotic stability of T-periodic solutions we prove the following.

Theorem 2.3. *Suppose that all conditions of theorem 2.2 are satisfied. Let t_0 be any positive number and let $\{u, V\}$ be any solution of (2.2) on $[t_0, \infty)$. Then there exists a T-periodic solution $\{\overline{u}, \overline{V}\}$ of (2.2) on R such that*

$$\int_\Omega u(t,x)dx + \int_\Gamma V(t,x)d\Gamma = \int_\Omega \overline{u}(t,x)dx + \int_\Gamma \overline{V}(t,x)d\Gamma \quad \text{for any } t \geq t_0,$$

$$u(t) - \overline{u}(t) \to 0 \quad \text{weakly in } L^2(\Omega) \text{ as } t \to \infty,$$
$$V(t) - \overline{V}(t) \to 0 \quad \text{weakly in } L^2(\Gamma) \text{ as } t \to \infty,$$

and

$$\beta(u(nT + \cdot)) \to \beta(\overline{u}) \quad \text{in } L^2(0,T;Y) \text{ as } n \to \infty.$$

References

[1] A. Visintin, Sur le problème de Stefan avec flux non linéaire, Boll. U.M.I., Anal. Funz. Appl. Serie V, 18(1981), 63–86.

[2] M. Niezgodka, I. Pawlow and A. Visintin, Remarks on the paper by A. Visintin "Sur le problème de Stefan avec flux non linéaire", Boll. U.M.I., Anal. Funz. Appl. Serie V, 18(1981), 87–88.

[3] M. Niezgodka and I. Pawlow, A generalized Stefan Problem in Several space variables, Applied Math. Opt., 9(1983), 193–224.

[4] J. R. Cannon, *The one-dimensional heat equation*, Encyclopedia of Mathematics and its Applications, Vol. 23, Addison-Wesley Publishing Company, 1984.

[5] T. Hintermann, Evolution equation with dynamic boundary condition, Proc. Roy. Soc. Edinburgh, Ser. A, 113(1989), 43–60.

[6] M. Grobbelaar and V. Dalsen, On B-evolution theory and dynamic boundary conditions on a portion of the boundary, Appl. Anal., 40(1991), 151–172.

[7] M. Primicerio and J. F. Rodrigues, The Hele-Shaw problem with nonlocal injection condition, to appear in Proceeding of the meetnig "Nolinear Mathematical Problems in Industry,", Iwaki, Japan, 1992.

[8] T. Aiki, Two-phase Stefan problems with dynamic boundary conditions, to appear in Advances in Mathematical Sciences and Applications.

[9] T. Aiki, Multi-dimensional Stefan problems with dynamic boundary conditions, Tech. Rep. Math. Sci., Chiba Univ., No. 18, 1992.

[10] A. Haraux and N. Kenmochi, Asymptotic behavior of solutions to some degenerate parabolic equations, Funk. Ekvac., 34(1991), 19–38.

[11] T. Aiki, N. Kenmochi and J. Shinoda, Periodic stability for some degenerate parabolic equations with nonlinear flux conditions, Nonlinear Anal. TMA., 17(1991), 885–902.

[12] T. Aiki, N. Kenmochi and J. Shinoda, Periodic stability for Stefan problems with flux boundary conditions, *Free boundary problems involving solids*, Vol. 281, *Pitman*, New York, pp 158 – 163, 1993.

[13] T. Aiki, Periodic stability of solutions to some degenerate parabolic equations with dynamic boundary conditions, Tech. Rep. Ser. SAMS.

[14] T. Aiki, Periodic stabilities of solutions for multi-dimensional Stefan problems with nonlinear dynamic boundary conditions, in preprint,.

4th Int. Coll. on Differential Equations, pp. 11-20
D. Bainov and V. Covachev (Eds)

OVERDETERMINED PROBLEMS FOR ELLIPTIC EQUATIONS

DEJENE ALEMAYEHU[1] and GIOVANNI PORRU[2]
[1]*Mathematics Department, A.A. University, Addis Ababa, Ethiopia.*
[2]*Dipartimento di Matematica, Via Ospedale 72, 09124 Cagliari, Italy.*

ABSTRACT. By using a sophisticated maximum principle, three overdetermined problems for elliptic equations in two variables are discussed.

Keywords. Free Boundary Problems, Maximum Principles.

1. INTRODUCTION

We consider classical solutions $u = u(x)$ of the second order equation

$$(g(q)u_i)_i = h(q) \tag{1.1}$$

in a smooth domain $\Omega \subset R^N$. Here g and h are two smooth real functions defined on $(0,\infty)$. In order for equation (1.1) to be elliptic we assume $g(q) > 0$ and

$$g(q) + g'(q)q > 0 \tag{1.2}$$

on solutions of (1.1). As usual, the subindex i $(i = 1, ..., N)$ denotes partial differentiation with respect to x^i, the summation convention (from 1 to N) over repeated indices is in effect and $q^2 = u_i u_i$. For the classical theory of second order elliptic equations including (1.1) we refer to [1,2].

In [3] J. Serrin proved that if there exists a positive smooth solution $u(x)$ of equation (1.1) in a bounded connected smooth domain $\Omega \subset R^N$ and if $u = 0$ and $u_n =$ constant on the boundary of Ω then Ω is a ball and $u(x)$ is radially symmetric. Here and in the sequel, u_n denotes external differentiation. For proving his result Serrin used the moving plane method and a sophisticated version of the Hopf boundary maximum principle. H. Weinberger [4], in a note following Serrin's article, gave another proof of the previous result in case $g = 1$ and $h = -1$. Weinberger's method is based on an interior maximum principle for the function $v(x) = q^2/2 + u$ as well as a Rellich type equality. Although Weinberger's paper deals with a very special equation, the method works under very weak assumptions on the boundary of Ω. In [5], N. Garofalo and J. Lewis generalized Weinberger's argument to equation (1.1) when $h(q) = -1$, g satisfies (1.2) as well as some more conditions, and $u(x)$ is a weak solution.

The methods pointed out by Serrin and Weinberger do not seem to work in the following situation. Let $\Omega = \Omega_0 \setminus \Omega_1$, where Ω_0 and Ω_1 are bounded domains in R^N such that Ω_1 is strictly contained in Ω_0. Let Γ_0 and Γ_1 be the boundaries of Ω_0 and Ω_1, respectively. We are interested in the existence of a solution $u(x)$ of equation (1.1), satisfying the (overdetermined) boundary conditions:

(1.3) $$u|_{\Gamma_0} = 0, \qquad u|_{\Gamma_1} = -1,$$

(1.4) $$u_n|_{\Gamma_0} = q_0, \quad u_n|_{\Gamma_1} = -q_1,$$

where q_0 and q_1 are two positive free constants and the subindex n denotes external differentiation. The expected result is that such a problem is solvable only when Γ_0 and Γ_1 are concentric N-spheres and q_0, q_1 satisfy suitable restrictions. In case Ω_0 and Ω_1 are convex, the above problem has been solved by L. Payne and G. Philippin in [6] for the equation

$$\left((q^{N-2})u_i\right)_i = 0$$

by using a maximum principle for the function $v(x) = q\exp(\lambda u)$ (λ is any real parameter) as well as Rellich's equality. For the equation

$$\left((q^{p-2})u_i\right)_i = 0 \quad (p > 1,\ p \neq N)$$

the problem has been discussed by G. Porru and F. Ragnedda in [7] by using a maximum principle for the function $v(x) = q(\lambda u + 1)^\gamma$, $\gamma = (N-1)/(p-N)$ and a Rellich type equality. For the case $p = 2$ see also [8]. For more general equations, the previous arguments do not work. In the second section of this paper we propose a new argument which applies to a wide class of equations when N=2. We will make use of the following maximum principle obtained in [9]. We consider in a plane domain Ω the partial differential equation (1.1). We also consider the ordinary differential equation (associated to (1.1))

(1.5) $$G(\phi(t))\phi'(t) + \frac{1}{t}g(\phi(t))\phi(t) = h(\phi(t)),$$

where

(1.6). $$G(s) = g(s) + g'(s)s = \frac{d}{ds}(g(s)s).$$

The following theorem holds.

THEOREM 1.1. *Let $\phi(t)$, $t_0 < t < t_1$, be a positive and strictly monotone solution of equation (1.5). Let $\psi(s)$ be the inverse function of $\phi(t)$. If $u(x)$ is a classical solution of equation (1.1) in a plane domain Ω such that all values of $q(x)$ belong to the range of $\phi(t)$, then the function*

(1.7) $$v(x) = \int_{q_0}^{q(x)} \psi'(s)s\,ds - u(x)$$

attains its maximum and its minimum value on the boundary of Ω. If $v(x)$ is not a constant then the external derivative $v_n(x)$ is strictly positive at each point in $\partial\Omega$ where $v(x)$ attains its maximum value. Moreover, if $v(x)$ is a constant in Ω then $u(x)$ must be a radial function.

Proof. In the proof of Theorem 2.2 of [9] it is shown that $v(x)$ satisfies an elliptic equation. Hence, the first part of our theorem follows by classical maximum principles for elliptic equations ([10]). The last assertion follows by Remark 2.1 and Lemma 2.2 of [9].

2. MAIN RESULTS

Throughout this section we assume Ω to be a smooth plane ringshaped domain bounded externally by a curve Γ_0 and internally by a curve Γ_1. Suppose Γ_0 and Γ_1 enclose the connected domains Ω_0 and Ω_1, respectively. We consider the equation

$$(g(q)u_i)_i = 0 \tag{2.1}$$

in Ω, and the boundary conditions

$$u|_{\Gamma_0} = 0, \qquad u|_{\Gamma_1} = -1, \tag{2.2}$$

$$u_n|_{\Gamma_0} = q_0, \quad u_n|_{\Gamma_1} = -q_1, \tag{2.3}$$

where q_0 and q_1 are two positive free constants. Assume condition (1.2), which may be reformulated as follows

$$G(s) = \frac{d}{ds}(g(s)s) > 0. \tag{2.4}$$

We consider only regular solutions of equation (2.1), that is smooth functions $u(x)$ whose gradient is non vanishing in Ω.

THEOREM 2.1. *Under condition (2.4), problem (2.1), (2.2), (2.3) has a regular solution $u(x)$ in Ω only when Γ_0 and Γ_1 are concentric circles with radii r_0, r_1, respectively, satisfying*

$$g(q_0)q_0r_0 = g(q_1)q_1r_1. \tag{2.5}$$

Proof. First let us show that if problem (2.1), (2.2), (2.3) is solvable then one must have $q_1 \neq q_0$. By contradiction, let $q_1 = q_0$. If $u(x)$ is a solution of our problem, let us define

$$a^{ij} = \delta^{ij} + h(q)u_iu_j, \tag{2.6}$$

where δ^{ij} is the Kronecker delta and $h(q) = g'(q)/g(q)$. By assumption (2.4) it follows that the matrix $[a^{ij}]$ is positive definite. With notations (2.6) in mind, equation (2.1) may be rewritten as

$$a^{ij}u_{ij} = 0. \tag{2.7}$$

By $q^2 = u_ku_k$ it follows that

$$qq_{ij} = u_{ijk}u_k + u_{ik}u_{jk} - q_iq_j.$$

Hence,

$$qa^{ij}q_{ij} = a^{ij}u_{ijk}u_k + a^{ij}u_{ik}u_{jk} - a^{ij}q_iq_j.$$

By using equation (2.7) and the positivity of the matrix $[a^{ij}]$, the latter equality implies

$$qa^{ij}q_{ij} \geq -(a^{ij})_k u_{ij}u_k - a^{ij}q_iq_j. \tag{2.8}$$

Since $u_j u_{ij} = q q_i$, by (2.6) it follows

$$(a^{ij})_k u_{ij} u_k = h' q_k u_i u_j u_{ij} u_k + 2h u_{ik} u_j u_{ij} u_k =$$

$$= (h' q_k u_k q u_i + 2h q^2 q_i) q_i.$$

Insertion of the last equality into (2.8) leads to

$$q a^{ij} q_{ij} + b^i q_i \geq 0, \tag{2.9}$$

where

$$b^i = h' q_k u_k q u_i + 2h q^2 q_i + a^{ij} q_j.$$

Since $u(x)$ is regular, q is positive in Ω. The strong maximum principle for the elliptic inequality (2.9) implies that either q is a constant or $q(x) < q_1 (= q_0)$ throughout Ω. In the latter case, by the boundary maximum principle ([8]) we must have $q_n > 0$ at each point in Γ_1. Let us write equation (2.1) in normal coordinates on the (closed) level line Γ_1. We find

$$G(q_1) q_\nu + g(q_1) K q_1 = 0, \tag{2.10}$$

where K denotes the curvature and ν is the internal normal ($\nu = -n$). If q is a constant then $q_\nu = 0$ and, by (2.10), K vanishes at each point in Γ_1, impossible. If $q_n > 0$ on Γ_1 then $q_\nu < 0$ and by (2.10) it follows that one must have $K > 0$ at each point in Γ_1. But then, Ω_1 is convex and, consequently,

$$|\Gamma_0| > |\Gamma_1|, \tag{2.11}$$

where $|\Gamma_0|$ and $|\Gamma_1|$ denote the lengths of Γ_0 and Γ_1, respectively. On the other hand, integration of equation (2.1) over Ω and use of Green's Theorem as well as conditions (2.3) lead to

$$g(q_0) q_0 |\Gamma_0| = g(q_1) q_1 |\Gamma_1|. \tag{2.12}$$

Being $q_0 = q_1$, (2.12) gives $|\Gamma_0| = |\Gamma_1|$, a contradiction to (2.11). Hence one must have $q_0 \neq q_1$.

Equation (1.5) corresponding to (2.1) reads

$$G(\phi(t)) \phi'(t) + \frac{1}{t} g(\phi(t)) \phi(t) = 0. \tag{2.13}$$

By setting $\phi(t) = s$, $t = \psi(s)$ equation (2.13) becomes:

$$\frac{G(s)}{\psi'(s)} + \frac{g(s)s}{\psi(s)} = 0. \tag{2.14}$$

Recalling the expression of $G(s)$ given by (2.4), equation (2.14) may be rewritten as

$$\frac{d}{ds}(g(s) s \psi(s)) = 0.$$

By integration we find

$$\psi(s) = \alpha\big(g(s)s\big)^{-1}, \tag{2.15}$$

where α is a real parameter. Observe that for $\alpha \neq 0$ the domain of $\psi(s)$ is $(0,\infty)$, hence, the range of the function $\phi(t)$ is $(0,\infty)$. Let $u(x)$ be a regular solution of problem (2.1), (2.2), (2.3). By Theorem 1.1, the function

$$v(x) = -\alpha \int_{q_0}^{q(x)} \big(g(s)s\big)^{-2} G(s)s\,ds - u(x) \tag{2.16}$$

for any fixed non vanishing real number α attains its maximum and its minimum value on $\Gamma_0 \cup \Gamma_1$. By using conditions (2.2), (2.3) we find:

$$v(x)|_{\Gamma_1} = -\alpha \int_{q_0}^{q_1} \big(g(s)s\big)^{-2} G(s)s\,ds + 1. \tag{2.17}$$

Let us take α so that

$$\alpha^{-1} = \int_{q_0}^{q_1} \big(g(s)s\big)^{-2} G(s)s\,ds. \tag{2.18}$$

If we choose this value of α in (2.16) then (2.17) yields $v(x)|_{\Gamma_1} = 0$. By using (2.2) and (2.3) we also find $v(x)|_{\Gamma_0} = 0$. Hence, by Theorem 1.1 it follows that $v(x) = 0$ in Ω and that $u(x)$ is a radial function. By using conditions (2.2), we infer that Γ_0 and Γ_1 must be concentric circles. The radii r_0 and r_1 are:

$$r_0 = \psi(q_0), \qquad r_1 = \psi(q_1), \tag{2.19}$$

with $\psi(s)$ defined by (2.15) with the value of α given by (2.18). Equality (2.5) follows easily by the equation

$$g(s)s\psi(s) = \text{constant}$$

and by (2.19). The theorem is proved.

Now let us discuss the equation

$$\big(g(q)u_i\big)_i = 2 \tag{2.20}$$

under the boundary conditions (2.2), (2.3). Assume condition (2.4) as well as the following restriction on q_1, q_0

$$\int_{q_1}^{q_0} G(\tau)\tau\,d\tau \geq 1. \tag{2.21}$$

THEOREM 2.2. *Under conditions (2.4) and (2.21), problem (2.20), (2.2), (2.3) has a regular solution $u(x)$ in Ω only if*

$$\int_{q_1}^{q_0} G(\tau)\tau\,d\tau < 2. \tag{2.22}$$

and if Γ_0, Γ_1 are concentric circles with radii r_0, r_1, respectively, satisfying

$$(r_0 - g(q_0)q_0)r_0 = (r_1 - g(q_1)q_1)r_1. \tag{2.23}$$

Proof. Equation (1.5) corresponding to (2.20) reads

$$G(\phi(t))\phi'(t) + \frac{1}{t}g(\phi(t))\phi(t) = 2.$$

By setting, as before, $\phi(t) = s$, $t = \psi(s)$ the last equation becomes:

$$\frac{G(s)}{\psi'(s)} + \frac{g(s)s}{\psi(s)} = 2.$$

This equation may be written as

$$\frac{d}{ds}(g(s)s\psi(s)) = \frac{d}{ds}(\psi(s))^2.$$

By integration we find

$$g(s)s\psi(s) = (\psi(s))^2 - \alpha^2$$

and

$$2\psi(s) = g(s)s + \left(\left(g(s)s\right)^2 + 4\alpha^2\right)^{\frac{1}{2}}, \tag{2.24}$$

where α is a real parameter. Notice that, for any real value of α, the domain of $\psi(s)$ is $(0,\infty)$, hence the range of $\phi(t)$ is $(0,\infty)$. First, let us take $\alpha = 0$ in (2.24). We find

$$\psi(s) = g(s)s.$$

Let $u(x)$ be a regular solution of problem (2.20), (2.2), (2.3). Then, by Theorem 1.1 the function

$$v(x) = \int_{q_1}^{q(x)} G(s)s\,ds - u(x) \tag{2.25}$$

attains its maximum (and its minimum) value on $\Gamma_0 \cup \Gamma_1$.

By equation (2.20) we have

$$\frac{1}{2}\left(x^j u_j g u_i - g q^2 x^i + g u u_i\right)_i = -\frac{1}{2}x^i q(gq)_i + x^i u_i + u.$$

Integration of this equation over Ω and use of Green's formula as well as conditions (2.2), (2.3) lead to

$$\frac{1}{2}\int_{\Gamma_1} g u u_n ds = -\frac{1}{2}\int_\Omega x^i\left(\int_{q_1}^{q(x)} G(\tau)\tau\,d\tau\right)_i dx + \int_\Omega x^i u_i dx + \int_\Omega u\,dx, \tag{2.26}$$

where the equation $(gq)_i = G(q)q_i$ has been used. Integrating by parts and using again conditions (2.2), (2.3) we obtain

$$-\frac{1}{2}\int_\Omega x^i\left(\int_{q_1}^{q(x)} G(\tau)\tau\,d\tau\right)_i dx = -\int_{q_1}^{q_0} G(\tau)\tau\,d\tau|\Omega_0| + \int_\Omega\left(\int_{q_1}^{q(x)} G(\tau)\tau\,d\tau\right) dx$$

and

$$\int_{\Omega} x^i u_i dx = \int_{\Gamma_1} x^i n^i u\, ds - 2\int_{\Omega} u\, dx = 2|\Omega_1| - 2\int_{\Omega} u\, dx.$$

By inserting the last two results into (2.26), after some simplifications we find

$$\frac{1}{2} g(q_1) q_1 |\Gamma_1| = -\int_{q_1}^{q_0} G(\tau)\tau\, d\tau |\Omega_0| + 2|\Omega_1| + \int_{\Omega} v\, dx, \tag{2.27}$$

where $v = v(x)$ is the function given by (2.25).

We have:

$$v(x)|_{\Gamma_1} = 1, \qquad v(x)|_{\Gamma_0} = \int_{q_1}^{q_0} G(\tau)\tau\, d\tau.$$

Hence, by condition (2.21) and Theorem 1.1 one must have

$$v(x) \leq \int_{q_1}^{q_0} G(\tau)\tau\, d\tau$$

throughout Ω. Therefore, by (2.27) it follows:

$$\frac{1}{2} g(q_1) q_1 |\Gamma_1| \leq -\int_{q_1}^{q_0} G(\tau)\tau\, d\tau |\Omega_0| + 2|\Omega_1| + \int_{q_1}^{q_0} G(\tau)\tau\, d\tau |\Omega|.$$

Since $|\Omega| = |\Omega_0| - |\Omega_1|$ the last inequality implies

$$\frac{1}{2} g(q_1) q_1 |\Gamma_1| \leq \Big(2 - \int_{q_1}^{q_0} G(\tau)\tau\, d\tau\Big) |\Omega_1|,$$

from which (2.22) follows.

Now let us consider the function $\psi(s)$ defined by (2.24) with a non negative α. We find

$$2\psi'(s) = G(s) + g(s)sG(s)\Big(\big(g(s)s\big)^2 + 4\alpha^2\Big)^{-\frac{1}{2}}. \tag{2.28}$$

Let us define

$$F(\alpha) = \int_{q_1}^{q_0} \psi'(s)s\, ds = \frac{1}{2}\int_{q_1}^{q_0} G(s)s\, ds + \frac{1}{2}\int_{q_1}^{q_0} g(s)sG(s)s\Big(\big(g(s)s\big)^2 + 4\alpha^2\Big)^{-\frac{1}{2}} ds.$$

For $\alpha \geq 0$, $F(\alpha)$ is strictly decreasing. Moreover, one has

$$F(0) = \int_{q_1}^{q_0} G(s)s\, ds$$

and

$$\lim_{\alpha \to \infty} F(\alpha) = \frac{1}{2}\int_{q_1}^{q_0} G(s)s\, ds.$$

Hence, by using conditions (2.21) and (2.22) we infer that there exists a unique value of α for which the following equation holds:

$$\int_{q_1}^{q_0} \psi'(s)s\, ds = 1. \tag{2.29}$$

Let us consider the function

$$v(x) = \int_{q_1}^{q(x)} \psi'(s) s\, ds - u(x)$$

where $\psi(s)$ is given by (2.24) with the value of α satisfying (2.29). By (2.2), (2.3) and (2.29), $v(x) = 1$ on Γ_0 as well as on Γ_1. By Theorem 1.1 it follows that $v(x) = 1$ throughout Ω and that $u(x)$ is a radial function. By using again conditions (2.2), we infer that Γ_0 and Γ_1 must be concentric circles. The radii r_0 and r_1 are:

$$2r_0 = g(q_0)q_0 + \left(\left(g(q_0)q_0 \right)^2 + 4\alpha^2 \right)^{\frac{1}{2}}$$

and

$$2r_1 = g(q_1)q_1 + \left(\left(g(q_1)q_1 \right)^2 + 4\alpha^2 \right)^{\frac{1}{2}}.$$

Elimination of α from the last two equations leads to equality (2.23). The theorem is proved.

Finally, let us discuss the equation

$$(g(q)u_i)_i = -g(q)q \tag{2.30}$$

again under the boundary conditions (2.2), (2.3). On $g(s)$ we assume hypothesis (2.4) as well as the following conditions

$$\lim_{s\to 0} g(s)s = 0, \quad \text{and} \quad \lim_{s\to\infty} g(s)s = \infty. \tag{2.31}$$

The following theorem holds.

THEOREM 2.3. *Under conditions (2.4) and (2.31), problem (2.30), (2.2), (2.3) has a regular solution $u(x)$ in Ω if and only if*

$$\int_{q_0}^{q_1} \frac{G(s)}{g(s)} ds > 1 \tag{2.32}$$

and Γ_0, Γ_1 are concentric circles with radii r_0, r_1, respectively, satisfying

$$g(q_0)q_0 r_0 \exp(r_0) = g(q_1)q_1 r_1 \exp(r_1). \tag{2.33}$$

Proof. Equation (1.5) corresponding to (2.30) reads

$$G(\phi(t))\phi'(t) + \frac{1}{t} g(\phi(t))\phi(t) = -g(\phi(t))\phi(t).$$

The last equation can be rewritten as follows

$$\frac{d}{dt}\left(t g(\phi(t))\phi(t) \right) = -t g(\phi(t))\phi(t).$$

Integrating we get

$$g(\phi(t))\phi(t) = \alpha \frac{\exp(-t)}{t}, \tag{2.34}$$

where α is an arbitrary constant. If $t = \psi(s)$ denotes the inverse function of $s = \phi(t)$ then equation (2.34) reads

$$\psi(s)\exp(\psi(s)) = \frac{\alpha}{g(s)s}. \tag{2.35}$$

Because of conditions (2.4) and (2.31), the function $\psi(s)$ defined by equation (2.35), for any positive fixed value of α, is positive, decreasing on $(0,\infty)$ and its range is $(0,\infty)$.

Let $u(x)$ be a regular solution of problem (2.30), (2.2), (2.3). By Theorem 1.1 the function

$$v(x) = \int_{q_1}^{q(x)} \psi'(s)s\,ds - u(x) \tag{2.36}$$

(for any fixed positive α) assumes it maximum (and its minimum) value on $\Gamma_0 \cup \Gamma_1$.

Let us prove (2.32) by contradiction. Assume

$$\int_{q_0}^{q_1} \frac{G(s)}{g(s)}ds \le 1. \tag{2.37}$$

By equation (2.35) it follows:

$$\psi'(s)s = -\frac{G(s)}{g(s)}\frac{\psi(s)}{1+\psi(s)}. \tag{2.38}$$

Inserting (2.38) into (2.36) and using (2.2), (2.3)) we obtain

$$v(x)|_{\Gamma_0} = \int_{q_0}^{q_1} \frac{G(s)}{g(s)}\frac{\psi(s)}{1+\psi(s)}ds. \tag{2.39}$$

From the latter equality we infer that if $q_1 \le q_0$ then $v(x)|_{\Gamma_0} \le 0$. If $q_1 > q_0$ then by (2.39) and (2.37) we find $v(x)|_{\Gamma_0} < 1$. By using (2.2) and (2.3) we obtain $v(x)|_{\Gamma_1} = 1$. Hence, by the maximum principle one must have $v(x) < 1$ in Ω and $v_\nu < 0$ on Γ_1 (ν is the internal normal). Therefore, by (2.36) it follows:

$$\psi'(q_1)q_1q_\nu - q_1 < 0.$$

The last inequality and (2.38) imply

$$-G(q_1)\frac{\psi(q_1)}{1+\psi(q_1)}q_\nu < g(q_1)q_1 \tag{2.40}$$

at each point in Γ_1. On the other hand, equation (2.30) in normal coordinates on Γ_1 reads

$$G(q_1)q_\nu + Kg(q_1)q_1 = -g(q_1)q_1.$$

The last equation and inequality (2.40) imply

$$(1+K)\frac{\psi(q_1)}{1+\psi(q_1)} < 1. \tag{2.41}$$

The function $\psi(s)$ which appears above is the solution of equation (2.35), so it also depends on α. As $\alpha \to 0$ then $\psi(s) \to 0$, and as $\alpha \to \infty$ then $\psi(s) \to \infty$ uniformly on the interval (q_0, q_1). Therefore, letting $\alpha \to \infty$, (2.41) implies that $K \leq 0$ at each point in Γ_1, impossible. Inequality (2.32) is proved.

Now, by using (2.32) and (2.38) it is easy to verify that the equation

$$\int_{q_1}^{q_0} \psi'(s) s \, ds = 1 \tag{2.42}$$

has a unique (positive) solution α. Consider the function $v(x)$ defined by (2.36) where $\psi(s)$ is obtained by (2.35) with the value of α satisfying (2.42). It results $v(x)|_{\Gamma_0} = v(x)|_{\Gamma_1} = 1$. Arguing as in the previous cases, we find that $v(x) = 1$ in Ω, that $u(x)$ is radial and that Γ_0 and Γ_1 are concentric circles. We have $q_0 = \phi(r_0)$ and $q_1 = \phi(r_1)$. By equation (2.34) with $t = r$ we find

$$g(\phi(r))\phi(r) r \exp(r) = \text{constant}$$

from which (2.33) follows. The proof of the theorem is complete.

ACKNOWLEDGEMENT. This research was partially supported by Regione Autonoma della Sardegna.

REFERENCES

1. D. Gilbarg and N.S. Trudinger, Elliptic partial differential equations of second order. Springer-Verlag, Berlin, Heidelberg, New York (1977).
2. B. Kawohl, Rearrangements and Convexity of level sets in PDE. Lectures Notes in Math. 1150, Springer-Verlag, Berlin, Heidelberg, New York (1985).
3. J. Serrin, Arch. Rat. Mech. Anal. 43, 304, 318 (1971).
4. H.F. Weinberger, Arch. Rat. Mech. Anal. 43, 319, 320 (1971).
5. N. Garofalo and J.L. Lewis, Am. J. Math. 111, 9, 33 (1989).
6. G.A. Philippin and L.E. Payne, in Symposia Math. XXX, Academic Press, London and New York (1989), p. 119.
7. G. Porru and F. Ragnedda, Applicable Anal. 37, 1, 18 (1990).
8. G.A. Philippin, Math. Methods in Appl. Sc. 12, 387, 392 (1990).
9. G. Porru, A. Tewodros and S. Vernier-Piro, To appear.
10. M.H. Protter and H.F. Weinberger, Maximum Principles in differential equations. Springer-Verlag, Berlin, Heidelberg, New York (1984).

4th Int. Coll. on Differential Equations, pp. 21-29
D. Bainov and V. Covachev (Eds)

Mixed Finite-Element Methods for the Equations of Flow in Porous Media

Myron B. Allen
Department of Mathematics
University of Wyoming
Laramie, Wyoming 82071, U.S.A.
allen@uwyo.edu

Abstract

The equations governing steady flow of fluid in a two-dimensional porous medium form a system of first-order partial differential equations that is equivalent to a second-order elliptic equation $-\nabla \cdot (K\nabla p) = f$. The coefficient K, known as the hydraulic conductivity or mobility, typically varies in space depending upon the lithology of the medium. This fact causes numerical difficulties in schemes for discretizing the equations.

Among the most attractive of these schemes are mixed finite-element methods. Properly formulated, mixed methods yield approximations of comparable accuracy for the pressure p and the Darcy velocity $\mathbf{v} = -K\nabla p$. This paper reviews the advantages of mixed methods and examines iterative schemes for solving the resulting linear algebraic systems in the presence of fine numerical grids and highly variable coefficients K.

Keywords: Mixed finite-elements, iterative methods, groundwater flow, porous media.

Introduction

Modeling fluid flow and transport in porous media is an important scientific task, not only because of its practical importance in groundwater hydrology and petroleum engineering, but also because of the many difficult mathematical problems that arise. Among these problems is the need to compute accurate fluid velocities. For this task, one starts with known fluid and rock properties and solves the differential equations of mass and momentum balance. These equations determine the pressure (or head) and fluid velocity. Since geologic effects typically cause the domains and coefficients for the equations to be highly variable, it is almost always necessary to use numerical methods in this context. This paper discusses the use of mixed finite-element methods to solve the equations of flow in porous media.

During the past decade, mixed finite-element methods have attracted a great deal of attention among computational hydrologists and petroleum engineers. The methods' main attraction is their ability to produce approximate velocities that are comparable in accuracy to computed pressures. However, difficulties associated with the linear algebraic systems that the methods generate has hindered their widespread application. For large problems, it is essential to use iterative solution schemes that converge rapidly, are relatively insensitive to geologic variability, and possess natural parallelism. After reviewing the physics of the problem and comparing the "classical" and mixed-method formulations, we examine iterative schemes that address these desiderata.

Physics

The equations governing steady fluid flow in a saturated, two-dimensional porous medium Ω have the following forms:

$$\begin{aligned} \nabla \cdot \mathbf{u} &= f \quad \text{in } \Omega, \\ \mathbf{v} &= -K\,\nabla p \quad \text{in } \Omega. \end{aligned} \tag{1}$$

The first equation is the mass balance, while the second is a version of the momentum balance known as Darcy's law. Here $\mathbf{v} = (v_1, v_2)$, p, and f represent the Darcy velocity, pressure (or head, depending upon the formulation), and source term, respectively. The coefficient $K(x, y)$, identified in various disciplines as the hydraulic conductivity or the mobility, depends upon the properties of the porous rock skeleton as well as those of the fluid occupying the pores.

In natural porous media, K varies in space owing to changes in the local lithology. It can also have a tensor character, if the geologic formation of interest has a fabric that gives rise to preferential flow directions. For this paper, we assume that K is a scalar function of spatial position, that it is bounded above, and that $\inf K(x, y) > 0$.

The spatial variability in K causes difficulties in numerical models. One important difficulty that this paper does not examine is that of determining $K(x, y)$ from physical measurements, which are often sparse. There are other difficulties that would remain even if we knew K exactly. One is the need to use fine spatial grids to resolve the variations in the medium and the resulting variations in p and $\mathbf{v}$. Fine grids lead to poorly conditioned algebraic systems and therefore to slow convergence of iterative schemes. Another difficulty arises from the variability in K itself, which affects the matrix entries of the linear system and again leads to poor conditioning.

The "classical" approach

The following approach to solving Equations (1) is fairly standard, having been used by hydrologists and petroleum engineers for over two decades. First, combine the mass balance and Darcy's law to get the second-order elliptic equation

$$-\nabla \cdot (K \nabla p) = f \quad \text{in } \Omega.$$

Then discretize the equation using finite differences or Galerkin finite elements on a grid having mesh size h. The resulting linear algebraic system is symmetric and positive definite, owing to the ellipticity of the differential operator, so one can solve for the approximate pressure p_h using a variety of numerical schemes. Having computed p_h, one can differentiate numerically to obtain $\mathbf{v}_h(\mathbf{x}) = -K(\mathbf{x})\nabla p_h(\mathbf{x})$.

There are problems with this approach that are easiest to describe in the context of Galerkin methods. Standard finite-element theory yields optimal-order error estimates for the pressure, having the form $\|p - p_h\|_0 \leq C h^s \|p\|_s$. Here, $\|\cdot\|_s$ denotes the usual norm associated with the Sobolev space $H^s(\Omega)$, and C is a positive constant independent of p and h. The exponent s depends on the trial space being used; for example, when p_h is a piecewise bilinear function on the grid, $s = 2$. The most important problem with this error estimate is that it implies that the approximate velocity is less accurate: $\|\mathbf{v} - \mathbf{v}_h\|_0 \leq (C \sup_\Omega K) h^{s-1} \|p\|_s$. This loss of accuracy is undesirable in problems, such as those involving underground contaminant transport, where one needs accurate velocities. A related but somewhat more technical problem is that the numerical velocities $\mathbf{v}_h = -K\nabla p_h$ are typically discontinuous across the boundaries of cells in the grid.

Mixed finite-element methods

Mixed finite-element methods use a different approach to discretization. Here, one solves the simultaneous first-order system

$$\begin{aligned} K^{-1}\mathbf{v} + \nabla p &= 0 \quad \text{in } \Omega, \\ \nabla \cdot \mathbf{v} &= f \quad \text{in } \Omega. \end{aligned} \tag{2}$$

For concreteness, we assume that $p = 0$ on $\partial\Omega$, although it is possible to impose nonzero Dirichlet boundary conditions as well as flux conditions of the form $\mathbf{v} \cdot \mathbf{n} = \gamma$ on $\partial\Omega$.

To arrive at a discrete approximation, consider two Hilbert spaces. The first is the "velocity space,"

$$\mathcal{V} = H(\operatorname{div}, \Omega) := \left\{ \mathbf{u} \in L^2(\Omega) \times L^2(\Omega) \quad : \quad \nabla \cdot \mathbf{u} \in L^2(\Omega) \right\}.$$

The second is the pressure space,

$$\mathcal{P} := \left\{ q \in L^2(\Omega) \quad : \quad q = 0 \quad \text{on} \quad \partial\Omega \right\}.$$

Equations (2) have the following variational formulation:

$$\begin{aligned} \textstyle\int_\Omega K^{-1}\mathbf{v} \cdot \mathbf{u} \quad - \int_\Omega p \nabla \cdot \mathbf{u} &= 0 \qquad \forall\, \mathbf{u} \in \mathcal{V}, \\ \textstyle\int_\Omega q \nabla \cdot \mathbf{v} \qquad\qquad &= \textstyle\int_\Omega q f \quad \forall\, q \in \mathcal{P}. \end{aligned} \tag{3}$$

Mixed finite-element schemes impose Equations (3) on finite-dimensional subspaces $\mathcal{P}_h$ and $\mathcal{V}_h$ of $\mathcal{P}$ and $\mathcal{V}$, respectively. Among the simplest choices of subspaces are the lowest-order Raviart-Thomas spaces [1] on rectangular grids. To be specific, let Δ be a quasi-regular rectangular grid of mesh size h on Ω. This grid partitions Ω into cells $\Omega_{i,j} := (x_{i-1}, x_i) \times (y_{j-1}, y_j)$. For the pressure space, let

$$\mathcal{P}_h := \left\{ q \in \mathcal{P} \quad : \quad q|_{\Omega_{i,j}} \quad \text{is constant} \right\}$$

be the space of functions that are piecewise constant with respect to Δ. The velocity space $\mathcal{V}_h$ is slightly more complicated. Call

$$\mathcal{V}_h^x := \left\{ u_1 \in L^2(\Omega) \quad : \quad u_1|_{\Omega_{i,j}} \text{ is linear in } x, \text{ constant in } y \right\}$$

and

$$\mathcal{V}_h^y := \left\{ u_2 \in L^2(\Omega) \quad : \quad u_2|_{\Omega_{i,j}} \text{ is constant in } x, \text{ linear in } y \right\}.$$

Then $\mathcal{V}_h := \mathcal{V}_h^x \times \mathcal{V}_h^y$. In other words, x-velocities are piecewise linear in x and piecewise constant in y, while y-velocities are piecewise constant in x and piecewise linear in y. It is crucial for the error estimates cited later that the divergence of any function in $\mathcal{V}_h$ belong to $\mathcal{P}_h$ [1].

The groundwater flow equations (2) then have the following finite-element formulation: Find $(p_h, \mathbf{v}_h) \in \mathcal{P}_h \times \mathcal{V}_h$ such that

$$\begin{aligned} \underbrace{\textstyle\int_\Omega K^{-1}\mathbf{v}_h \cdot \mathbf{u}}_{A(\mathbf{u},\mathbf{v}_h)} \quad - \underbrace{\textstyle\int_\Omega p_h \nabla \cdot \mathbf{u}}_{B(\mathbf{u},p_h)} &= 0 \qquad \forall\, \mathbf{u} \in \mathcal{V}_h, \\ \underbrace{\textstyle\int_\Omega q \nabla \cdot \mathbf{v}_h}_{B(\mathbf{v}_h,q)} \qquad\qquad &= \underbrace{\textstyle\int_\Omega q f}_{F(q)} \quad \forall\, q \in \mathcal{P}_h. \end{aligned}$$

As the underbraces in these equations suggest, we can express these equations in terms of bilinear forms $A: \mathcal{V}_h \times \mathcal{V}_h \to \mathbb{R}$, $B: \mathcal{V}_h \times \mathcal{P}_h \to \mathbb{R}$ and a linear functional $F: \mathcal{P}_h \to \mathbb{R}$ as follows:

$$\begin{aligned} A(\mathbf{u}, \mathbf{v}_h) \quad -B(\mathbf{u}, p_h) &= 0 \qquad \forall\, \mathbf{u} \in \mathcal{V}_h, \\ B(\mathbf{v}_h, q) \qquad\qquad &= F(q) \quad \forall\, q \in \mathcal{P}_h. \end{aligned} \tag{4}$$

We discuss the linear algebraic aspects of this system shortly.

This approach to discretization is easily extended. For example, it is possible to use finite-element subspaces involving higher-degree piecewise polynomials, to develop analogous spaces for triangular grids, and to generalize the construction to three space dimensions.

Many researchers have devoted considerable effort to analyzing the convergence of mixed finite-element methods. Brezzi [2], Raviart and Thomas [1], Falk and Osborn [3], and Douglas and Roberts [4] prove the following optimal-order error estimates:

THEOREM. *For the lowest-order Raviart-Thomas spaces,*

$$\begin{aligned} \|p - p_h\|_0 &\leq Ch\|p\|_1, \\ \|\mathbf{v} - \mathbf{v}_h\|_0 &\leq Ch\|p\|_2. \end{aligned}$$

In addition, Ewing, Lazarov, and Wang [5] prove the following superconvergence results:

THEOREM. *Along the "Gauss lines,"*

$$\left.\begin{aligned} &|p(x_i, y_j) - p_h(x_i, y_j)| \\ &|v_1(x_{i+1/2}, y) - v_{1_h}(x_{i+1/2}, y)| \\ &|v_2(x, y_{j+1/2}) - v_{2_h}(x, y_{j+1/2})| \end{aligned}\right\} \leq Ch^2(\|p\|_2 + \|\mathbf{v}\|_3).$$

Thus the computed pressures and velocities have the same order of accuracy globally, and along nodal lines of the grid these variables have the same order of accuracy as the pressures computed in standard finite-difference formulations.

By now the prospect of computing pressures and velocities with comparable accuracy has led to many applications of mixed methods. Early among the applications to porous-media flows are the papers of Douglas, Ewing, and Wheeler [6] and Ewing, Russell, and Wheeler [7], who examine miscible displacement in oil reservoirs; Ewing and Heineman [8], who model multiphase, multicomponent flows in oil reservoirs; and Ewing, Koebbe, Gonzales, and Wheeler [9] and Allen, Ewing, and Koebbe [10], who apply the mixed method to groundwater flows. One of the barriers to more widespread application has been the need for efficient iterative methods for solving the linear algebraic equations (4). We turn now to this problem.

Linear algebra

Using standard nodal bases for $\mathcal{P}_h$ and $\mathcal{V}_h$ and adopting a lexicographic ordering of equations and unknowns, we reduce Equations (4) to a linear system having the following block structure:

$$\begin{bmatrix} A & B \\ B^T & 0 \end{bmatrix} \begin{bmatrix} V \\ P \end{bmatrix} = \begin{bmatrix} 0 \\ F \end{bmatrix}. \tag{5}$$

The vector V contains nodal values of the velocities v_1 and v_2, associated with cell edges in the grid, and P contains nodal pressures, associated with cell centroids. The matrix A is symmetric and positive definite and has the block structure

$$A = \begin{bmatrix} A_1 & 0 \\ 0 & A_2 \end{bmatrix}.$$

The blocks A_1 and A_2 are tridiagonal, their entries having the form $\int_\Omega K^{-1}\varphi_k\varphi_\ell$, where φ_k, φ_ℓ are functions belonging to the basis for $\mathcal{V}_h$.

The matrix B has the block structure

$$B = \begin{bmatrix} B_1 \\ B_2 \end{bmatrix},$$

where B_1 and B_2 mimic the usual difference approximations to $\partial/\partial x$ and $\partial/\partial y$. The vector F contains integrals involving the source function f. For a detailed specification of the entries in this linear system, we refer readers to Allen, Ewing, and Lu [11].

The system (5) has several interesting properties. First, the matrix A is tridiagonal, with positive eigenvalues bounded away from zero by a constant independent h. Second, the matrix on the left side of Equation (5) is nonsingular and symmetric, but it is indefinite. To see this, observe that

$$\begin{bmatrix} A & B \\ B^\top & 0 \end{bmatrix} = \begin{bmatrix} I & 0 \\ B^\top A^{-1} & I \end{bmatrix} \begin{bmatrix} A & 0 \\ 0 & -B^\top A^{-1}B \end{bmatrix} \begin{bmatrix} I & A^{-1}B \\ 0 & I \end{bmatrix}.$$

The spectrum of A lies in $(0,\infty)$, while the spectrum of $-B^\top A^{-1}B$ lies in $(-\infty,0)$. Hence, by Sylvester's law of inertia, the matrix on the left has eigenvalues to the left and right of the origin.

Third, blockwise row reduction of Equation (5) yields the equations

$$\begin{aligned} (B^\top A^{-1}B)P &= F, \\ V &= -A^{-1}BP. \end{aligned}$$

The Schur complement $B^\top A^{-1}B$ of A is symmetric and positive definite. However, solving the system in this form is impractical in large problems, since $B^\top A^{-1}B$ is not sparse.

Iterative schemes

Several researchers have examined efficient iterative schemes for solving the system (5) and similar systems arising from other mixed methods. Ewing, Lazarov, Lu, and Vassilevski [12], for example, analyze a class of preconditioners for solvers based on the conjugate-gradient method. Beckie, Wood, and Aldama [13] examine a multigrid scheme for higher-order mixed methods based on the use of Lagrange multipliers to enforce the appropriate continuity conditions across cell boundaries. We devote the remainder of this paper to a discussion of two matrix splitting schemes introduced by Allen, Ewing, and Lu [11].

The first splitting has the form

$$\begin{bmatrix} \omega I & B \\ B^\top & 0 \end{bmatrix} \begin{bmatrix} V \\ P \end{bmatrix}^{(k+1)} = \begin{bmatrix} 0 \\ F \end{bmatrix} + \begin{bmatrix} \omega I - A & 0 \\ 0 & 0 \end{bmatrix} \begin{bmatrix} V \\ P \end{bmatrix}^{(k)},$$

where I is the identity and ω is a parameter depending on the spectral radius $\rho(A)$ of A. Computationally, this scheme has the following form:

(i) $G^{(k)} \leftarrow F + \omega^{-1}B^{\mathsf{T}}(\omega I - A)V^{(k)}$.

(ii) Solve $\omega^{-1}B^{\mathsf{T}}BP^{(k+1)} = G^{(k)}$.

(iii) $V^{(k+1)} \leftarrow \omega^{-1}(\omega I - A)V^{(k)} - BP^{(k+1)}$.

The scheme has two attractive features. First, the overall iteration matrix

$$M_1 := \begin{bmatrix} \omega I & B \\ B^{\mathsf{T}} & 0 \end{bmatrix}^{-1} \begin{bmatrix} \omega I - A & 0 \\ 0 & 0 \end{bmatrix}$$

allows one to construct schemes that are insensitive to the fineness of the grid Δ:

THEOREM. *If* $\omega \geq \rho(A)$, *then*

$$\rho(M) \leq 1 - \mathcal{O}\left(\frac{\inf(K)}{\sup(K)}\right). \tag{6}$$

Thus, for example, if we make the computationally convenient choice $\omega = \|A\|_\infty \geq \rho(A)$, then the overall convergence rate of the scheme is independent of the mesh size h.

Second, the matrix $B^{\mathsf{T}}B$ is just the five-point difference approximation to the Laplace operator. Therefore, we can execute step *(ii)* of the algorithm, which is the costly step, by using any fast scheme for the five-point approximation to the Poisson equation.

There is one disadvantage apparent in the estimate (6): The convergence rate of the iterative scheme deteriorates when $\inf(K) << \sup(K)$. Allen, Ewing, and Lu [11] report results of several numerical experiments that illustrate this fact. The problem is a serious one in models of natural porous media, in which spatial changes in depositional environment can cause K to vary over several orders of magnitude.

The second scheme uses a different matrix splitting:

$$\begin{bmatrix} D & B \\ B^{\mathsf{T}} & 0 \end{bmatrix} \begin{bmatrix} V \\ P \end{bmatrix}^{(k+1)} = \begin{bmatrix} 0 \\ F \end{bmatrix} + \begin{bmatrix} D - A & 0 \\ 0 & 0 \end{bmatrix} \begin{bmatrix} V \\ P \end{bmatrix}^{(k)}. \tag{5}$$

Here, D is a diagonal matrix, the simplest effective structure for which is the diag (A). Computationally, the scheme has the following form:

(i) $G^{(k)} \leftarrow -F + B^{\mathsf{T}}D^{-1}(D - A)V^{(k)}$.

(ii) Solve $B^{\mathsf{T}}D^{-1}BP^{(k+1)} = G^{(k)}$.

(iii) $V^{(k+1)} \leftarrow D^{-1}(D - A)V^{(k)} - D^{-1}BP^{(k+1)}$.

This scheme has iteration matrix

$$M_2 := \begin{bmatrix} D & B \\ B^{\mathsf{T}} & 0 \end{bmatrix}^{-1} \begin{bmatrix} D - A & 0 \\ 0 & 0 \end{bmatrix},$$

and its convergence rate is independent of mesh size h and of variations in K. In fact,

THEOREM. *If* $D = \operatorname{diag}(A)$, *then* $\rho(M) \leq \frac{1}{2}$.

Allen, Ewing, and Lu [11] report computational experiments that corroborate this estimate. The price that we pay for this benefit is the need to solve a matrix equation involving $B^{\mathsf{T}}D^{-1}B$ in step *(ii)* of the algorithm. This matrix is essentially a five-point difference approximation to the operator $-\nabla\cdot(K\nabla)$, for which constructing a fast solver is problematic.

Instead of executing step *(ii)* exactly, we use several V-cycles of a multigrid scheme to solve the five-point system approximately. Thus the matrix splitting serves as an "outer" iteration, while the multigrid cycles constitute an "inner" iteration that generates an approximate value for $P^{(k)}$. In our implementation, each V-cycle involves two Gauss-Seidel iterations at each level in a nest $\Delta = \Delta_0 \supset \Delta_1 \supset \cdots \supset \Delta_L$ of successively coarser grids, the mesh size of Δ_k being $2^k h$. For the intergrid transfers, we use full weighting as a restriction operator and bilinear interpolation as a prolongation operator [14].

One attractive feature of the multigrid scheme is its amenability to parallel processing. Curran and Allen [15] discuss one approach to parallelizing the algorithm. Broadly speaking, one can arrange for each processor to manage a rectangular region, or *patch*, of the original fine grid. The relaxation sweep on any patch requires some values of latest iterates from the nearest-neighbor patches. Therefore, before executing a relaxation sweep, a processor must trade information about a "boundary layer" of nodal values with the processor that manages the nearest-neighbor patch. The parallel implementation thus requires communication between processors before each sweep of the Gauss-Seidel smoother.

To assess the scheme's parallelism, Curran and Allen [15] examine execution times on a 1024-processor nCube 2 having a hypercube archtecture. In particular, they present *scaled speedups* by examining execution times on subcubes having dimension 0 (1 processor), 1 (2 processors), ..., 10 (1024 processors), running problems of proportionately larger size on larger subcubes. Each subcube is a set of processors linked by the shortest possible physical paths in the machine.

Table 1 shows timings for a sequence of runs involving a 32 × 32-cell grid on the one-processor subcube, a 64 × 32-cell grid on the two-processor subcube, a 64 × 64-cell grid on the four-processor subcube, and so forth, up to a 512 × 512 grid on a 512-processor cube. A perfectly parallel algorithm would require the same execution time for all runs. In practice, interprocessor communication and computational overhead disrupt this ideal relationship. However, Table 1 suggests that this overhead is quite small in practice.

Table 1: RUNTIMES (SECONDS) FOR SCALED FLOW PROBLEMS ON THE NCUBE 2.

Number of processors	Setup time	Communication time	Total time
1	0.178	0.205	161.516
2	0.184	1.784	159.274
4	0.201	3.690	157.260
8	0.214	4.639	158.484
16	0.251	5.775	159.967
32	0.319	5.832	160.231
64	0.530	5.908	160.620
128	0.782	5.950	160.985
256	1.396	5.979	161.673
516	2.691	6.030	163.005

Discussion

A variety of extensions are needed to make these iterative schemes useful in modeling realistic porous-media flows. The most obvious need is to extend the scheme for the flow equation to time-dependent, three-dimensional settings. The extension to time-dependent problems is straightforward. Indeed, the resulting matrix equations are better conditioned than those for the steady-state case, since the accumulation term enhances diagonal dominance. The extension to three dimensions is also straightforward conceptually, but it results in more computationally intensive codes. Still, the principles that allow parallelization remain intact.

More interesting is the need to extend the method to problems involving tensor conductivities [16]. It is in the context of tensor conductivities that the two-level iterative scheme for the mixed-method equation has the greatest potential for practical use. Shen [17], through delicate analysis, shows that one can lump the matrix A in the mixed-method system and preserve global accuracy in the scalar case. Thus, when K is a scalar, one can dispense with the outer iterations used here. However, the analysis does not appear to extend to the case when the conductivity K is a tensor. In this case, the inner-outer iteration scheme still offers reasonable prospects for an effective, robust solution procedure.

Acknowledgments

The U.S. National Science Foundation provided support through grant number EHR-910-8774.

References

1. Raviart, P.A., and Thomas, J.M., "A mixed finite element method for second order elliptic problems," in *Mathematical Aspects of Finite Element Methods,* Lecture Notes in Mathematics vol. 606, ed. by I. Galligani and E. Magenes, pp. 292-315, Springer-Verlag, Berlin, 1977.

2. Brezzi, F., "On the existence, uniqueness, and approximation of saddle point problems arising from Lagrangian multipliers," *R.A.I.R.O. Analyse Numérique 2* (1974), 129-151.

3. Falk, R.S., and Osborn, J.E., "Error estimates for mixed methods," *R.A.I.R.O. Analyse Numérique 14* (1980), 249-277.

4. Douglas, J., and Roberts, J.E., "Global estimates for mixed methods for second order elliptic equations," *Math. Comp. 444* (1985), 39-52.

5. Ewing, R.E., Lazarov, R.D., and Wang, J., "Superconvergence of the velocities along the Gauss lines in mixed finite-element methods," *SIAM J. Numer. Anal.*, to appear.

6. Douglas, J., Ewing, R.E., and Wheeler, M.F., "The approximation of the pressure by a mixed method in the simulation of miscible displacement," *R.A.I.R.O. Analyse Numérique 17* (1983), 17-33.

7. Ewing, R.E., Russell, T.F., and Wheeler, M.F., "Convergence analysis of an approximation of miscible displacement in porous media by mixed finite elements and a modified method of characteristics," *Comp. Meth. Appl. Mech. Eng. 47* (1984), Chapter 3.

8. Ewing, R.E., and Heineman, R.F., "Mixed finite element approximation of phase velocities in compositional reservoir simulation," *Comp. Meth. Appl. Mech. Eng.* *47* (1984), Chapter 7.

9. Ewing, R.E., Koebbe, J.V., Gonzalez, R., and Wheeler, M.F., "Computing accurate velocities for fluid flow in porous media," *Proceedings, Fifth Annual Symposium on Finite Elements and Flow Problems,* Austin, Texas, January 23-26, 1984.

10. Allen, M.B., Ewing, R.E., and Koebbe, J.V., "Mixed finite element methods for computing groundwater velocities," *Numer. Meth. P.D.E. 3* (1985), 195-207.

11. Allen, M.B., Ewing, R.E., and Lu, P., "Well conditioned iterative schemes for mixed finite-element models of porous-media flow," *SIAM Jour. Sci. Stat. Comp. 13:3* (1992), 794-814.

12. Ewing, R.E., Lazarov, R.D., Lu, P., and Vassilevski, P.S., "Preconditioning indefinite systems arising from mixed finite element discretization of second-order elliptic systems," *Proceedings of the Conference on Preconditioned Conjugate Gradient Methods,* Nijmegen, The Netherlands, June 15-17, 1989.

13. Beckie, R., Wood, E.F., and Aldama, A.A., "Mixed finite element simulation of saturated groundwater flow using a multigrid accelerated domain decomposition technique," *Water Resour. Res. 29:9* (1993), 3145-3157.

14. Briggs, W., *A Multigrid Tutorial,* SIAM, Philadelphia, Pennsylvania, 1987.

15. Curran, M.C. and Allen, M.B., "A multigrid-based solver for mixed finite-element approximations to groundwater flow," in *Computational Methods in Water Resources IX, Vol. I: Numerical Methods in Water Resources,* ed. by T.F. Russell et al., Elsevier Applied Science, London, 1992, 579-585.

16. Koebbe, J.V., "A computationally efficient modification of mixed finite element methods for flow problems with full transmissivity tensors," *Numer. Meth. P.D.E. 9:4* (1993), 339-356.

17. Shen, J., " Mixed Finite Element Methods: Analysis and Computational Aspects," Ph.D. dissertation, University of Wyoming, Laramie, Wyoming, 1992.

4th Int. Coll. on Differential Equations, pp. 31-38
D. Bainov and V. Covachev (Eds)

Limit sets of impulsive dynamical systems

D. D. BAINOV AND S. I. NENOV

Technical University - Plovdiv, Plovdiv, BULGARIA;
Technological University of Sofia, Sofia, BULGARIA

Abstract. The aim of the present paper is to define C^2-smooth impulsive flows on 2-dimensional smooth manifold as a class of partially continuous and locally C^2-smooth flows. An theorem analogue of the Poincaré - Bendixon's theorem is proved.

Key words and phrases: compatible triple, impulsive flow, minimal set, limit set.
1991 AMS Subject classification: 34A37, 34C35, 58F25.

1 Introduction

The article deals with Poincaré - Bendixon's type theorems.

The classical Poincaré - Bendixon's theorem for C^2-smooth flow on 2-dimensional manifold was generalized by A. J. Schwarz in 1963 [11]. The main goal of the present work is to extend the same theorem for a class of partially continuous and locally C^2-smooth flows on 2-dimensional smooth manifold. The flows in the class under consideration will be named impulsive flows by means of analogy with the notion of the impulsive equation.

In 1960 V. Milman and A. Mishkis have defined the object "impulsive systems" [8], [9] as differential systems whose solution are subject to "instant" perturbations.

At the last time the theory of these systems has developed intensively because of their numerous applications in mechanics - shock mechanical systems, control theory - "fast" outside actions on controlled systems, population dynamics, applied robotics ets.

The impulsive flows on smooth manifold are characterized by the next properties: determination, partial continuity and local regularity. Each impulsive flow is defined by its impulsive sets — the union of 2-dimensional submanifolds with boundary at 2-dimensional phase manifold and by its impulsive mapping - a difeomorphism between the impulsive manifolds and a smooth flow in the phase manifold. The evolutional portraid of any impulsive flow in arbitrary domain not intersecting the impulsive set does not differ from the evolutional portraid of the smooth flow. When the moving point of the impulsive flow "reaches" the boundary of the impulsive set it "jumps" (under the action of the impulsive mapping) in eventually another connected component of the impulsive set. After that the evolution goes on under the action of the smooth flow.

1.1 Acknowledgment

We owe special gratitude to Prof. A. Dishliev for his careful reading of the manuscript.

2 Statement of the problem

Throughout the paper will be used the following notations: $N = \{1, 2, \ldots\}$, $R_- = (-\infty, 0)$, $R_+ = (0, \infty)$. If M is a 2-dimensional manifold and A_1, A_2 are submanifolds of M we denote by ∂A_1 the boundary of A_1 in M; by $\mathrm{Diff}^\kappa(A_1, A_2)$ – the space of all C^κ - diffeomorphisms from A_1 onto A_2, $\kappa \in N$. If $\Phi : M \longrightarrow M$ then $\Phi|A_1$ is the trace of Φ over A_1, and $Fix\ \Phi|A_1$ stands for the set of all fixed points of the mapping $\Phi|A_1$.

Hereafter it will be assumed that the 2-dimensional manifold M is embedded in R^4 as a closed subset. Moreover, we suppose that $\rho(\cdot,\cdot)$ is a complete metric in M generated by this embedding [6]. Let $m_0 \in M$ and $r \in R_+$. We set $\mathbf{B}(m_0, r) = \{m \in M : \rho(m, m_0) < r\}$, $\mathbf{B}(A_1, r) = \cup_{a \in A_1} B(a, r)$, $\rho(A_1, A_2) = inf\{\rho(a_1, a_2) : a_1 \in A_1,\ a_2 \in A_2\}$.

Let A is a submanifold of M such that $\{A_s : s \in S\}$ is the set of all connected components of A where $S = S_- \cup S_0 \cup S_+ \subseteq N$ is an index set, $S_- \cap S_0 = S_+ \cap S_0 = S_- \cap S_+ = \emptyset$ and the potency of sets A_1 and A_2 are equal. Setting $A_\sigma = \cup\{A_s : s \in S_\sigma\}$, $\sigma \in \{-, 0, +\}$ and suppose that $\Phi : M \longrightarrow M$ and $T : R \times M \longrightarrow M$ are respectably a map and a flow on M.

We will say that the conditions (H1) are hold if:

$(H1.1)$ *M is a Hausdorff's, paracompact, conected, $2-$dimensional, C^2-smooth manifold withouth boundary.*

$(H1.2)$ *A_s is a $2-$dimensional, C^2- smooth manifold with boundary $\partial A_s \neq \emptyset$ for each $s \in S$ and $M \setminus A \neq \emptyset$.*

$(H1.3)$ *For ewery compact set $K \subset M$ there exist a number $\delta_0 \in R_+$ such that $\rho(A_{s_1}, A_{s_2}) \geq \delta_0$, $s_1, s_2 \in S$ and $s_1 \neq s_2$.*

$(H1.4)$ *$\Phi|\partial A$ is an involution; $\Phi|A_- \in Diff^2(A_-, A_+)$ and $\Phi|A_s \in Diff^2(A_s, A_s)$ as $s \in S_0$.*

$(H1.5)$ *T is a C^2-smooth flow on M.*

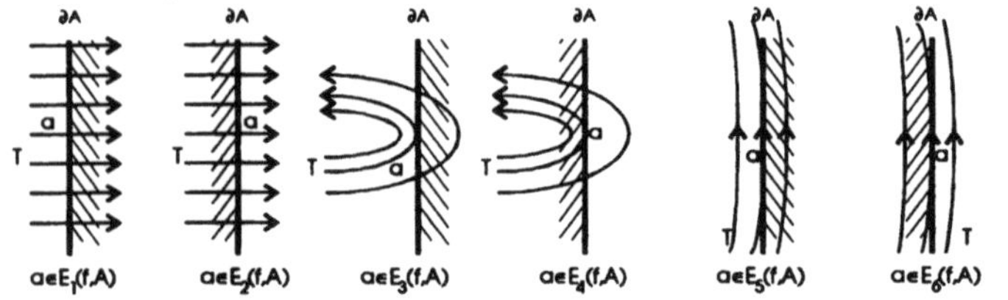

fig 1.a fig 1.b fig 1.c

Assuming now the conditions (H1) to be valid. The point $a \in \partial A$ is a point of entrance (exit) of the flow T into (from) the manifold A if there exist $\theta \in R_+$ such that $\gamma_1 = \{T(t, a) : -\theta < t < 0\} \subset M \setminus A$ and $\gamma_2 = \{T(t, a) : 0 < t < \theta\} \subset A \setminus \partial A$ ($\gamma_1 \subset A \setminus \partial A$, $\gamma_2 \subset M \setminus A$). The set of all points of entrance (exit) of T into (from) A will be denoted by $E_1(T, A)$ ($E_2(T, A)$) (figure 1.a). The point $a \in \partial A$ is called a point of outher (inner) contact of the flow T with ∂A if there exist $\theta \in R_+$ such that the arc $\gamma = \{T(t, a) : -\theta < t < \theta\}$ contacted ∂A at the point a and $\gamma \subset \overline{M \setminus A}$ ($\gamma \subset A$). The set of all points of outher (inter) contact of T with ∂A will be denoted by $E_3(T, A)$ ($E_4(T, A)$)

(figure 1.b). The point $a \in \partial A$ is outher (inter) common point of the flow T and ∂A if there exist $\theta_1, \theta_2, \theta_3 \in R_+$ such that the implications $\gamma_1 \subset M \setminus A$, $\gamma_2 \subset \partial A$, $\gamma_3 \subset M \setminus A$, $(\gamma_1 \subset A \setminus \partial A$, $\gamma_2 \subset \partial A$, $\gamma_3 \subset A \setminus \partial A)$ are valid for the arcs $\gamma_1 = \{T(t,a) : -\theta_1 - \theta_3 < t < -\theta_1\}$, $\gamma_2 = \{T(t,a) : -\theta_1 \leq t \leq \theta_2\}$, $\gamma_3 = \{T(t,a) : \theta_2 < t < \theta_2 + \theta_3\}$. The set of all outher (inner) common points of T and A will denoted by $E_5(T,A)$ $(E_6(T,A))$ (figure 1.c).

Definition 1 *Let the conditions (H1) be satisfied. The ordered triple (T, A, Φ) will be named compatible triple on the manifold M if:*

1. *The set of all points in $E_3(T,A) \cup E_4(T,A)$ and all disjunctive arcs in $E_5(T,A) \cup E_6(T,A)$ does not have a limit point in ∂A.*
2. *The equalities*

$$\partial A = \cup_{\sigma=1}^{6} E_\sigma(T,A), \quad \Phi(E_1(T,A)) = E_2(T,A);$$

$$\Phi(E_\sigma(T,A)) = E_\sigma(T,A), \sigma \in \{3,\dots,6\}.$$

are valid.□

Remark 1 *The first condition in Definition 1 is equivalent to the next one: for each compact set $K \subset M$ there exist $\delta_0 \in R_+$ such that $\rho(a_1,a_2) \geq \delta_0$, $\rho(l_1,l_2) \geq \delta_0$ and $\rho(a_1,l_1) \geq \delta_0$ (sf. (H1.3)) for arbitrary point $a_1, a_2 \in E_3(T,A) \cup E_4(T,A)$, $a_1 \neq a_2$ and arbitrary arcs $l_1, l_2 \in E_5(T,A) \cup E_6(T,A)$, $l_1 \cap l_2 = \emptyset$.* □

Remark 2 *It follows from the Definition 1 that $E_1(T,A)$ and $E_2(T,A)$ are open, disjunctive sets in ∂A; Fix $(\Phi|A) \subset A \setminus (E_1(T,A) \cup E_2(T,A))$.* □

Before giving of precise definition of the term impulsive flow P on the manifold M generated by the compatible triple (T, A, Φ) we will illustrate the motion of the mapping point $P(t, m_0)$ of P.

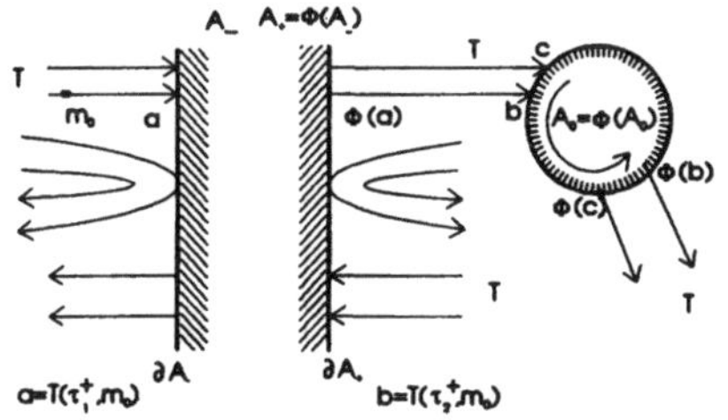

fig 2

Let us suppose that $m_0 \in M \setminus A$ and $t \in R_+$. Then $P(0, m_0) = m_0$ at the initial moment $t = 0$. The mapping point $P(t, m_0)$ starts from the point m_0 and moving along the trajectory of the flow T defined by m_0 to the first moment $\tau_1^+(m_0) \in R_+$ such that $T(\tau_1^+(m_0), m_0) \in \partial A_s$ for some $s \in S$. The mapping point $P(t, m_0)$ "jumps" from the point $P(\tau_1^+(m_0), m_0) = T(\tau_1^+(m_0), m_0)$ to the point $P(\tau_1^+(m_0) + 0, m_0) = \Phi(P(\tau_1^+(m_0), m_0)) \in \Phi(A_s)$ at the moment $\tau_1^+(m_0)$. After that $P(t, m_0)$ continues its motion along the trajectory of the flow T defined by the point $P(\tau_1^+(m_0) + 0, m_0)$ to first moment $\tau_2^+(m_0) \in (\tau_1^+(m_0), \infty)$ when $T(\tau_2^+(m_0) - \tau_1^+(m_0), \Phi(T(\tau_1^+, m_0))) \in \partial A$ and etc. (figure 2).

In the case $t \in R_-$ the dynamical picture of the process under consideration is analogous.

Definition 2 *Let (T, A, Φ) be a compatible triple on M and $m_0 \in M \setminus A$. The interval $(\alpha, \beta) \subset R$, $\alpha < 0 < \beta$ and the mapping*

$$P(\cdot, m_0) : (\alpha, \beta) \longrightarrow M, \quad t \longrightarrow P(t, m_0)$$

will be named an existence interval of trajectory and a trajectory of the impulsive flow P on M (defined by the point m_0) respectively, if:

1. *If τ^+ is the first moment when $T(\tau^+, m_0) \in \partial A$ then $P(t, m_0) = P(t, m_0)$ as $t \in [0, \tau^+]$ and $P(\tau^+ + 0, m_0) = \Phi(P(\tau^+, m_0))$.*

2. *If τ^- is the first moment when $T(\tau^-, m_0) \in \partial A$ then $P(t, m_0) = P(t, m_0)$ as $t \in (\tau^-, 0]$ and $P(\tau^-, m_0) = \Phi(P(\tau^- + 0, m_0))$.*

3. *The group equality*
$$P(t_1 + t_2, m_0) = P(t_2, P(t_1, m_0)),$$
holds for each two numbers $t_1 \in (\alpha, \beta)$, $t_2 \in R$ such that $t_1 + t_2 \in (\alpha, \beta)$. □

Maximal existence interval $(\omega^-(m_0), \omega^+(m_0))$ of the trajectory $P(\cdot, m_0)$ will be named the unique interval for which the next condition holds: there exist no extension of the trajectory $P(\cdot, m_0)$ on an interval containing $(\omega^-(m_0), \omega^+(m_0))$.

The mapping P resulting as the point $P(t, m_0)$ for each $m_0 \in M \setminus A$ and each number $t \in (\omega^-(m_0), \omega^+(m_0))$ will be called C^2-smooth impulsive flow on M generated by the compatible triple (T, A, Φ).

The flow P is said to be complete if for every point $m_0 \in M \setminus A$ we have $\omega^-(m_0) = -\infty$ and $\omega^+(m_0)) = \infty$.

Let us denote by $\mathcal{O}(m_0) = \{P(t, m_0) : t \in (\omega^-(m_0), \omega^+(m_0))\}$ the trajectory of P defined by m_0.

If $\mathcal{O}(m_0) = \{m_0\}$ then m_0 is called stationary point of impulse flow P. It is evident that $m_0 \in M \setminus A$ is stationary point of P if m_0 is a stationary point of T.

The trajectory $\mathcal{O}(m_0)$ is periodical one provided there exist $\omega \in R_+$ such that $P(t + \omega, m_0) = P(t, m_0)$ for all $t \in R$.

3 Preliminary notes

The next lemma proves that each point in $M \setminus A$ determines in a unique way maximal existence interval of a impulsive flow.

Lemma 1 *Let P be an impulsive flow generated by the compatible triple (T, A, Φ). Then:*

1. *Each point $m_0 \in M \setminus A$ determines a unique maximal existence interval $(\omega^-(m_0), \omega^+(m_0))$ of the trajectory $P(t, m_0)$.*

2. *If T is a complete flow on M then*

$$\omega^+(m_0) = \begin{cases} \lim\limits_{i \longrightarrow \infty} \tau_i^+, & p^+(m_0) = \infty, \\ +\infty, & p^+(m_0) < \infty; \end{cases}$$

$$\omega^-(m_0) = \begin{cases} \lim\limits_{i \longrightarrow \infty} \tau_i^-, & p^-(m_0) = \infty, \\ -\infty, & p^-(m_0) < \infty. \end{cases}$$

where $\{\tau_j^\pm = \tau_j^\pm(m_0) : j = 1, \ldots, p^\pm(m_0)\}$ are successive moments when the trajectory $P(t, m_0)$, $t \in R_\pm$ reaches ∂A, i.e. $P(\tau_j^\pm(m_0), m_0) \in \partial A$. □

Lemma 2 *Let P be an impulsive flow on M generated by the compatible triple (T, A, Φ), $m_0 \in M \setminus A$ and $\mathcal{O}(m_0) \cap \partial A \subseteq E_1(f, A) \cup E_2(f, A)$. Then for each number $i = 1, \dots, p^+(m_0)$ ($j = 1, \dots, p^-(m_0)$) there exist a neighborhood V of m_0 in $M \setminus A$ such that the mappings*

$$\tau_i^+ : V \longrightarrow R_+,\ m \longrightarrow \tau_i^+(m) \qquad (\tau_j^- : V \longrightarrow R_-,\ m \longrightarrow \tau_j^-(m))$$

are C^2-smooth. □

Definition 3 *1. The set $\mathcal{O}(m_0; \alpha, \beta) = \{P(t, m_0) : t \in (\alpha, \beta)\}$ will be called arc of the trajectory $\mathcal{O}(m_0)$ of P, $\omega^-(m_0) \leq \alpha \leq \beta \leq \omega^+(m_0)$.*

2. *The set $\mathcal{O}^*(m_0) = \cup\{\overline{\mathcal{O}(m_0; \alpha, \beta)} : (\alpha, \beta) \subseteq (\omega^-(m_0), \omega^+(m_0))\}$ will be called $\mathcal{O}^*$-trajectory of P defined by the point m_0.*

3. *The set $\mathcal{O}_*(m_0) = \mathcal{O}^*(m_0) \setminus \partial A$ will be called $\mathcal{O}_*$-trajectory of P defined by the point m_0.*

4. *The set $\Lambda \subset M$ will be called invariant set of P ($\mathcal{O}_*$-invariant set) if $\Lambda = \cup\{\mathcal{O}(m) : m \in \Lambda \setminus A\}$ ($\Lambda = \cup\{\mathcal{O}_*(m) : m \in \Lambda \setminus A\}$).*

5. *The set $\Lambda \subset M$ will be called $\mathcal{O}^*$-invariant set of P if $\Lambda = \cup\{\mathcal{O}^*(m) : m \in \Lambda \setminus A\}$ or the set $(M \setminus Int\ A) \setminus \Lambda$ is $\mathcal{O}^*$-invariant set of P.* □

Remark 3 *We need to introduce the notations of $\mathcal{O}^*$-invariant sets of impulsive flow because of the fact that a periodical trajectory is not a closed set in general.* □

Remark 4 *We introduce the separate definition of $\mathcal{O}^*$-invariant set of impulsive flow P because in general case the set $M \setminus Int\ A$ contents points from sets $E_\sigma(T, A)$, $\sigma \in \{3, \dots, 6\}$ and the impulsive flow P isn't defined in this sets. Clearly, the set $M \setminus Int\ A$ is an $\mathcal{O}^*$-invariant set of P.*

It is evident that any invariant set Λ of the impulsive flow P may be represented as $\Lambda = \Lambda_1 \cup \Lambda_2$ where Λ_1 is a closed set in $M \setminus A$, $\mathcal{O}_*$-invariant set of P, i.e. $\Lambda_1 = \Lambda \cap (M \setminus A)$, $\Lambda_2 = \Lambda \cap \partial A$. As far as the main goal of the present work is to characterize a class of invariant sets of P it follows that we need to study the $\mathcal{O}_*$-invariant sets of P only.

Lemma 3 *Assume Λ_1 and Λ_2 are $\mathcal{O}_*$-invariant sets of the impulsive flow P. Then the sets $\Lambda_1 \cup \Lambda_2$, $\Lambda_1 \cap \Lambda_2$, $\Lambda_1 \setminus \Lambda_2$, $Int\ \Lambda_1$ and $\partial \Lambda_1$ are $\mathcal{O}_*$-invariant sets of P too.* □

Definition 4 *The set Λ, closed in $M \setminus A$, will be named a minimal set of the impulsive flow P if:*

1. *Λ is a $\mathcal{O}_*$-invariant set of P.*

2. *Λ contains no a nonempty, own, closed in $M \setminus A$, $\mathcal{O}_*$-invariant set of P.* □

The last definition and Zorn's lemma lead immediately to the conclusion that every nonempty, compact and $\mathcal{O}_*$-invariant set of P contains a minimal set.

Lemma 4 *Suppose that Λ is a nonempty, compact and minimal set of the complete flow P, generated by the compatible triple (T, A, Φ). Then either:*

1. *Λ is nowhere nondense in $M \setminus A$ subset; or*

2. *Λ consists of faintly many connected components of $M \setminus A$.* □

Remark 5 *The proofs of the results, which was included in this subsection, was given in [2].* □

Let us recall that the invariant set Λ of the impulsive flow P approximates the point $a \in \Lambda$ if there exist $\delta \in R_+$ such that the equality $\overline{\Lambda \setminus \gamma} = \Lambda$ holds for the arc $\gamma = \{T(t,a) : t \in (-\delta, \delta)\} \cap B(a, \delta)$.

We impose an additional condition now:

($H2$) *Every nonempty, compact and minimal set Λ of P wich contains no a conected components of $M \setminus A$, does not approximate the points $\overline{\Lambda} \cap E_\sigma(T, A)$, $\sigma \in \{3, \ldots, 6\}$.*

Finally, we will note of two facts. Let P be an impulsive flow in R^2, generated by the compatible triple (T, A, Φ) and $E_\sigma(T, A) = \emptyset$ as $\sigma \in \{3, \ldots, 6\}$. Then:

If T is a complete flow then the flow P is complete too.

The factor space $M = R^2/(\Phi|\partial A)$ is a 2-dimensional smooth manifold and the impulsive flow P generates a smooth flow R on M module a diffeomorphism.

4 Main results

4.1 An analogue of the Schwarz theorem

The aim of the present subsection is to introduce the next result.

Theorem 1 *Assume the condition (H2) is hold and let Λ is a nonempty, compact and minimal set of the complete C^2-smooth flow P.*

Then either:

1. *Λ is a stationary point of P, or*

2. *Λ is a periodical trajectory of P, or*

3. *Λ consists of finite many connected components of $M \setminus A$.* □

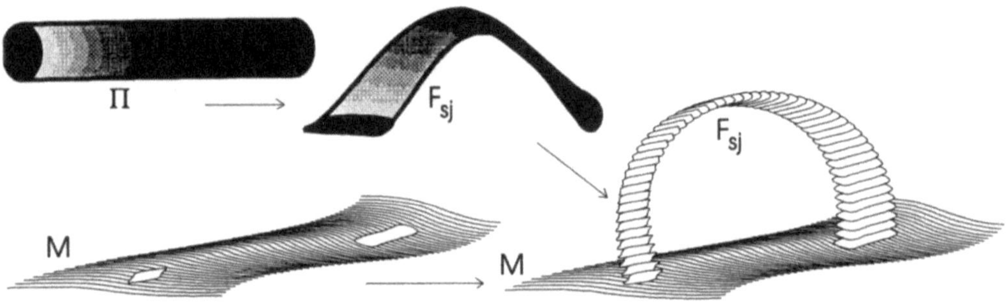

fig 3

The basic idea in the proof is to add handles to the phase manifold M connect two corresponding (under the impulsive mapping) components of the impulsive set A. We extended the impulsive flow P to C^2-smooth flow on the smooth manifold along the handles. Finally, the classical Schwarz' theorem is applied to the so-constructed flow (figure 3).

Remark 6 *The proof of theorem is given in [2].* □

4.2 Limit sets of impulsive flow

The next definition is an analogue of classical definition for limit set of dynamical systems.

Definition 5 *Let P be an impulsive flow in M, $m_0 \in M \backslash A$. The point $m \in M \backslash Int\ A$ will be called ω-limit (α-limit) point of the trajectory $\mathcal{O}(m_0)$ if there exist sequence $t_1 < t_2 < \ldots$ such that $\lim t_i = +\infty$ and $\lim P(t_i, m_0) = m$ ($\lim P(-t_i, m_0) = m$). The set of all ω-limit (α-limit) points of the trajectory $\mathcal{O}(m_0)$ will be denoted by $\omega(m_0)$ ($\alpha(m_0)$).* □

It is evident that $\omega(m_0)$ is $\mathcal{O}^*$-invariant set of P and $\omega(m_0)$ is not connected subset on $M \setminus Int\ A$, in general case.

Theorem 2 *Let the following condition hold:*

1. *$M/(\Phi|\partial A)$ is orientational manifold.*
2. *P is complete C^2-smooth flow on M generated by the compatible triple (T, A, Φ).*
3. *$m_0 \in M \setminus A$, $\omega(m_0)$ is a nonempty, compact set and $\omega(m_0)$ contains no a connected components of $M \setminus A$.*
4. *The condition (H2) hold.*

Then:

1. *There exist function $f : C \longrightarrow M \setminus Int\ A$ and points $\{c_1, c_2, \ldots, c_n\}$, $n \in N$ such that:*
   (a) *$\omega(m_0) = f(C)$.*
   (b) *$f|C \setminus \{c_1, c_2, \ldots, c_n\}$ is a continuous function.*
   (c) *$\lim_{c \longrightarrow c_i} f(c) \in \partial A$, $i \in \{1, \ldots, n\}$.*
2. *For every neighborhood U of $\omega(m_0)$ in $M \setminus Int\ A$ there exist $t_0 \in R_+$ such that $\mathcal{O}(m_0; t_0, \infty) \subset U$.* □

References

[1] D.D. Bainov and P. Simeonov, *Theory of differencial equations with impulse effects*, Ellias Horwood, 1989.

[2] D.D. Bainov and S. Nenov, *Impulsive dinamical systems. An analogue of the Poincaré - Bendixson theorem* (to appear).

[3] A. Denjoy, *Sur les courbes definees les equations differentialles a la suface du tore*, J. Math. Pures Appl. 11(9), 1932, pp. 333 - 373.

[4] A. Fomenko, *Differential geometry and topology. Supplementary shapters.* Moscow State Univ. 1983 (in Russian).

[5] F. Hartman, *Ordinary differential equations*, John Wiley & Sons, 1964.

[6] M. Hirsch, *Differential topology – Graduate texts in mathematics*, 33, Springer - Verlag, 1976.

[7] V. Lakshmikantham, D. D. Bainov and P. Simeonov, *Theory of impulsive differential equations*, World Scientific, 1989.

[8] V. Mil'man and A. Myshkis, *On the stability of motion in the presence of impulses*, Sib. Math. J. 1 (2), 1960, pp. 233 - 237 (in Russian).

[9] ———, *Random impulses in linear dynamical systems. Approx. methods for solving differential equations* Acad. Sci. of USSR Inst. M. preprint, 1963, pp. 64 - 81.

[10] J. Pallis and W. DeMelo, *Geometric theory of dynamical systems. An introduction*, Springer - Verlag, 1982.

[11] A. J. Schwarz, *A generalization of a Poincaré – Bendixson theorem to closed two-dimensional manifolds*, Amer. J. Math. 85, 1963, pp. 453 - 458.

[12] E. R. Van Kampen, *Remarks on systems of ordinary differential equations*, Amer. J. Math. 59, 1937, pp. 144 - 152.

4th Int. Coll. on Differential Equations, pp. 39-43
D. Bainov and V. Covachev (Eds)

Non-classical dynamic programming in optimal control.

S.A. Belbas
Mathematic Department, University of Alabama
Tuscaloosa, AL 35487-0350, U.S.A.
Abstract. We outline some aspects of dynamic programming for semi-linear Goursat systems.
Keywords. Dynamic programming, optimal control, Goursat systems.

1. Introduction.

Traditional dynamic programming has been associated with systems governed by one of the following: ordinary differential equations, functional differential equations, parabolic partial differential equations, and Cauchy problems for hyperbolic partial differential equations.

The ideas of dynamic programming can be applied, with appropriate adjustments, to other types of problems, notably to problems in which the "time" parameters do not have the physical meaning of time but rather correspond to appropriate parametrizations of the system dynamics. The present author has applied dynamic programming to two types of problems which traditionally had not been considered as falling within the scope of dynamic programming. These problems are: Goursat problems for hyperbolic systems, and nonlinear elliptic systems. In this paper we present an outline of some aspects of dynamic programming for semi–linear Goursat systems.

The problem of optimal control of Goursat systems has been treated in [1, 2] and in several works cited in [2], using methods of the type of Pontryagin's extremum principles, and in [3, 4] using dynamic programming. Goursat systems are used to describe physical situations involving exchange of heat or mass between a stationary medium and a moving medium; such models may be found in many books on the equations of mathematical physics.

The material of this paper follows our talk at this Colloquium, with one qualification: in this paper, we emphasize one aspect of dynamic programming that has not been presented in our previous publications.

2. Semilinear Goursat systems.

A semilinear controlled Goursat system has the form

$$\frac{\partial^2 y}{\partial s\, \partial t} = f(s,t,\, y(s,t),\, a(s,t)) \tag{2.1}$$

where $a(\cdot,\cdot)$ is the control function. This system is accompanied by boundary conditions along characteristics (or, in more generality, along families of curves that are transversal to the characteristics). The simplest case of boundary conditions has the form

$$y(\sigma,\, t) = x_2(t - \tau);\;\; y(s,\tau) = x_1(s - \sigma);$$

$$y(\sigma,\, t) = x_0 \qquad (\text{for } s \geq \sigma,\, t \geq \tau) \tag{2.2}$$

where $(\sigma,\, \tau)$ is an initial point on the st–plane. If we insist on continuous solutions for (2.1), we postulate the consistency condition $x_1(0) = x_2(0) = x_0$.

From now on, we restrict (s,t) to an interval $[0,A] \times [0,B]$. Following the terminology of multi–dimensional systems theory [5] and certain parts of theoretical physics [6], we shall call the pair (s,t) "two–dimensional time", even though, in applied problems, (s,t) may not have the physical dimensions of time.

An optimal control problem for (2.1–2) concerns the minimization of a cost functional J, given by

$$J = \int_0^A \int_0^B F(s,t,y(s,t),\, a(s,t))\, dtds + \int_0^A F_1(s,y(s,B),\, a(s,B))\, ds +$$

$$+ \int_0^B F_2(t,y(A,t),\, a(A,t)\, dt \; + \; F_0(y(A,B) \tag{2.3}$$

It is convenient to consider a parametrized functional K, as follows:

$$K = \int_{\sigma}^{A} \int_{\tau}^{B} F(s,t,y(s,t), a(s,t))\, dtds + \int_{\sigma}^{A} F_1(s,y(s,B), a(s,B))\, ds +$$

$$+ \int_{\tau}^{B} F_2(t,y(A,t), a(A,t))\, dt + F_0((y(A,B)) \tag{2.4}$$

In equation (2.4), y(s,t) depends on the initial two–dimensional time (σ, τ), the boundary data $\tilde{x} \equiv (x_0, x_1(\cdot), x_2(\cdot))$, and the control function $a(\cdot,\cdot)$. When we wish to emphasize these dependencies, we shall write $y(s,t;\ \sigma, \tau, \tilde{x}, \alpha)$ instead of merely y(s,t).

As explained in [3], the form (2.1) of a Goursat system is not suitable for the application of dynamic programming arguments, and an extended state $\tilde{y}$ must be introduced. Roughly, $\tilde{y} = (y_0, y_1(\cdot), y_2(\cdot))$, where y_0 is the value of y at any given two–dimensional time, and $y_1(\cdot)$, $y_2(\cdot)$ are the values of y along the two characteristics passing through any given two–dimensional time. The details of constructing $\tilde{y}$, and the relevant evolution equations, have been presented in [3] and will not be repeated here. We mention only that $\tilde{y}(s,t;\ \sigma, \tau, \tilde{x}, a)$ satisfies a Goursat equation with an infinite–dimensional state–space:

$$\frac{\partial^2\, \tilde{y}(s,t)}{\partial s\ \partial t} = \tilde{f}(s,t,\tilde{y}(s,t), \tilde{a}(s,t));\ \tilde{y}(\sigma, \tau) = \tilde{x} \tag{2.5}$$

where $\tilde{a}$ is an extension of the control function $a(\cdot,\cdot)$, defined in a way analogous to the extension $\tilde{y}$ of the state y. In [3], we have presented two types of dynamic programming for a controlled Goursat system: dynamic programming using one–dimensional variations along characteristics, and dynamic programming using two–dimensional "Stieltjes–like" variations. Here, we present another type of dynamic programming, using variations along an increasing path that intersects all characteristics transversally.

Consider a smooth curve γ, given by

$$s = h(\xi),\quad t = k(\xi),\quad 0 \le \xi \le C\,;$$

$$h(0) = k(0) = 0,\quad h(C) = A,\ k(C) = B \tag{2.6}$$

The value function on γ is

$$W(\xi, \tilde{x}) = \inf_{\tilde{a}(\cdot,\cdot)}\ K(\xi, \tilde{x}, \tilde{a}) \tag{2.7}$$

where $K(\xi, \tilde{x}, \tilde{a})$ represents the functional K of (2.4), calculated with $\sigma = h(\xi)$, $\tau = k(\xi)$, $\tilde{y}(h(\xi), k(\xi)) = \tilde{x}$, and extended control function $\tilde{a}$; the infimum in (2.7) is taken over an appropriate set of admissible extended controls.

If we consider two points Q, $\hat{Q}$ on γ, with $Q = (h(\xi), k(\xi))$ and $\hat{Q} = (h(\xi) + h'(\xi)\delta\xi + o(\delta\xi), k(\xi) + k'(\xi)\delta\xi + o(\delta\xi))$, with $\delta\xi > 0$, then

$$\tilde{y}(\hat{Q}) - \tilde{y}(Q) = \left\{h'(\xi)\frac{\partial\tilde{y}}{\partial s} + k'(\xi)\frac{\partial\tilde{y}}{\partial t}\right\}\delta\xi + o(\delta\xi) \tag{2.8}$$

The derivatives $\frac{\partial\tilde{y}}{\partial t}, \frac{\partial\tilde{y}}{\partial t}$ can be expressed in terms of the operators P_1, P_2 introduced in [3]. If a_0, $a_1(\cdot)$, $a_2(\cdot)$ are the components of an extended control $\tilde{a}$, then

$$\frac{\partial\tilde{y}(\sigma, \tau)}{\partial s} = P_2^{\sigma,\tau}(\tilde{x}, a_2(\cdot))\ ; \frac{\partial\tilde{y}(\sigma, \tau)}{\partial t} = P_1^{\sigma,\tau}(\tilde{x}, a_1(\cdot) \tag{2.9}$$

We set

$$P_\gamma^\xi(\tilde{x}, \tilde{a}) = h'(\xi)\, P_2^{\sigma,\tau}(\tilde{x}, a_2(\cdot)) + k'(\xi)\, P_1^{\sigma,\tau}(\tilde{x}, a_1(\cdot))\Big|_{\substack{\sigma = h(\xi)\\ \tau = k(\xi)}} \tag{2.10}$$

Then we have

$$W(\xi + \delta\xi, \tilde{y}(\hat{Q}) - W(\xi, \tilde{x}) = \frac{\partial W(\xi, \tilde{x})}{\partial\xi} + \left\langle\frac{\partial W(\xi, \tilde{x})}{\partial\tilde{x}}, P_\tau^\xi(\tilde{x}, \tilde{a})\right\rangle$$

(2.11) Now, turning to the integral terms in (2.4), we may calculate the derivative $\frac{\partial K}{\partial\xi}$ as follows:

$$\frac{\partial K}{\partial\xi} = -h'(\xi)\int_\tau^B F(\tau, t, y(\sigma, t), a(\sigma, t))dt - k'(\xi)\int_\sigma^A F(s, \tau, y(s, \tau), a(s, \tau))ds -$$

$$-h'(\xi)\, F_1(\sigma, y(\sigma, B), a(\sigma, B)) - k'(\xi)\, F_2(\tau, y(A, \tau), a(A, \tau))\Big|_{\substack{\sigma = h(\xi)\\ \tau = k(\xi)}} =$$

$$= -h'(\xi)\int_\tau^B F(\sigma, t, x_1(t-\tau), a_1(t-\tau))dt - k'(\xi)\int_\sigma^A F(s, \tau, x_2(s-\sigma),$$

$$a_2(s-\sigma))ds - h'(\xi)\, F_1(\sigma, x_1(B-\tau), a_1(B-\tau)) - k'(\xi)\, F_2(\tau, x_2(A-\sigma),$$

$$a_2(A-\sigma))\Big|_{\substack{\sigma = h(\xi)\\ \tau = k(\xi)}} \equiv -\Phi_\gamma(\xi, \tilde{x}, \tilde{a}) \tag{2.12}$$

Given the calculations of (2.11 – 12), a dynamic programming argument applied to the variation of K from Q to $\hat{Q}$ yields the following optimality condition:

$$\inf_{\tilde{a}}\left\{\frac{\partial W(\xi, \tilde{x})}{\partial\xi} + \left\langle\frac{\partial W(\xi, \tilde{x})}{\partial\tilde{x}}, P_\gamma^\xi(\tilde{x}, \tilde{a})\right\rangle - \Phi_\gamma(\xi, \tilde{x}, \tilde{a})\right\} = 0 \tag{2.13}$$

Boundary conditions at the "final" two–dimensional time (A, B) may be obtained from (2.4):

$$W(C, \tilde{x}) = F_0(x_0) \qquad (2.14)$$

We end this section with three remarks. First, the equations of dynamic programming over an increasing path can be extended to include functionals with discount factors and non–local terms. (The appropriate forms of discount factors for controlled Goursat systems have been established in [3].) Second, the equations (2.13 – 14) can be solved approximately using finite–difference approximations; the corresponding work for dynamic programming along characteristics has been carried out in [4]. Third, the equations of dynamic programming deal primarily with the (usually, approximate) calculation of the value function; the actual synthesis of optimal controls will, in general, require an extended version of the Goursat system. One such extension has been studied in [7].

References.

1. A.I. EGOROV, On optimal control in certain systems with distributed parameters (in Russian), Prikl. Mat. i Mekh., No. 4, 688–696 (1963).

2. V.O. VASILYEV, V.A. SROCHKO, V.I. TERLETSKII, Methods of optimization and their applications, part 2: optimal control (in Russian), Nauka, Novosibirk (1990).

3. S.A. BELBAS, Dynamic programming approach to the optimal control of systems governed by Goursat–Darboux equations, Int. J. Control, Vol. 51, 1279–1294 (1990).

4. S.A. Belbas, Dynamic programming and maximum principle for discrete Goursat systems, J. Math. Anal. Appl., Vol. 161, 57–77 (1991).

5. N.K. BOSE, Applied multi–dimensional systems theory, Van Nostrand Reinhold, New York (1982).

6. N. KALITZIN, Multitemporal theory of relativity, Bulgarian Academy of Sciences, Sofia (1975).

7. A.N. VITYUK, On R–solutions, associated with hyperbolic differential inclussions (in Russian), Abstracts of this Colloquium, p. 263, Plovdiv (1993).

4th Int. Coll. on Differential Equations, pp. 45-53
D. Bainov and V. Covachev (Eds)

Painlevé analysis of nonlinear reaction-diffusion

GUO BEN-YU
Shanghai University of Science and Technology, Shanghai, China

Abstract. Painlevé analysis becomes a powerful tool for studying nonlinear reaction-diffusion systems. It is used for finding periodic solutions, travelling waves and the expressions of the coalescence between waves.

Keywords: Nonlinear reaction-diffusion, Painlevé analysis.

1. INTRODUCTION

Many phenomena in biology, chemistry and other fields can be described by various nonlinear reaction-diffusion systems, see [1-2]. People looked for their solutions by using many precise techniques. But most of them are not constructive. On the other hand, many attractive techniques were proposed in studying integrable systems, see [3-4]. As we know, it is not easy to generalize them to nonintegrable systems. But Painlevé analysis has been used successfully for nonlinear reaction-diffusion systems, see [5-8]. Especially Guo Ben-yu and Chen Zhi-xiong obtained periodic solutions, travelling waves and coalescence between waves, and studied chaotic behaviors of several systems via this approach, see [9-12].

2. PAINLEVÉ ANALYSIS AND ARS TEST

In 1887, Picard [13] considered the equation

$$\frac{d^2w}{dz^2} = F(z, w, \frac{dw}{dz})$$

where z is a complex variable, F is analytic in z, algebraic in w and rational in $\frac{dw}{dz}$. The question is whether or not all branch points and essentially singular points of its solutions are fixed. This problem led to the begining of singularly analysis, see [14-15]. A singular point is said to be movable, if its position depends on integral constant.

Definition 1. (Painlevé Property for Ordinary Differential Equations.) An ordinary differential equation (or a system of equations) in the complex domain is said to have the Painlevé property if the only movable singular points that its solutions can exhibit, are poles.

Clearly, Picard problem is connected with the question whether or not a second-order ordinary differential equation has the Painlevé property. Many theoretical results can be found in [16-18].

By the end of the seventies, Ablowitz, Ramani, Segur and other authors studied nonlinear evolution equations which can be solved by the inverse scatering transformation (IST). It was found that the resulting equations reduced by similarity transformation have Painlevé property see [19]. Therefore they guessed that it is valid for all equations solvable by IST. This is refered to as ARS conjecture. There are two parts in ARS method. The first is the group reduction. The another is the judgement on Painlevé property of the resulting equations. To do this, we can use α-method by Painlevé (see [20]), or ARS test (see [19]). But both of them are only necessary conditions for Painlevé property. ARS test includes three steps. For the detail, we consider the equation

$$F(z, w, \frac{dw}{dz}, \cdots, \frac{d^n w}{dz^n}) = 0, \tag{1}$$

where F is analytic in z, and rational in $\frac{d^j w}{dz^j}, 0 \leq j \leq n$. Let the solution be the form of Laurent series

$$w = \sum_{j=0}^{\infty} a_j (z - z_0)^{j+p}$$

where z_0 is a pole depending on integral constant, and p is an undetermined negative integer. By substituting it into (1) and comparing the coefficients of terms $(z - z_0)^{j+p}$, we get a series of recursive equations for a_j. The values of p and a_0 can be evaluated by letting $j = 0$. But p may have several possible values as well as a_0. If p is not an integer, then the solution of (1) has algebraic branch point and thus (1) has no Painlevé property. If p is an integer but not negative, then the solution either has essential singularity or has no singularity. Finally if p is really a negative integer, then we turn to the next step. In this case, we have the recursive formulas as follows

$$Q(j)a_j = R_j(z_0, a_0, \cdots, a_{j-1}), \quad j \geq 1,$$

where $Q(j)$ is a polynomial, R_j is rational in all variables. The roots of $Q(j)$ are so called resonance points. Furthermore if there exists a non-negative integer j_0 such that $Q(j_0) = 0$ and $R_{j0} \neq 0$, then it is contradictory and so (1) has no Painlevé property. If $Q(j_0) = 0$ and $R_{j_0} = 0$, then we obtain an identity and so a_{j_0} is arbitrary, which is an integral constant. By Cauchy-Kowalevskaya Theorem, if the number of integral constants is less than n, then the above series is not general solution of (1). Therefore it has no Painlevé property.

It is believed that there is a deep connection between the integrability and Painlevé property. Certainly it would be better to extend ARS method to partial differential equation (PDE) directly.

Definition 2. (see [21]). Let S be an analytic and non-characteristic complex hypersurface in the n-demensional complex space C^n. If all solutions of a PDE or a system of equations are analytic on C^n/S and meromorphic on C^n, then we say that it has Painlevé property.

The above definition is a direct generalization of Definition 1. But it is hard to be used.

Definition 3 (see [22]). A PDE or a system of equations defined on C^n is said to have Painlevé property, if the solutions are single-valued about the movable noncharacteristic singularity manifold.

We now explain the above definition. Let $\phi = 0$ be the movable, noncharacterstic (i.e., $\phi_x \phi_t \neq 0$) singular manifold of solutions. Assume that

$$U = \frac{1}{\phi^p} \sum_{j=0}^{\infty} U_j \phi^j \tag{2}$$

where ϕ and U_j are analytic functions in a neighborhood of the manifold $\phi = 0$. Putting the above expansion (2) into the equation and analyzing the leading part, we get the value of p and a series of recursive relations for U_j. We say that this equation or a system of equations has Painlevé property, if the following three conditions are satisfield:

(i) p is a positive integer,

(ii) the recursive relations are consistent for all U_j,

(iii) there are enough free functions in the sense of Cauchy-Kowalevskaya Theorem.

Clearly Definition 3 is a natural generalization of ARS method. Weiss, Tabor and Carnevale guessed that every equation with complete integrability has Painlevé property in this sense. It is refered to as WTC conjecture (see [23]). But Clarkson [24] found that WTC conjecture is neither sufficient nor necessary. Nevertheless, it is still powerful as a detector for the integrability. On the other hand, the procedure of this method brings many equations which often lead to the Bäcklund transformation, Lax pairs, symmetry operators, bilinear forms and rational solutions.

3. EXPLICIT SOLUTIONS OF NONLINEAR REACTION-DIFFUSION EQUATION

WTC method is also a powerful tool for constructing explicit solutions of dynamical systems, no matter they are integrable or not. Especially, it has been applied successfully to nonlinear reaction-diffusion systems. To show this technique, let (see [25])

$$S = \frac{\phi_{xxx}}{\phi_x} - \frac{3\phi_{xx}^2}{2\phi_x^2}, \quad C = -\frac{\phi_t}{\phi_x}, \quad Q = -\frac{\phi_{xx}}{2\phi_x}.$$

S is called Schwartzian derivative, while C has the dimension of velocity.

We now consider Fisher equation in biology (see [2])

$$U_t = U_{xx} + aU(1-U), \quad a > 0. \tag{3}$$

For convenience of analysis, let $u = \frac{1}{6}U$ and $\tau = 5t$. Then (3) changes into

$$5u_\tau = u_{xx} + au(1-6u). \tag{4}$$

By Painlevé analysis, we have $p = 2, u_0 = \frac{1}{a}\phi_x^2$, and

$$\begin{aligned}(j-6)(j+1)u_j\phi_x^2 &= 5(u_{j-2,\tau} + (j-3)u_{j-1}\phi_\tau) - u_{j-2,xx} \\ &-2(j-3)u_{j-1,x}\phi_x - (j-3)u_{j-1}\phi_{xx} - au_{j-2} + 6a\sum_{k=1}^{j-1} u_{j-k}u_k, \quad \forall j \geq 2.\end{aligned} \tag{5}$$

Clearly, the resonance points are $j = -1$ and $j = 6$. For $j = 1$, we obtain

$$u_1 = \frac{1}{a}(\phi_\tau - \phi_{xx}).$$

By Kruskal prescription (see [22]), we know that (5) is inconsistent at the resonance point $j = 6$ and so (3) has no Painlevé property. But this failure does not stop us to derive interesting results. Since $p = 2$, we set

$$u = \frac{u_0}{\phi^2} + \frac{u_1}{\phi} + u_2. \tag{6}$$

Therefore, (5) leads to

$$u_2 = \frac{1}{12} + \frac{1}{a}[-\frac{1}{4}(\frac{\phi_{xx}}{\phi_x})^2 + \frac{1}{3}(\frac{\phi_{xxx}}{\phi_x}) - \frac{\phi_{x\tau}}{\phi_x} + \frac{1}{2}\frac{\phi_{xx}\phi_\tau}{\phi_x^2} - \frac{1}{12}(\frac{\phi_\tau}{\phi_x})^2], \quad (7)$$

$$\frac{\phi_{xx} - \phi_\tau}{\phi_x}u_2 = \frac{1}{12}(\frac{\phi_{xx} - \phi_\tau}{\phi_x}) + \frac{1}{a}[-\frac{1}{2}\frac{\phi_{xx\tau}}{\phi_x} + \frac{5}{12}\frac{\phi_{\tau\tau}}{\phi_x} + \frac{1}{12}\frac{\phi_{xxxx}}{\phi_x}] \quad (8)$$

and

$$5u_{2,\tau} = u_{2,xx} + au_2(1 - 6u_2). \quad (9)$$

The last equation is the same as (4). Thus (6) is the auto-Bäcklund transformation. We are going to solve (7)-(9). By computation, we get

$$u_2 = \frac{1}{12} + \frac{1}{a}(C_x - \frac{1}{12}C^2 + \frac{1}{3}S - CQ + Q^2),$$
$$5C_\tau = C_x - 7CC_x + 6C_{xx} + 2SC + C^3$$

and

$$80C_{xxx} - 248CC_{xx} + 16S_{xx} - 8S_xC + 40SC_x + 76C^2C_x - 20C_x^2$$
$$= 76SC^2 + 3C^4 - 4S^2 + a^2.$$

We now focus on the simplest case. It means that C and S are constants. Let $\varepsilon = \pm 1$ and $a = 6b^2$. We find the following four cases:

(i) $S = -\frac{b^2}{2}, C = \varepsilon b$,
(ii) $S = \frac{b^2}{2}, C = \varepsilon ib$,
(iii) $S = -3b^2, C = 0$,
(iv) $S = 3b^2, C = 0$.

The first case corresponds to

$$\phi = \frac{\overline{A}\exp(\frac{b}{2}\xi_1) + \overline{B}\exp(-\frac{b}{2}\xi_1)}{A\exp(\frac{b}{2}\xi_1) + B\exp(-\frac{b}{2}\xi_1)}, \quad \xi_1 = x - \varepsilon b\tau$$

where $A, B, \overline{A}$ and $\overline{B}$ are arbitrary constants provided $\overline{A}B - A\overline{B} = \frac{1}{b}$. Finally

$$U_2 = \frac{1 - \varepsilon}{2} + \frac{\varepsilon}{1 + \frac{A}{B}\exp(b\eta_1)} - \frac{AB}{(A\exp(\frac{b}{2}\eta_1) + B\exp(-\frac{b}{2}\eta_1))^2}$$

with $\eta_1 = x - 5\varepsilon bt$. It can be checked that for $\varepsilon = 1, U_2 \to 1$ as $x \to -\infty$ and $U_2 \to 0$ as $x \to +\infty$. For $\varepsilon = -1$, we get an x-reversed solution. Thus both of them are heteroclinic solutions. It is interesting that when a goes from 0 to $+\infty$, the wave speed takes all postive numbers for $\varepsilon = 1$ and all negative numbers for $\varepsilon = -1$. In the second case, we get a complex solution which tends to $\frac{1+\varepsilon}{2}$ oscillately, as $t \to +\infty$. On the other hand, the third and the fourth cases lead to homoclinic solutions including periodic solutions, see [10]. We can construct new solutions by the auto-Bäcklund transformation (6). But we get the same solutions in turn. Indeed, this is one of essential features of nonintegrable systems.

4. THE COALESCENCE BETWEEN WAVES

Painlevé analysis provides also other kinds of solutions. We now consider the Nagumo equation (see [26])

$$U_t = U_{xx} + U(U-a)(1-U) \tag{10}$$

where a is a constant and $-1 \le a < 1$. By an argument as before, we have $p=1, U_0 = \sqrt{2}\varepsilon\phi_x, \varepsilon = \pm 1$ and

$$U_1 = \frac{a+1}{3} + \frac{\sqrt{2}}{2}\varepsilon\left(\frac{\phi_t}{3\phi_x} - \frac{\phi_{xx}}{\phi_x}\right).$$

(10) has no Painlevé property. But we still set

$$U = \frac{\phi_0}{\phi} + U_1 - \sqrt{2}\varepsilon(\ln\phi)_x + U_1. \tag{11}$$

Then

$$\frac{\phi_{xt}}{\phi_x} = \frac{\phi_{xxx}}{\phi_x} - \frac{3}{2}\left(\frac{\phi_t}{3\phi_x} - \frac{\phi_{xx}}{\phi_x}\right)^2 + \frac{1}{3}(a^2 - a + 1)$$

and

$$U_{1,t} = U_{1,xx} + U_1(U_1 - a)(1 - U_1). \tag{12}$$

By (12), (11) is the auto-Bäcklund transformation.

We now look for ϕ. It can be verified that

$$4C_{xx} - 4CC_x + \frac{4}{9}C^3 - \frac{2}{3}(a^2 - a + 1)C - \frac{\sqrt{2}\varepsilon}{9}(a+1)(a-2)(2a-1) = 0.$$

By Cole-Hopf transformation $C = -\frac{3f_x}{f}$, we obtain

$$f_{xxx} - \frac{1}{6}(a^2 - a + 1)f_x + \frac{\sqrt{2}\varepsilon}{108}(a+1)(a-2)(2a-1)f = 0$$

and so

$$C = \frac{\sqrt{2}\varepsilon}{2}\frac{\alpha(2a-1)e^{\xi_1} - \beta(a-2)e^{\xi_2} - \gamma(a+1)e^{\xi_3}}{\alpha e^{\xi_1} + \beta e^{\xi_2} + \gamma e^{\xi_3}}$$

where γ, β and γ are arbitrary constants, and

$$\begin{aligned}
\xi_1 &= -\tfrac{\sqrt{2}\varepsilon}{6}(2a-1)x - \tfrac{(2a-1)^2}{6}t,\\
\xi_2 &= \tfrac{\sqrt{2}\varepsilon}{6}(a-2)x - \tfrac{(a-2)^2}{6}t,\\
\xi_3 &= \tfrac{\sqrt{2}\varepsilon}{6}(a+1)x - \tfrac{(a+1)^2}{6}t.
\end{aligned}$$

Obviously, if $\tilde{\phi}$ is a special solution of the equation $\phi_t + C\phi_x = 0$, then for any function $\Phi, \Phi(\tilde{\phi})$ is also a solution. Since $S(\Phi(\tilde{\phi})) = S(\tilde{\phi})$, we get

$$\Phi(\tilde{\phi}) = \frac{A + B\tilde{\phi}}{m + n\tilde{\phi}},$$

where A, B, m and n are arbitrary constants, but $An \neq Bm$. Finally we obtain

$$U = \frac{a+1}{3} - \frac{\sqrt{2}\varepsilon}{6}C - \frac{\sqrt{2}\varepsilon}{2}\frac{\tilde{\phi}_{xx}}{\tilde{\phi}_x} + \frac{\sqrt{2}B\varepsilon\tilde{\phi}_x}{A + B\tilde{\phi}}.$$

We now consider some special cases. Firstly, if $\alpha \neq 0$, and $\beta = \gamma = 0$, then

$$C = C_1 = \frac{\sqrt{2}\varepsilon}{2}(2a-1).$$

If in addition

$$\tilde{\phi} = \exp\left(\frac{\sqrt{2}\varepsilon}{2}(x - C_1 t)\right),$$

then we get

$$U = \frac{B}{B + A\exp\left(-\frac{\sqrt{2}\varepsilon}{2}(x - C_1 t)\right)}.$$

For $\varepsilon = 1, U \to 0$ as $x \to -\infty$ and $U \to 1$ as $x \to +\infty$. Thus it is a heteroclinic solution. For $\varepsilon = -1$, we get an x-reversed solution. Similarly, we get other heteroclinic solutions with the speed C_2 or C_3,

$$C_2 = -\frac{\sqrt{2}\varepsilon}{2}(a-2), \quad C_3 = \frac{\sqrt{2}\varepsilon}{2}(a-1).$$

We next consider the case with $\alpha = 0$ and $\beta\gamma \neq 0$. Then

$$C = -\frac{\sqrt{2}\varepsilon}{2}\frac{(a-2)\beta e^{\eta_2} + (a+1)\gamma e^{\eta_3}}{\beta e^{\eta_2} + \gamma e^{\eta_3}}$$

with

$$\eta_2 = -\tfrac{1}{2}[\sqrt{2}\varepsilon a x + (a^2 - 2a)t],$$
$$\eta_3 = -\tfrac{1}{2}[\sqrt{2}\varepsilon(a-1)x + (a^2 - 1)t].$$

If we take

$$\tilde{\phi} = \frac{\beta}{a}e^{\eta_2} + \frac{\gamma}{a-1}e^{\eta_3},$$

then

$$U = \frac{Aa^2(a-1) + Ba\gamma e^{\eta_3}}{Aa(a-1) + B\beta(a-1)e^{\eta_2} + B\gamma a e^{\eta_3}}.$$

This solution expresses the coalescence of two waves. In particular, if $0 < a < 1$, then this solution goes from 0(or 1) to 1(or 0) as x increases from $-\infty$ to $+\infty$ for $\varepsilon = 1$(or -1). Similarly, if $\gamma = 0$ and $\alpha\beta \neq 0$, then we get another solution.

Now, let $a = -1$. Then (10) becomes Newell-Whitehead (or KPP) equation. We have

$$C = \frac{3\sqrt{2}\varepsilon}{2}\frac{-\alpha e^{\frac{\sqrt{2}}{2}\varepsilon x} + \beta e^{-\frac{\sqrt{2}}{2}\varepsilon x}}{\alpha e^{\frac{\sqrt{2}}{2}\varepsilon x} - \beta e^{-\frac{\sqrt{2}}{2}\varepsilon x} + \gamma e^{\frac{3}{2}t}}.$$

It is interesting that $C \to 0$ and U tends a steady state as $t \to +\infty$. By choosing $\tilde{\phi}$ suitably, we get the solution as in [8].

Finally, (10) becomes Huxley equation for $a = 0$. We get

$$C = \frac{\sqrt{2}\varepsilon}{2}\frac{2\beta - (\alpha+\gamma)\exp(\frac{\sqrt{2}}{2}\varepsilon x + \frac{t}{2})}{\beta + (\alpha+\gamma)\exp(\frac{\sqrt{2}}{2}\varepsilon x + \frac{t}{2})}$$

and

$$U=\frac{B\beta+B(\alpha+\gamma)\exp(\frac{\sqrt{2}}{2}\varepsilon x+\frac{t}{2})}{A+\frac{1}{2}B\beta(\sqrt{2}\varepsilon x-2t)+B(\alpha+\gamma)\exp(\frac{\sqrt{2}}{2}\varepsilon x+\frac{t}{2})}.$$

Obviously $C\to -1$ as $t\to+\infty$. Besides, it also provides the special solutions of some porous medium problems (see [11]).

5. THE SYSTEM OF NONLINEAR REACTION-DIFFUSION

Painlevé approach could be generalized to systems of nonlinear reaction-diffusion. As an example, we consider the Noeys-Field system for Belousov-Zhabotinskii reaction

$$\begin{cases} U_t = DU_{xx}+U(1-U-rV), \\ V_t = V_{xx}-bUV \end{cases} \tag{13}$$

where D, r and b are positive constants. This system has no Painlevé property. Now, let

$$U=\sum_{j=0}^{\infty}U_j\phi^{j-p},\quad V=\sum_{j=0}^{\infty}V_j\phi^{j-q}.$$

We find that $p=2$ and $q=1$ or 2.

We first take $p=q=2$ and $bD\neq 3$. Let

$$U=\frac{6}{b}u,\quad V=\frac{6(bD-1)}{rb}v,\quad t=\frac{1}{5(3-bD)}\tau.$$

Then

$$\begin{cases} 5(3-bD)u_\tau = Du_{xx}+u(1-\frac{6}{b}u-\frac{6}{b}(bD-1)v), \\ 5(3-bD)v_\tau = v_{xx}-6uv. \end{cases} \tag{14}$$

Set

$$u=\frac{u_0}{\phi^2}+\frac{u_1}{\phi}+u_2,\quad v=\frac{v_0}{\phi^2}+\frac{v_1}{\phi}+v_2.$$

We have from (14) that

$$\begin{cases} u_0=v_0=\phi_x^2, \\ u_1=-\phi_{xx}-(3bD-2b-3)\phi_\tau, \\ v_1=-\phi_{xx}+(2bD-3b+3)\phi_\tau, \end{cases}$$

The auto-Bäcklund transformation is

$$\begin{cases} u=[-\frac{\partial^2}{\partial x^2}-(3bD-2b-3)\frac{\partial}{\partial\tau}]\ln\phi+u_2, \\ v=[-\frac{\partial^2}{\partial x^2}+(2bD-3b+3)\frac{\partial}{\partial\tau}]\ln\phi+v_2. \end{cases}$$

If C and S are constants, then we get

$$C^2=-\frac{b}{2(3bD-2b-3)(2b^2D^2+2b^2D-21bD+6b+9)}$$

and

$$2b(2bD-3)(bD+b-6)(bD^2+bD+9D-15)=0$$

which implies the three consistent conditions for the existence of travelling waves. In the first case, $b = \frac{3}{2D}$. We obtain that for $D \in (0, \frac{3}{4}) \bigcup (2, +\infty)$,

$$C_1 = \frac{\varepsilon}{3}\sqrt{\frac{D}{(2-D)(3-4D)}}.$$

If we take $\varepsilon = -1$, then

$$\begin{cases} U_2 = \frac{A^2(2-D)e^{2k_1\eta_1} + AB(3D-1)}{(2-D)(Ae^{k_1\eta_1} + Be^{-k_1\eta_1})^2}, \\ V_2 = \frac{B^2(3-4D)e^{-2k_1\eta_1}}{2r(2-D)(Ae^{k_1\eta_1} + Be^{-k_1\eta_1})^2} \end{cases}$$

where A, B are arbitrary constants,

$$k_1 = \frac{1}{4}\sqrt{\frac{3-4D}{D(2-D)}}, \quad \eta_1 = x - \frac{5\varepsilon}{2}\sqrt{\frac{D}{(2-D)(3-4D)}}t.$$

In particular, if $r = \frac{4D-3}{2(D-2)}, \varepsilon = -1$ and $A = B = 1$, then $U_2 \to 0, V_2 \to 1$ as $x \to -\infty$, whereas $U_2 \to 1, V_2 \to 0$ as $x \to +\infty$ or $t \to +\infty$. In the other two cases, $b = \frac{6}{D+1}$ or $b = \frac{3(5-3D)}{D(D+1)}$. We can derive travelling waves with the speeds C_2 and C_3 respectively (see [12]).

We now turn to the second Painlevé expansion with $p = 2$ and $q = 1$. It can be verified that the corresponding recursive equations are consistent if and only if $bD = \frac{1}{3}$. Thus let $b = \frac{1}{3D}$ and

$$U = \frac{U_0}{\phi^2} + \frac{U_1}{\phi} + U_2, \quad V = \frac{V_0}{\phi} + V_1.$$

We know that $U_0 = 6D\phi_x^2$, and V_0 is arbitrary. For simplicity, let $V_0 = \frac{10mD\phi_x}{r}$, m being a constant, and look for travelling waves. Let C and S are constants. Then we get

$$[C + m][(15D-6)C^2 + m(15D-12)C - 6m^2 + \frac{1}{2D}] = 0. \tag{15}$$

Clearly $C_4 = -m$ is one of the roots of (15). Furthermore, we get that for $|m| > \frac{1}{5D}$,

$$\begin{cases} U_2 = \frac{6AB(1-25m^2D^2)}{(Ae^{k_4\eta_4} + Be^{-k_4\eta_4})^2}, \\ V_1 = \frac{25D^2m^2}{r} + \frac{10mDk_4(Ae^{k_4\eta_4} - Be^{-k_4\eta_4})}{r(Ae^{k_4\eta_4} + Be^{-k_4\eta_4})} \end{cases}$$

with

$$k_4 = \frac{1}{2}\sqrt{\frac{25m^2D^2 - 1}{D}}, \quad \eta_4 = x + 5Dmt.$$

We point out that the previous statements show that the system (13)with $bD = \frac{1}{3}$ possesses the travelling waves with any speed $m, |m| > \frac{1}{5D}$. It means the chaotic behavior in some sense. On the other hand, for $|m| < \frac{1}{5D}$ and $A = B$, we get double periodic solutions describing the oscillations. We can derive other solutions related to the other two roots of (15), see [12].

References

1. J. Smoller, Shock Waves and Reaction-Diffusion Equations, Springer-Verlag, Berlin (1980).

2. J. D. Murray, Mathematical Biology, Springer-Verlag, Berlin (1989).
3. M. J. Ablowitz and H. Segur, Solitons and the Inverse Scattering Transform, SIAM, Philadelphia (1981).
4. R. K. Dodd, J. C. Eilbeck, J. D. Gibbon and H.C. Morris, Solitions and Nonlinear-Wave Equations, Academic Press, London (1982).
5. M. J. Ablowitz and A. Zeppetella, Bull. Math. Biol., 41, 835 (1979).
6. R. Conte, Phys. Lett. A., 134, 100 (1988).
7. R. Conte and M. Musette, J. Phys. A:Math. Gen., 22, 169 (1989).
8. F. Cariello and M. Tabor, Physica D, 39, 77 (1989).
9. Chen Zhi-xiong and Guo Ben-yu, J. Phys. A., 22, 5187 (1989).
10. Guo Ben-yu and Chen Zhi-xiong, J. Phys. A., 24, 645 (1991).
11. Chen Zhi-xiong and Guo Ben-yu, IMA J. Appl. Math., 48, 107 (1992).
12. Guo Ben-yu and Chen Zhi-xiong, Analytic solutions of Noyes-Field system for Belousov-Zhabotinskii reaction (unpublished).
13. E. Picard, C. R. Acad. Sc. Paris, 104, 41 (1887).
14. P. Painlevé, C. R. Acad. Sc. Paris, 116, 88 (1893).
15. B. Gambier, Acta Math., 33, 1 (1909).
16. W-H. Steeb and N. Euler, Nonlinear Evolution Equations and Painlevé Test, world Scientific, Singapore (1988).
17. J. Chazy, Acta Math., 34, 317 (1911).
18. S. Kowalevskaya, in: A Russian Childhood, B. Stillman, (Ed.), Springer-Verlag, New York (1987).
19. M. J. Ablowitz, A. Ramani and H. Segur, J. Math. Phys., 21, 715 (1980).
20. E. L. Ince, Ordinary Differential Equations, Dover, New York (1956).
21. R. S. Ward, Phys. Lett. A, 102, 279 (1984).
22. J. Weiss, M. Tabor and G. Carnevale, J. Math. Phys., 24, 522 (1983).
23. J. Weiss, Phys. Lett. A, 102, 329 (1984).
24. P. A. Clarkson, Physica D, 18, 209 (1986).
25. R. Conte, in:Inverse Problems in Action, P. C. Sabatier, (Ed.), Springer-Verlag, Berlin (1990).
26. J. Nagumo, S. Arimoto and S. Yoshizawa, Proc. IRE., 50, 2061 (1962).

4th Int. Coll. on Differential Equations, pp. 55-70
D. Bainov and V. Covachev (Eds)

Weighted interpolation and Hardy inequalities with some spectral theoretic applications

R. C. Brown[1] and D. B. Hinton[2]
[1]Department of Mathmatics University of Alabama Tuscaloosa, AL 35487-0350, USA
[2]Department of Mathematics University of Tennessee Knoxville, TN 37996-1300, USA

Abstract - In this paper we survey some recent results concerning interpolation inequalities involving derivatives of sum and product form in weighted Lebesgue spaces as well as applications to the spectral theory of ordinary differential operators. The weighted Hardy-type inequality is also discussed.

Key-Words: Weighted normed inequalities; Hardy inequality; spectral theory.

1. INTRODUCTION

We are interested in the inequalities

$$\int_I w|u^{(j)}|^q \le K_1 \left\{ \epsilon^{-\phi} \left(\int_I v_0 |u|^p \right)^{q/p} + \epsilon^{\theta} \left(\int_I v_1 |u^{(m)}|^r \right)^{q/r} \right\} \tag{1.1}$$

or

$$\int_I w|u^{(j)}|^q \le K_2 \left(\int_I v_0 |u|^p \right)^{(q/p)\lambda} \left(\int_I v_1 |u^{(m)}|^r \right)^{(q/r)(1-\lambda)} . \tag{1.2}$$

If $j = 0$ and either $q = p$ and $v_0 = w$ or $\lambda = 0$ then (1.2) yields the Hardy-type inequality

$$\int_I w|u|^q \le K_3 \left(\int_I v_1 |u^{(m)}|^r \right)^{q/r} . \tag{1.3}$$

Here w, v_0, v_1 are positive measurable functions or "weights", $1 \le p, q, r \le \infty$, ϕ, θ, λ are certain parameters, and ϵ (in (1.1)) is an arbitrarily small positive number in the fixed interval $(0, \epsilon_0)$ where $\epsilon_0 \le \infty$. Also $I = (a, b)$, $[a, b)$, or $(a, b]$ where in the first case $-\infty \le a < b \le \infty$, and in the latter two a or b respectively are finite. In (1.1) and (1.2) u comes from one of the domains:

$$D^{p,r}(v_0, v_1; I) := \left\{ u : u \in AC^m(I); \int_I v_0 |u|^p, \int_I v_1 |u^{(m)}|^r < \infty \right\}$$

$$D_L^{p,r}(v_0, v_1; I) := \{ u \in D^{p,r}(v_0, v_1; I) : \lim_{t \to a+} u^{(i)}(t) = 0, \ i = 0, \dots, m-1 \}$$

$$D_R^{p,r}(v_0, v_1; I) := \{ u \in D^{p,r}(v_0, v_1; I) : \lim_{t \to b-} u^{(i)}(t) = 0, \ i = 0, \dots, m-1 \}$$

$$D_{LR}^{p,r}(v_0, v_1; I) := \{ u \in D_L^{p,r}(v_0, v_1; I) \cap D_R^{p,r}(v_0, v_1; I) \} .$$

In the above definitions $AC^m(I)$ designates the functions u such that u is $m-1$ times continuously differentiable and $u^{(m-1)}$ is absolutely continuous on every compact subinterval of I. For the Hardy-type inequality (1.3) we require that $u \in AC^m(I)$,

$\int_I v_1|u^{(m)}|^r < \infty$, and that u together with its first $m-1$ derivatives vanish at one or both of the endpoints of I; by analogy with the previous definitions we write these domains as $D_L^r(v_1;I)$, $D_R^r(v_1;I)$, or $D_{LR}^r(v_1;I)$. The constants K_1, K_2, or K_3 may depend on the weights, the interval I, and the other parameters but not on u. Note also that the sum inequality (1.1) can be written in the more familiar form

$$\int_I w|u^{(j)}|^q \leq K(\epsilon)\left(\int_I v_0|u|^p\right)^{q/p} + \epsilon\left(\int_I v_1|u^{(m)}|^r\right)^{q/r}. \tag{1.4}$$

(To see this substitute δ for ϵ in (1.1) and then set $\epsilon = \delta^\theta$ in (1.4) with the result that $K(\epsilon) = K_1\epsilon^{-\phi/\theta}$.)

The inequalities (1.1)–(1.3) can be generalized in several ways. For instance we may replace I by a domain $\Omega \subset \mathbf{R}^n$ where $n > 1$ and the ordinary derivatives by those of multi-index distributional type; further the Lebesgue norms can be replaced by weighted Sobolev, Orlicz, or Hölder norms.

Exactly when do such inequalities hold? In other words, what conditions on w, v_0, v_1 and p, q, r are necessary and sufficient for (1.1)–(1.3). Except for the unweighted case ($w = v_0 = v_1 = 1$), a few examples of (1.1), (1.2), or for the Hardy-type inequality (1.3) (see Theorem 4.1) satisfactory answers are not known. However a vast amount of information concerning special cases is available. Our intention here is to survey a representative selection of recent results. Because of space limitations, complete proofs will be given only if they are brief or apply to results not in the literature. Wherever possible however we will try to indicate the method of proof and to give complete references. In the final section of the paper we will sketch some significant applications relating to the spectral theory of ordinary differential operators.

Because of its technical simplicity our emphasis throughout will be on the real variable case; however many of the results presented here can be extended to the multidimensional case with some of the norms mentioned above. Full details may be found in [5],[6], and [10].

2. A BASIC INTERPOLATION LEMMA

A large part of the one variable theory of inequalities (1.1) and (1.2) depends on the following simple result.

Lemma 2.1. *Let $J = [c, d]$ be a finite subinterval of $\mathbf{R}$, and let $d - c = L$. Then for all $u \in AC^m(J)$ the inequality*

$$\|u^{(j)}\|_{\infty,J} \leq K\left\{L^{-(j+1)}\int_J |u| + L^{m-j-1}\int_J |u^{(m)}|\right\}. \tag{2.1}$$

holds.

Proof. It is enough to take $J = [0,1]$. For if $u \in AC^m([0,1])$ then $\hat{u}(t) := u((t-c)/L) \in AC^m(J)$. Then if

$$\|u^{(j)}\|_{\infty,[0,1]} \leq K\left\{\int_{[0,1]} |u| + \int_{[0,1]} |u^{(m)}|\right\}, \tag{2.2}$$

$\hat{u}$ satisfies (2.1) as we can see by changing variables; since if $s := (t-c)/L$ then

$$\hat{u}^{(k)}(t) = u^{(k)}(s)L^{-k},$$
$$ds = L^{-1}dt.$$

There are several ways to obtain (2.2). The simplest is to let

$$\mathcal{G} := \left\{(u, u^{(m)}) : u \in AC^m([0,1])\right\},$$

and define

$$\mathcal{T} := \mathcal{G} \longrightarrow AC^m([0,1]) \subseteq L^\infty[0,1]$$

by

$$\mathcal{T}(u, u^{(m)}) = u^{(j)}.$$

It is enough to show that $\mathcal{T}$ is continuous. We will do this via the Closed Graph Theorem. We first show that $\mathcal{G} \subset L^1[0,1] \times L^1[0,1]$ is closed. Let $\{u_k\}$ be a sequence of functions in $AC^m([0,1])$ such that $u_k \to \alpha$ and $u_k^{(m)} \to \beta$, the convergence in both cases being in $L^1[0,1]$. Then each u_k can be written as

$$u_k = \tilde{u}_k + c_{0k} + c_{1k}t + \cdots + c_{(m-1)k}t^{(m-1)} \tag{2.3}$$

where

$$\tilde{u}_k := \int_0^t \frac{(t-s)^{m-1}}{(m-1)!} u_k^{(m)}(s)\, ds. \tag{2.4}$$

Hölder's inequality applied to (2.4), and the L^1 convergence of $u_k^{(m)}$ shows that the $\tilde{u}_k$ converge uniformly to the function

$$\tilde{\alpha} := \int_0^t \frac{(t-s)^{m-1}}{(m-1)!}\beta(s)\, ds$$

which is in $AC^m([0,1])$. Call the polynomial part of (2.3) π_k^{m-1}. Then it follows that $\pi_k^{m-1} \to \alpha - \tilde{\alpha}$ in L^1. Since the set of all polynomials on $[0,1]$ of degree at most $m-1$ is a finite dimensional Banach space with respect to the L^1 norm which is equivalent to $\mathbf{C}^m$, the mapping f from this subspace of polynomials to $\mathbf{C}^m$ defined by $f(\pi_k^{m-1}) = (c_{0k}, \ldots, c_{(m-1)k})$ is continuous; consequently the vectors $(c_{0k}, \ldots, c_{(m-1)k})$ converge in the Euclidean norm to a vector $(c_0, \ldots, c_{m-1})$ which in turn implies that π_k^{m-1} converges uniformly to $\pi^{m-1} := c_0 + c_1 t + \ldots c_{m-1}t^{m-1}$. Thus $\alpha = \tilde{\alpha} + \pi^{m-1}$ which is in $AC^m([0,1])$ and furthermore $\alpha^{(m)} = \beta$. In exactly the same way we can show that the operator $u \to u^{(j)}$ has closed graph in $L^1[0,1] \times L^\infty[0,1]$. From these facts it follows that $\mathcal{T}$ has closed graph and has a Banach space domain. By the Closed Graph Theorem we have immediately that $\mathcal{T}$ is continuous.

Since this method is nonconstructive, we give a second proof of (2.2) which gives upper bounds on K. Let $\mathcal{F}_j$ denote the class of functions $\lambda_j \in AC^m([0,1])$ having essentially bounded mth derivative and satisfying the interpolation conditions

$$\lambda_j^{(i)}(1) = \begin{cases} 1, & \text{if } i = m-j-1 \\ 0, & \text{otherwise,} \end{cases} \qquad \lambda_j^{(i)}(0) = 0 \text{ for } i = 0, \ldots, m-1. \tag{2.5}$$

We will show that

$$K \leq \inf_{\lambda_j \in \mathcal{F}_j} \max \left\{ \|\lambda_j\|_\infty, 2^{j+1} \|\lambda_j^{(m)}\|_\infty \right\}, \tag{2.6}$$

or alternatively that

$$K \leq \inf_{\lambda_k \in \mathcal{F}_k} \max \left\{ \left(\frac{1}{(m-j-1)!} + \sum_{k=j}^{m-1} \|\lambda_k\|_\infty \right), \sum_{k=j}^{m-1} \|\lambda_k^{(m)}\|_\infty \right\}. \tag{2.7}$$

To proceed with the proof: let e be the midpoint of J. Suppose $t \in [e, d]$. Let L_t be the length of $[c, t]$. Define

$$z_j(s) = \lambda_j((s-c)/L_t) L_t^{m-j-1} (-1)^{m-j-1}.$$

Then it is clear that

$$z_j^{(m)}(s) = \lambda_j^{(m)}((s-c)/L_t) L_t^{-(j+1)} (-1)^{m-j-1},$$

and that

$$z_j^{(i)}(t) = \begin{cases} (-1)^{m-j-1}, & \text{if } i = m-j-1 \\ 0, & \text{otherwise,} \end{cases} \qquad z_j^{(i)}(c) = 0 \text{ for } i = 0, \ldots, m-1. \tag{2.8}$$

If $u \in AC^m(J)$ integration by parts gives the formula

$$\begin{aligned} \int_c^t u^{(m)} z_j - \int_c^t u(-1)^m z_j^{(m)} &= \sum_{i=0}^{m-1} (-1)^i u^{(m-i-1)} z_j^{(i)} \Big|_{c+}^{t-} \\ &= u^{(j)}(t) \end{aligned} \tag{2.9}$$

Therefore using (2.8) and the fact that $L/2 \leq L_t \leq L$ we obtain the inequality

$$|u^{(j)}(t)| \leq \|\lambda_j\|_\infty L^{m-j-1} \int_J |u^{(m)}| + 2^{j+1} \|\lambda_j^{(m)}\|_\infty L^{-(j+1)} \int_J |u| \tag{2.10}$$

which establishes (2.6). If $t \in [c, e)$ we redefine z_j so that

$$z_j(s) = -\lambda_j((d-s)/L_t) L_t^{m-j-1}.$$

Integration by parts as in (2.9) gives

$$\int_t^d u^{(m)} z_j - \int_t^d u(-1)^m z_j^{(m)} = u^{(j)}(t),$$

which in turn yields (2.10).

The derivation of the alternate bound (2.7) is equally straightforward. We start with the identity

$$u^{(j)}(t) = \int_c^t \frac{(t-s)^{m-j-1}}{(m-j-1)!} u^{(m)}\, ds + u^{(j)}(c^+) + \cdots + \frac{(t-c)^{m-j-1} u^{(m-1)}(c^+)}{(m-j-1)!}. \tag{2.11}$$

By retracing the proof of (2.10) we see that

$$|u^{(k)}(c)| \le \|\lambda_k^{(m)}\|_\infty L^{-(k+1)} \int_J |u| + \|\lambda_k\|_\infty L^{m-k-1} \int_J |u^{(m)}| \qquad \text{for } k = j, \ldots, m-1.$$

Since $t - s \le L$ (2.7) follows by substituting these bounds on $u^{(k)}(c)$ into (2.11).

In the case $j = 1$, $m = 2$ we can derive an optimal value for K by letting λ_1 be the quadratic "bang-bang" spline

$$\lambda_1(t) := \begin{cases} 2t^2, & \text{if } 0 \le t < 1/2 \\ -2(t-1)^2 + 1, & \text{if } 1/2 \le t < 1, \end{cases}$$

so that $\lambda_1''(t) = 4$ for $t \in [0, 1/2]$ and $\lambda_1''(t) = -4$ for $t \in (1/2, 1]$. Application of (2.7) yields that $K \le 4$. On the other hand, taking $J = [0, 1]$ and $u = t - 1/2$ in (2.2) shows that $K = 4$. We do not know however if (2.6), (2.7) give good bounds on K for other m, j.

For still another proof of Lemma 2.1 using divided differences see Brown and Hinton [2].

3. WEIGHTED SUM INEQUALITES

Let us see how the basic lemma gives a weighted sum inequality (1.1). Suppose $I = [a, \infty)$ and f is a positive continuous function. Let $J_t := [t, t + \epsilon f(t)]$ and

$$S_1(\epsilon_0) := \sup_{\substack{t \in I \\ 0 < \epsilon \le \epsilon_0}} \left\{ f^{-\phi} \left(\frac{\int_{J_t} w}{\epsilon f} \right) \left(\frac{\int_{J_t} v_0^{-p'/p}}{\epsilon f} \right)^{q/p'} \right\} < \infty,$$

$$S_2(\epsilon_0) := \sup_{\substack{t \in I \\ 0 < \epsilon \le \epsilon_0}} \left\{ f^{\theta} \left(\frac{\int_{J_t} w}{\epsilon f} \right) \left(\frac{\int_{J_t} v_1^{-r'/r}}{\epsilon f} \right)^{q/r'} \right\} < \infty \qquad (3.1)$$

where $\phi := q(j + 1/p - 1/q)$ and $\theta := q(m - j - 1/r + 1/q)$. For simplicity let $q = p = r$. Then if $t_0 := a$, $t_{i+1} := t_i + \epsilon f(t_i)$, etc., and $\Delta_i := [t_i, t_{i+i}]$, a calculation (Hölder's inequality) shows that

$$\int_{\Delta_i} w|u^{(j)}|^p \le K \left\{ \epsilon^{-jp} S_1(\epsilon_0) \int_{\Delta_i} v_0 |u|^p + \epsilon^{(m-j)p} S_2(\epsilon_0) \int_{\Delta i} v_1 |u^{(m)}|^p \right\}.$$

Adding we get the weighted sum inequality

$$\int_I w|u^{(j)}|^p \le K_1 \left\{ \epsilon^{-jp} \left(\int_I v_0 |u|^p \right) + \epsilon^{(m-j)p} \left(\int_I v_1 |u^{(m)}|^p \right) \right\}.$$

This argument also works if $q \ge \max\{p, r\}$.

If f can be taken to be nondecreasing, a variation on the above argument (cf. [3]) shows (with $q \ge \max\{p, r\}$) that the conditions

$$R_1(\epsilon_0) := \sup_{\substack{t \in I \\ 0 < \epsilon \le \epsilon_0}} f^{-\phi} w(t) \left(\frac{\int_{J_t} v_0^{-p'/p}}{\epsilon f} \right)^{q/p'} < \infty,$$

$$R_2(\epsilon_0) := \sup_{\substack{t \in I \\ 0 < \epsilon \le \epsilon_0}} f^{\theta} w(t) \left(\frac{\int_{J_t} v_1^{-r'/r}}{\epsilon f} \right)^{q/r'} < \infty \qquad (3.2)$$

are sufficient for the inequality.

Other conditions resembling (3.1) can be stated if $q < \min\{p, r\}$, $p \leq q < r$, or $r \leq q < p$. See [4] for details.

How are such conditions verified?

Definition 3.1. Let z be a positive measurable function on I. Define functions $C_z(t)$ and $D_z(t)$ by

$$C_z(t) := \inf_{s \in J_t} \frac{z(s)}{z(t)},$$
$$D_z(t) := \sup_{s \in J_t} \frac{z(s)}{z(t)}.$$

If there are positive constants C_z, D_z such that $C_z(t) \geq C_z$ and $D_z(t) \leq D_z$ for all $t \in I$. We call z ***f*-bounded above and below on *I***. (Recall that $J_t := [t, t + \epsilon f(t)]$ so that these functions and constants may depend on ϵ.)

Since $z(t)C_z(t) \leq z(s) \leq D_z(t)z(t)$ for $s \in [t, t + \epsilon f(t)]$, the previous definition allows us to bound the the definitions of $S_1(\epsilon_0), S_2(\epsilon_0)$ by the simpler expressions:

$$S_1(\epsilon_0) \leq \sup_{t \in I} f(t)^{-\phi} D_w(t) C_{v_0}(t)^{-q/p} w(t) v_0(t)^{-q/p},$$
$$S_2(\epsilon_0) \leq \sup_{t \in I} f(t)^{\theta} D_w(t) C_{v_1}(t)^{-q/r} w(t) v_1(t)^{-q/r}.$$

Example 3.1. Let $q \geq \max\{p, r\}$, $w(t) = t^{\beta}$, $v_0(t) = t^{\gamma}$, $v_1(t) = t^{\alpha}$, $f(t) = t^{\delta}$ where $\delta \leq 1$, and $I = [1, \infty)$. Then

$$1 \leq \sup_{s \in J_t} \frac{s}{t} \leq 1 + \epsilon t^{\delta - 1} \leq 1 + \epsilon.$$

It follows that the weights are all f-bounded above and below so that $S_1(\epsilon_0), S_2(\epsilon_0) < \infty$ if

$$\beta - \frac{q}{p}\gamma - \delta\phi \leq 0,$$
$$\beta - \frac{q}{r}\alpha + \delta\theta \leq 0 \tag{3.3}$$

and ϵ is sufficiently small (say $\epsilon \in (0, 1]$). If $I = (0, b]$, $b < \infty$, then these inequalities are reversed. If $I = (0, \infty)$, there is equality . Since in (3.3)

$$\beta \leq \min\left\{\delta\phi + \frac{q}{p}\gamma, -\delta\theta + \frac{q}{r}\alpha\right\},$$

β will be as large as possible relative to α and γ if δ is chosen by "equality", i.e.,

$$\delta = \frac{(\alpha/r - \gamma/p)}{(m + 1/p - 1/r)} \leq 1. \tag{3.4}$$

Therefore

$$\frac{\beta}{q} \leq \left(\frac{\gamma}{p}\right)\left(\frac{\theta}{\phi + \theta}\right) + \left(\frac{\alpha}{r}\right)\left(\frac{\phi}{\phi + \theta}\right). \tag{3.5}$$

In particular if $q = p = r$ we get that

$$\beta m \le (m-j)\gamma + j\alpha . \tag{3.6}$$

It turns out that if (3.4) is assumed, then (3.5) and (3.6) are necessary as well as sufficient for the sum inequality to hold. (In general the condition $S_1(\epsilon_0) < \infty$ is necessary and sufficient for the inequality provided the weights are chosen so that the two integral averages $S_1(\epsilon_0)$, $S_2(\epsilon_0)$ in (3.1) are equivalent.)

Example 3.2. The above conditions also work for $w(t) = e^{\beta t}$, $v_0(t) = e^{\gamma t}$, $v_1(t) = e^{\alpha t}$, $f(t) = e^{\delta t}$ where $\delta \le 0$, and $I = [0, \infty)$.

Example 3.3. If I is as above, $w(t) = e^{\beta t^2}$, $v_0(t) = e^{\gamma t^2}$, $v_1(t) = e^{\alpha t^2}$, and $f(t) = e^{\delta t^2}, \delta \le 0$, then

$$1 \le \frac{e^{s^2}}{e^{t^2}} \le O(e^{2\epsilon t})$$

so that $C_w(t), D_{v_0}(t)$, etc. may be of exponential order. Nevertheless since these terms are of *lower* order than f, w, v_0 and v_1 we will get exactly the same results as before except that there is the *strict* inequality:

$$\frac{\beta}{q} < \left(\frac{\gamma}{p}\right)\left(\frac{m-j-1/r+1/q}{m+1/p-1/r}\right) + \left(\frac{\alpha}{r}\right)\left(\frac{j+1/p-1/q}{m+1/p-1/r}\right).$$

Remarks 3.1. As mentioned above no necessary and sufficient conditions are known for the sum inequality (1.1) which are valid for all p, q, r, j, m and weights w, v_0, and v_1. However in the case $m = 1$, $j = 0$, $p = r$, $q \ge p$, and $\epsilon = 1$, then a necessary and sufficient condition for (1.1) has been derived by Oĭnarov **[26]**. In this case the inequality is equivalent to the existence of an embedding of a weighted Sobolev space having the norm

$$\left(\int_I v_0|u|^p + \int_I v_1|u'|^p\right)^{1/p}$$

into the weighted Lebesgue space $L^q(w; I)$. Oĭnarov's condition is too complicated to be stated here but it requires the finiteness of an integral average resembling those in (3.1). Some other necessary conditions valid for general m, j and q, p, r may be found in **[4]**.

We also remark here that Kwong and Zettl **[15]**, **[17]** have shown in the case $q = p = r$ that the sum inequality (1.1) is sometimes valid when the local integrability of $v_0^{-p/p'}$ is replaced by weaker conditions, for example $v_0 \ge 0$, $\int_I v_0 > 0$, and I bounded.

4. PRODUCT INEQUALITIES

Brown and Hinton **[3]** have shown that if

$$\frac{m}{q} \le \frac{m-j}{p} + \frac{j}{r} \tag{4.1}$$

and

$$S_1 := \sup_{\substack{t\in I\\ 0<\epsilon<\infty}} \left\{ f^{-\phi} \left(\frac{\int_{J_t} w}{\epsilon f} \right) \left(\frac{\int_{J_t} v_0^{-p'/p}}{\epsilon f} \right)^{q/p'} \right\} < \infty,$$

$$S_2 := \sup_{\substack{t\in I\\ 0<\epsilon<\infty}} \left\{ f^{\theta} \left(\frac{\int_{J_t} w}{\epsilon f} \right) \left(\frac{\int_{J_t} v_1^{-r'/r}}{\epsilon f} \right)^{q/r'} \right\} < \infty \tag{4.2}$$

(recall that $\phi := q(j + 1/p - 1/q)$ and $\theta := q(m - j - 1/r + 1/q)$), then

$$\int_I w|u^{(j)}|^q \le K_2 \left(\int_I v_0|u|^p \right)^{(q/p)\lambda} \left(\int_I v_1|u^{(m)}|^r \right)^{(q/r)(1-\lambda)} \tag{4.3}$$

where

$$\lambda := \frac{m - j - 1/r + 1/q}{m + 1/p - 1/r},$$

$$1 - \lambda := \frac{j + 1/p - 1/q}{m + 1/p - 1/r} \tag{4.4}$$

A standard arithmetic-geometric mean inequality argument (cf. Mitrinovic [20]) shows that (4.3) is equivalent to the sum inequality

$$\int_I w|u^{(j)}|^q \le K_1 \left\{ \epsilon^{-\phi} \left(\int_I v_0|u|^p \right)^{q/p} + \epsilon^{\theta} \left(\int_I v_1|u^{(m)}|^r \right)^{q/r} \right\} \tag{4.5}$$

where $K_2 = K_1/(\lambda^\lambda(1-\lambda)^{(1-\lambda)})$. Conversely if (4.5) holds for all $\epsilon > 0$, then minimizing the right hand side as a function of ϵ (or using an arithmetic-geometric mean inequality) gives (4.3).

An interesting special case of (4.3) is to take $I = \mathbf{R}$, $w(t) = v_0(t) = v_1(t) = f(t) = 1$, and assume (4.1) (so that $S_1 = S_2 = 1$ for all ϵ). We get that

$$\int_{\mathbf{R}} |u^{(j)}|^q \le K_2 \left(\int_{\mathbf{R}} |u|^p \right)^{(q/p)\lambda} \left(\int_{\mathbf{R}} |u^{(m)}|^r \right)^{(q/r)(1-\lambda)}$$

This inequality is due to Gabushin [12] and is the one dimensional analogue of the famous interpolation inequality of Nirenberg [24], holding on $C_0^\infty(\mathbf{R}^n)$. A difficulty however with the strategy of obtaining (4.3) from (4.2) is that as we have seen in the previous section (4.2) may only hold for $0 < \epsilon \le \epsilon_0 < \infty$ (i.e., for sufficiently small ϵ). This is true for example in the case of power weights. However if $w(t) = t^\beta$, $v_0(t) = t^\gamma$, $v_1(t) = t^\alpha$, and $q = p = r$ where $I = [a, \infty)$, $a > 0$, and we use (3.2) instead of (3.1) as a sufficient condition for the sum inequality, then R_1 and R_2 will be independent of ϵ if α, γ are positive. If further $\delta \ge 0$ (so that f is nondecreasing), then the multiplicative inequality will be true if $\delta = (\alpha - \gamma)/m \le 1$ and

$$\beta m \le (m - j)\gamma + j\alpha$$

as before. Similarly one can take $w = v_0 = v_1 = \omega$, where ω is nondecreasing (Kwong and Zettl [16], [17]) so that we have for example the inequality

$$\int_I e^{ct}|u^{(j)}|^p \le K_2 \left(\int_I e^{ct}|u|^p \right)^{(m-j)/m} \left(\int_I e^{ct}|u^{(m)}|^p \right)^{j/m}$$

for $c > 0$.

In the case of power weights with $q = p = r$, $m = 2$, and $j = 1$, the following more general result may be found in [11].

Theorem 4.1. *Let $u \in D^{p,p}(t^\gamma, t^\alpha; (0,\infty))$, $1 \le p < \infty$, then the inequality*

$$\int_0^\infty t^\beta |u'|^p \le K_2 \left(\int_0^\infty t^\gamma |u|^p\right)^{1/2} \left(\int_0^\infty t^\alpha |u''|^p\right)^{1/2}$$

holds if and only if the following conditions are satisfied:

(i) $\{\alpha,\beta,\gamma\} \neq \{p-1,-1,-1-p\}$.

(ii) $\beta = \dfrac{\alpha+\gamma}{2}$.

(iii) $\lim\limits_{t\to 0+} u'(t) = 0$ *when* $\beta \le -1$ *and* $\beta > \alpha - p$.

(iv) $\lim\limits_{t\to\infty} u'(t) = 0$ *when* $\beta \ge -1$ *and* $\beta < \alpha - p$.

Also in the exceptional case $\{\alpha,\gamma\} = \{p-1,\ -1-p\}$ *the inequality*

$$\int_0^\infty t^{-1}|u'|^p \le K_2' \left\{ \left(\int_0^\infty t^{-1-p}|u|^p\right)^{1/2} \left(\int_0^\infty t^{p-1}|u''|\right)^{1/2} + \int_0^\infty t^{-1-p}|u|^p \right\}$$

is valid.

The above inequalities hold on $D^{p,r}(v_0, v_1; I)$. However if one is willing to allow u to vanish at the end points of I, many more multiplicative inequalities can be derived.

Assume that M is a real class $C^{(1)}$ function defined on $I := [a,\infty)$ such that

$$M(x) > 0 \text{ on } I. \tag{4.6}$$

$$\text{Either } M'(x) > 0 \text{ on } I \text{ or } M'(x) < 0 \text{ on } I \text{ and}$$
$$M' \text{ is a monotone function on } I. \tag{4.7}$$

$$M(a), M'(a) \text{ are positive.} \tag{4.8}$$

Set $f(x) := M(x)/|M'(x)|$. Assume that f is bounded below by a positive constant and additionally that:

$$\text{there is a } k > 0 \text{ such that } |f(s) - f(t)| \le k|s-t| \text{ for all } s,t \text{ in } [a,\infty). \tag{4.9}$$

For $\alpha, \beta, \gamma, \delta_2, \delta_3$ real numbers set

$$\begin{aligned} w(x) &:= f(x)^{q\beta}|M'(x)|^{q\delta_1} \\ v_0(x) &:= f(x)^{p\gamma}|M'(x)|^{p\delta_2} \\ v_1(x) &:= f(x)^{r\alpha}|M'(x)|^{r\delta_3} \end{aligned} \tag{4.10}$$

where

$$\delta_1 := \lambda\delta_3 + (1-\lambda)\delta_2\,. \tag{4.11}$$

In this setting it has been shown in [9] that:

Theorem 4.2. *Suppose (4.6)–(4.11) hold, $I = [a, \infty)$, $0 \le j \le m-1$, $1 \le p, q, r < \infty$, and $p \ge r$. Assume $q \ge q_0$ where q_0 is defined by*

$$\frac{m}{q_0} = \frac{m-j}{p} + \frac{j}{r}, \tag{4.12}$$

and $\lambda_0 \le \lambda \le 1$ where λ_0 is given by

$$\lambda_0 := \frac{j + p^{-1} - q^{-1}}{m - r^{-1} + p^{-1}}. \tag{4.13}$$

Define $\beta^\sharp$ by

$$\beta^\sharp + q^{-1} = \lambda(\alpha + r^{-1} - m + j) + (1-\lambda)(\gamma + p^{-1} + j). \tag{4.14}$$

and consider the conditions

(C1) $\alpha + r^{-1} > m$
(C2) $\alpha + r^{-1} < 1$
(C3) $\delta_3 \ge r^{-1} + (\alpha - 1)$
(C4) $\delta_3 \le r^{-1} + \alpha - m$

Then (1.2) is valid for all $u \in D_{LR}^{p,r}(v_0, v_1; I)$ with $\beta \le \beta^\sharp$ when:

A. *$M' < 0$, is nondecreasing, and either $\{(C1),(C3)\}$ or $\{(C2),(C4)\}$ hold.*
B. *$M' > 0$, is nondecreasing, and either $\{(C2),(C4)\}$ or $\{(C1),(C3)\}$ hold.*
C. *$M' > 0$, is nonincreasing, and either $\{(C2),(C3)\}$ or $\{(C1),(C4)\}$ hold.*

Corollary 4.1. *Let $0 \le j \le m-1$, $1 \le p, q, r < \infty$, $q \ge r$, and $I = [a, \infty)$. Assume that the hypotheses (4.6)–(4.14) of Theorem 4.1 hold, but in addition suppose that $f := M/|M'$ is bounded above and below by a positive constant. Then (1.2) will hold on $D_{LR}^{p,r}(v_0, v_1; I)$ with $\alpha = \beta = \gamma = 0$ if:*

A. *$M' < 0$, is nondecreasing, and $\delta_3 \ne 0$.*
B. *$M' > 0$, is nondecreasing, and $\delta_3 \ne 0$.*
C. *$M' > 0$ and is nonincreasing, and $\delta_3 \in \mathbf{R}$.*

Corollary 4.2. *Let $0 \le j \le m-1$, $1 \le p, q, r < \infty$, $q \ge r$, and $I = [a, \infty)$. Suppose $w = M^{q\beta}$, $v_0 = M^{p\gamma}$, and $v_1 = M^{r\alpha}$. Assume that the hypotheses (4.6)–(4.14) of Theorem 2.1 hold, but in addition let $|M'|$ is bounded above and below by positive constants on I. Then (1.1) holds for all $u \in D_{LR}^{p,r}(v_0, v_1; I)$ if either of the conditions:*

(i) *$\alpha + r^{-1} \notin \{1, \ldots, m\}$, and $\beta \le \beta^\sharp$,*
(ii) *$\alpha + r^{-1} \in \{1, \ldots, m\}$, and $\beta < \beta^\sharp$*

is satisfied.

Remark 4.1. If M satisfies (4.8) we say that a is *regular*. Otherwise a is *singular*. Similar results hold if $I = (a, b]$ where both a and b are finite but a is singular and b is regular. For simplicity assume that $a = 0$. In this case we need only reverse the inequalities involving β in the statements of Theorem 4.1 and Corollary 4.2, while also replacing the condition that f should be bounded below by a positive constant in all three results by the condition that $f(x) \to 0$ as $x \to 0$.

Remark 4.2. The case of two singular endpoints may be deduced from that of two one singular endpoint problems by use of Jensen's inequality.

Example 4.1. Let $M = t$ and $I = [1, \infty)$. By Corollary 4.2 (see also [**11**, Theorem 2.1]) we see that the inequality

$$\left(\int_I |t^\beta u^{(j)}|^q\right)^{1/q} \le K_2 \left(\int_I |t^\gamma u|^p\right)^{\lambda/p} \left(\int_i |t^\alpha u^{(m)}|^r\right)^{(1-\lambda)/r}$$

is valid on $D^{q,r}_{LR}(t^{q\gamma}, t^{r\alpha}; I)$ if either $\beta = \beta^\sharp$ and $\alpha + r^{-1} \notin \{1, \ldots, m\}$ or $\beta < \beta^\sharp$.

Example 4.2. Let $M = e^t$ and $I = [0, \infty)$. Since $f = 1$, this falls under Corollary 4.1. α, β, and γ play no role and we have the inequality

$$\left(\int_I |e^{\delta_1 t} u^{(j)}|^q\right)^{1/q} \le K_2 \left(\int_I |e^{\delta_2 t} u|^p\right)^{\lambda/p} \left(\int_I |e^{\delta_3 t} u^{(m)}|^r\right)^{(1-\lambda)/r}$$

which holds on $D^{p,r}_{LR}(e^{\delta_2 pt}, e^{\delta_3 rt}; I)$ if $\delta_3 \ne 0$. This inequality is also valid on $D^{p,r}_L(e^{\delta_2 pt}, e^{\delta_3 rt}; I)$ when $\delta_3 < 0$ and on $D^{p,r}_R(e^{\delta_2 pt}, e^{\delta_3 rt}; I)$ when $\delta_3 > 0$. (See [**9**, Example 2.2 and Remark 2.3]. However note that end point conditions are necessary; there can be no inequality on $D^{p,r}(e^{\delta_2 pt}, e^{\delta_3 rt}; I)$ when $\delta_3 < 0$, as is clear if u is a monomial of degree between j and $m - 1$.

Remarks 4.3. We have not discussed here the important problem of determining the optimal values of the constants K_1 or K_2 in the inequalities (1.1) or (1.2). (Indeed our policy has been to let K denote an arbitrary constant whose value may change from line to line.) For nontrivial weights these values are almost all unknown. In the unweighted case the multiplicative inequality for general p, q, and r includes many special cases due to Landau, Kolgomorov, and others for which the best constants are known. For a survey of results concerning this problem see [**17**].

5. WEIGHTED HARDY INEQUALITIES

The classical Hardy inequality (Hardy, Littlewood and Polya, [**13**, Theorem 330]) states that if $1 < p < \infty$, $c < -1$, $u \in AC((0, \infty))$, $u(0) = 0$, and

$$\int_0^\infty t^{c+p} |u'(t)|^p \, dt < \infty ,$$

then

$$\int_0^\infty t^c |u(t)|^p \, dt \le [p/|c+1|]^p \int_0^\infty t^{c+p} |u'(t)|^p \, dt .$$

If $c > -1$ the inequality is true if $u(\infty) = 0$.

There have been many generalizations of this inequality (for an interesting survey, see [**25**]). Most recently it has been shown (Stepanov, [**28**]) that

$$\int_0^\infty w|u|^q \le K_3 \left(\int_0^\infty v_1 |u^{(m)}|^r\right)^{q/r} .$$

where $q \ge r$ and $u(0) = 0$ holds if and only if

$$\sup_{t>0} \left[\int_t^\infty w(x)\, dx\right]^{1/q} \left[\int_0^t (t-x)^{(m-1)r'} v_1(x)^{-r'/r} dx\right]^{1/r'} < \infty ,$$

$$\sup_{t>0} \left[\int_t^\infty (x-t)^{(m-1)q} w(x)\, dx\right]^{1/q} \left[\int_0^t v_1(x)^{-r'/r} dx\right]^{1/r'} < \infty . \qquad \textbf{(5.1)}$$

(A slightly different but equivalent pair of conditions has also been given in [19].) When $m = 1$ the conditions (5.1) collapse into a single necessary and sufficient condition due to Muckenhoupt [21]. It seems hard however to verify conditions of this kind for specific weights.

Another approach is to show that Hardy's inequality holds for a large family of weights having a certain structure. For example one can obtain the following result (see [7]) by an essentially inductive technique:

Theorem 5.1. Let M be a positive locally absolutely continuous function on I such that either $M' > 0$ or $M' < 0$ a.e. on I and:

(i) *ω is a positive measurable function that is nonincreasing if $u^{(i)}(a) = 0$ and nondecreasing if $u^{(i)}(b) = 0$, $i = 0, \dots, m-1$;*

(ii) *we suppose that*

$$\beta = (q/r)(\alpha + 1) - mq - 1\,,$$
$$\beta_0 = (q/r)(\alpha_0 - 1) + mq + 1.$$

Then if M is increasing and concave up the inequality

$$\int_I M^\beta |M'|^{\beta_0} |u|^q \omega \le K_3 \left[\int_I M^\alpha |M'|^{\alpha_0} |u^{(m)}|^r \omega^{r/q} \right]^{q/r}$$

holds on $D_L^r(M^\alpha |M'|^{\alpha_0} \omega^{r/q}; I)$ if $\beta < -1 - (m-1)q$ and $\beta_0 \le 1$; and it holds on $D_R^r(M^\alpha |M'|^{\alpha_0} \omega^{r/q}; I)$ if $\beta > -1$ and $\beta_0 \ge 1 + (m-1)q$.

For other choices of M, e.g.: M increasing and concave down, M decreasing and concave up or down, the Hardy inequality will hold with other conditions on β_0, β. We note that for $m = 1$ the result essentially agrees with Love [18].

Corollary 5.1. Let M_j, M_j', $j = 1, \dots, l$, be functions satisfying the conditions of Theorem 5.1. Define

$$w := \left(\prod_{j=1}^{l} w_j^{1/\lambda_j} \right) \omega\,,$$
$$v_1 := \left(\prod_{j=1}^{l} v_{1j}^{1/\lambda_j} \right) \omega$$

where

$$w_j = M_j^{\beta_j} |M_j'|^{\beta_{j0}}\,,$$
$$v_{1j} = M_j^{\alpha_j} |M_j'|^{\alpha_{j0}}\,,$$

the λ_j are positive numbers such that $\sum_{j=1}^{l} \lambda_j^{-1} \ge 1$, and the other parameters are as above. Then the Hardy inequality

$$\int_I w|u|^q \le K_3 \left(\int_I v_1 |u^{(m)}|^r \right)^{q/r}$$

holds on $D^r_R(v_1;I)$ *or* $D^r_L(v_1;I)$.

Example 5.1. Let $M(t)=e^{-t^2}$, $q=r$, $I=[1,\infty)$. A calculation based on Theorem 5.1 gives:

$$\int_I e^{-ct^2}|u(t)|^q\,dt \le K_3\int_I e^{-ct^2}t^{-mq}|u(t)^{(m)}|^q\,dt\,.$$

If $M_1(t)=e^{-t^2}$, $M_2=t$, and $\lambda_1=\lambda_2=2$, a calculation based on Corollary 5.1 gives

$$\int_I e^{-ct^2/2}t^{(m-1)q/2}|u(t)|^q\,dt \le K_3\int_I e^{-ct^2/2}t^{(m-1)q/2}|u(t)^{(m)}|^q\,dt$$

where in both inequalities $c>(m-1)q$ and $u^{(i)}(1)=0$, $i=0,\dots,m-1$.

6. APPLICATIONS

6.1. Discreteness of the Spectrum. Our sum inequalities define mappings from the weighted "Sobolev-like" space $W^{p,r,m}(v_0,v_1;I)$ with norm

$$\|u\|_{v_0,v_1,I} := \left(\int_I v_0|u|^p\right)^{1/p} + \left(\int_I v_1|u^{(m)}|^r\right)^{1/r}$$

into $L^q(w;I)$.

Let $I=[a,\infty)$. Consider the "maximal" differential operator

$$L(u) := w^{-1}[(-1)^m\,(v_1u^{(m)})^{(m)} + v_0u]$$

defined in $L^2(w;I)$. It is known that L has discrete spectrum ([**23**, pp. 239–246]) if and only if

$$W^{2,2,m}(v_0,v_1;I) \hookrightarrow\hookrightarrow L^2(w;I)\,.$$

Set $w(t)=t^\beta$, $v_0(t)=t^\gamma$, $v_1(t)=t^\alpha$. If $f(t)=t^\delta$ and

$$\delta=\frac{(\alpha-\gamma)}{2m}\le 1$$

with $I=[a,\infty)$, $a>0$, it turns out from our theory of sum inequalities that the embedding holds if and only if $\beta\le\gamma$. Furthermore the embedding is compact if and only if $\beta<\gamma$. If $\delta=1$ but $\delta\ne\alpha-\gamma)/2m$, we can show compactness if and only if $(\alpha-\gamma)/2m>1$. This result was first proven by Hinton and Lewis, [**14**] using a different technique. If $m=1$ the necessary and sufficient condition of Oĭnarov for the embedding of $W^{r,1}(v_0,v_1;I)$ into $L^q(w;I)$ refered to in Section 3 has likewise been modified to give a necessary and sufficient condition for the discreteness of the spectrum of (6.1) defined in $L^q(w;I)$ when $m=1$ and $q=r$. See [**22**] for details. The resulting condition is however difficult to verify.

6.2 Nonoscillation Criteria. The differential operator

$$L(u)=\sum_{i=0}^{m}(-1)^i(p_iu^{(i)})^{(i)}$$

is said to be oscillatory at ∞ if for every $n>0$ there is an interval $[c,d]\subset(n,\infty)$ and a nontrivial solution y of $L(y)=0$ such that

$$y(c)=\cdots=y^{(m-1)}(c)=0=y(d)=\cdots=y^{(m-1)}(d).$$

THEOREM 6.1. *Let*

$$Q(u) := \sum_{i=0}^{m} \int_I p_i |u^{(i)}|^2 ,$$

and set

$$\mathcal{A}_m(c,d) := \left\{ u \in AC_{LR}^m(c,d) : u^{(m)} \in L^2(c,d) \right\} .$$

Then $L(u)$ is oscillatory at ∞ iff for every $n > 0$ there is an interval $[c,d] \subset (n,\infty)$ and a nontrivial $y \in \mathcal{A}_m(c,d)$ such that $Q(\tilde{y}) \le 0$ where $\tilde{y} := y$ on $[c,d]$ and 0 otherwise.

This known result (see [**27**]) and our inequalities can be used to give conditions that L be ***nonoscillatory***.

Example 6.1. Consider the two term operator

$$L(u) = (-1)^m (e^{ct} u^{(m)})^{(m)} + q(t) u .$$

Then given a weight w and an index $p > 2$ (to be determined later):

$$\begin{aligned} Q(u) &= \int_I e^{ct} |u^{(m)}|^2 + q|u|^2 \\ &\ge \int_I e^{ct} |u^{(m)}|^2 - |q||u|^2 \\ &\ge \int_I e^{ct} |u^{(m)}|^2 - \left(\int_I w^{-p/(p-2)} |q|^{p/(p-2)} \right)^{(p-2)/2} \\ &\times \left(\int_I w^{p/2} |u|^p \right)^{2/p} \end{aligned}$$

by Hölder's inequality. Suppose

$$\int_I w^{-p/(p-2)} |q|^{p/(p-2)} \le C ,$$

$$\int_I w^{p/2} |u|^p \le K_3 \left(\int_I e^{ct} |u^{(m)}|^2 \right)^{p/2} .$$

The first condition implies that for arbitrarily small η there is a sufficiently large n such that

$$\int_n^\infty w^{-p/(p-2)} |q|^{p/(p-2)} \le \eta .$$

Substitution of the last two conditions into the previous inequality gives

$$Q(u) \ge \int_n^\infty e^{ct} |u^{(m)}|^2 (1 - \eta K_3^{p/2}) .$$

It follows that $Q(u) > 0$ for *all* $u \in AC_{LR}^m([n,\infty))$. Hence L cannot be oscillatory. A calculation shows that the Hardy inequality

$$\int_I e^{pct/2} |u|^p \le K_3 \left(\int_I e^{ct} |u^{(m)}|^2 \right)^{p/2}$$

is valid on $D_R^{2,p}(e^{ct}; I)$. So we may take $w = e^{ct}$ and the condition for nonoscillation is that

$$\int_I e^{-cpt/(p-2)} |q|^{p/(p-2)} < \infty.$$

There is a similar condition for $c < 0$. Additional discussion of this method may be found in [**7**]; in the same way (see [**1**]) product inequalities can be used to derive nonoscillation criteria for on operators with middle terms.

REFERENCES

1. M. Ashbaugh, R. C. Brown, and D. B. Hinton, *Interpolation inequalities and nonoscillatory differential equations*, in "International Series of Numerical Mathematics, Vol 103," Proceedings of the Oberwolfach conference "General Inequalities 6", Birkhäuser Verlag, Basel, 1992, 243–255.
2. R.C. Brown and D. B. Hinton, *Sufficient conditions for weighted inequalities of sum form.*, J. Math. Anal. Appl. **112** (1985), 563–578.
3. ______, *Sufficient conditions for weighted Gabushin inequalities*, Časopis Pěst. Mat. **111** (1986), 113–121.
4. ______, *Necessary and sufficient conditions for one variable weighted interpolation inequalities*, Proc. London Math. Soc. **(2)35** (1987), 439–453.
5. ______, *Weighted interpolation inequalities of sum and product form in R^n*, J. London Math Soc. **(3)56** (1988), 261–280.
6. ______, *Weighted interpolation inequalities and embeddings in R^n*, Canad. J. Math. **47** (1990), 959–980.
7. ______, *A Weighted Hardy's inequality and nonoscillatory differential equations*, Quaes. Mathematicae **15** (1992), 197–212.
8. ______, *Interpolation inequalities with power weights for functions of one variable*, J. Math. Anal. Appl. **172** (1993), 233–242.
9. ______, *An interpolation inequality and its applications*, preprint.
10. R. C. Brown and B. Opic, *Embeddings of weighted Sobolev spaces into spaces of continuous functions*, Proc. R. Soc. Lond. A **439** (1992), 279–296.
11. R. C. Brown, D. B. Hinton, and M. K. Kwong, *A remark on a weighted Landau inequality of Kwong and Zettl*, preprint.
12. V. N. Gabushin, *Inequalities for norms of a function and its derivatives in L_p metrics*, Mat. Zam. **1** (1967), 291–298 (*Russian*).
13. G. H. Hardy, J. E. Littlewood and G. Pólya, "Inequalities," Cambridge University Press, Cambridge, 1959.
14. D. B. Hinton and R.T. Lewis, *Discrete spectra criteria for singular differential operators with middle terms*, Proc. Camb. Phil. Soc. **77** (1975), 337–346.
15. M. K. Kwong and A. Zettl, *Weighted norm inequalities of sum form involving derivatives*, Proc. Roy. Soc. Edinburgh **88A** (1981), 121–134.
16. ______, *Norm inequalities of product form in weighted L^p spaces*, Proc. Roy. Soc. Edinburgh **89A** (1981), 293–307.
17. ______, "Norm Inequalities for Derivatives and Differences," Lecture Notes in Mathematics 1536, Springer-Verlag Berlin, 1992.
18. E. R. Love, *Generalizations of Hardy's integral inequality*, Proc. Roy. Soc. Edinburgh **100A** (1985), 237–262.
19. F.J. Martin-Reyes and E. Sawyer, *Weighted inequalities for Riemann-Liouville fractional integrals of order one and higher*, Proc. A.M.S. **106** (1989), 727–733.
20. D.S. Mitrinovic, "Analytic Inequalities," Springer-Verlag, Berlin, 1970.
21. B. Muckenhoupt, *Hardy's inequality with weights*, Studia Math. **44** (1972), 31–38.
22. K. T. Mynbaev and M. O. Otelbaev, *Weighted function spaces and the spectrum of differential operators*, "Nauka", Moscow (1988), (*Russian*).
23. M. A. Naimark, "Linear Differential Operators, Part II," Frederick Ungar, New York, 1968.
24. L. Nirenberg, *On elliptic partial differential equations*, Annali della Scuola. Norm. Sup. Pisa, Ser III **16** (1958), 115–162.

25. B. Opic and A. Kufner, "Hardy-type Inequalities," Longman Scientific and Technical, Harlow, Essex, UK, 1990.
26. R. Oĭnarov, *On weighted norm inequalities with three weights*, J. London Math. Soc. **(2) 48** (1993), 103–116.
27. W. T. Reid, "Sturmian Theory for Ordinary Differential Equations," Springer- Verlag, Berlin, 1980.
28. V. Stepanov, *Two weighted estimates for Riemann-Liouville integrals*; Preprint No. 39, Mat. Ustav. Csav, Prague, 1988; Math. USSR Izvestiya 36 (1991), 669–681 (*Russian*) .

4th Int. Coll. on Differential Equations, pp. 71-82
D. Bainov and V. Covachev (Eds)

RELATIVITY AND GRAVITATION : PROBABILISTIC GERMS AND SURVEY

Jean-Pierre CAUDET
University of Poitiers
F - 86022 FRANCE

Abstract : Construction of the relativistic formalism and of its invariants on the basis of the combinatorial numbers via two limit procedures.

Key Words : Combinatorial principles, large numbers laws, brownian motion, semigroup property, Euclidean and Minkowskian metrics, Klein-Gordon equation, stochastic mechanics, relativity. Classification : 60J60, 03H05, 81P20, 83A05.

Il y a plus de science dans le fond d'un cornet que n'en a rêvé d'Alembert.
Alfred de Musset

1 - INTRODUCTION

This is a search for links between the geometric nature of the laws of physics, and their conjectured probabilistic germs, here tentatively developed for Relativity and Gravitation. In this case, a bifocal perspective situation appears between the macroworld of stars and the microworld of radiations. On the macroworld side, we are accustomed to the invariance associated with a classical perspective effect (in which the basic line is the straight line) : the geometric invariance of the so-called celestial sphere, or equivalently, of the constellations. Still, in a more secret way, this perspective effect also involves a dynamical invariance (Foucault's pendulum and Mach). On the microworld side, a brownian perspective effect (in which the basic line is the zigzag line) can be constructed, involving the fundamental invariance of the quadratic form $q^2 - c^2t^2$ reducible to the invariance of the combinatorial numbers $\binom{n}{k}$ through the Laplace and de Moivre theorem, with c^2 invariant by brownian perspective.

This brownian perspective effect appears, not at the level of the zigzag paths in a perfect gas, in which waves propagate according to the adiabatic equation of state, but below, at the non-standard level of (what we shall call modelized) radiations, as it appears when we search for the brownian kernel propagating a monochromatic wave conditionned by an isothermal equation of state : compare with the invariant celestial sphere at the level above the level of the moving Sun and planets.

From this brownian kernel constructed on the basis of the numbers $\binom{n}{k}$ can then be extracted successively gaussian Huygens' spheres, Poincaré-Cartan forms and Hermite polynomials, monochromatic waves, brownian perspective, relativistic brownian motion and Klein-Gordon equation, Lagrange's equations in the vicinity of the monochromatic state, and then, classically, geodesics, static ds^2 to be compared with monochromatic waves, basic Archimedes' lever identity (1 + 1 - 2 = 0) ruling curvature (E. Cartan, John A. Wheeler), Newtonian ds^2 and Schwarzchild, Poisson's equation and Einstein's equation. From this construction arched by the semigroup property arises the following conjecture naturally : Is gravitation an isostatic result of ergodism ?

Now, what is the origin and the real content of the laws of Nature ? It seems that two basic ideas are involved in these laws which are universal : As *the large numbers laws* also are universal, if they contain the germs of what we shall then have to call the so-called laws of Nature, which way are they going to give us these *geometrical laws* ?

In the particular case of Statics, and though it is based on experiment, symmetry is enough to explain this theory since it is founded upon Archimedes' lever, which arises from the symmetric figure associated with the balance, and which leads to Galileo's virtual velocities, d'Alembert's virtual displacements, and Lagrange's equations [13].

And Dynamics ? As Dynamics = Statics + the inertia forces + the relativistic formalism, we have to find out the germs of the inertia forces and of the relativistic formalism. Now, let us look at the large numbers laws carefully. The weak and strong laws lead to a degenerescence of the stochastic distribution, and the Laplace-de Moivre theorem does not, at the price of what we shall call the Laplace torsion it leads to the gaussian distribution, and more, this theorem implies the weak and strong laws. And more, the statement of the Laplace-de Moivre theorem begins with an initial whatever which results in the final gaussian distribution, which is indefinitely divisible and so, involves the semi-group property.

We here show that the inertia forces and the relativistic formalism come from the semi-group property, with all their germs coming from the combinatorial numbers $\binom{n}{k}$. In particular, the basic relativistic invariants come from the Laplace-de Moivre theorem, and then are reducible to the combinatorial numbers $\binom{n}{k}$, either directly, or from an involved brownian perspective effect. Or in brief, the large numbers laws conveniently normalized lead to the space-time dynamics.

2 - THE INERTIA FORCES FROM THE SEMI-GROUP PROPERTY

(Track 2) **Semi-group structure and brownian kernel.** Any semi-group Q of non-negative matrices Q_t^S defined on products $E_s \times E_t$ (s<t) of discrete spaces E_t $(s,t \in T)$ with $Q_t^s Q_u^t = Q_u^s$ $(s < t < u)$ and with $\sup_T \operatorname{Card} E_t < \infty$ involves :

(1) a semi-group Q^* of non-negative matrices, called the reduction of Q, of which the matrices Q_t^{*s} are defined on the discrete spaces $E^*_{s_\pm} \times E^*_{t_\pm}$ where $E^*_{t_-}$ (*resp.* $E^*_{t_+}$) is the exit (resp. entrance) boundary at time t

(2) an expansion of the time parameter set T according to Baire's theory

$$T = \sum_{o \leq \beta < \gamma} (T_\beta - T_{\beta+1})$$

with $\gamma \leq \sup_T \operatorname{Card} E^*_{t_\pm}$, and with

$$T_\beta - T_{\beta+1} = \sum_{0 \leq \alpha < \alpha(\beta)} (T_\beta^\alpha - T_\beta^{\alpha+1}) + (T_\beta^\Omega - T_{\beta+1}) ,$$

where $\alpha(\beta)$ is the first ordinal such that

$$T_\beta^{\alpha(\beta)} = T_\beta^{\alpha(\beta)+1} = \ldots = T_\beta^\Omega ,$$

with $T_0 = T$, $T_1 = T_1^0$ = the set of discontinuity of Q*, T_1^α = the derived set of order α of T_1 $(1 \leq \alpha \leq \Omega)$, and more generally with $T_\beta = T_\beta^0$ = the set of discontinuity of Q^* restricted to $T_{\beta-1}$, when β is of the second class, T_β^α = the derived set of order α of $T_\beta (1 \leq \alpha \leq \Omega)$.

On each connected component of $(T - T_\beta)$ intersected with $T_{\beta-1}^\Omega$, we find the germs of analyticity of class 1 for the associated random paths, the germs of connection between the boundary spaces, and the germs of continuity for the reduced semi-group $Q^* = \sum_{\eta \geq 0} {}^{(n)}Q^*$, where ${}^{(n)}Q^*$ is the transition matrix involving n random jumps exactly. And stationarity, reached for instance in the long run, allows these jump processes to approximate (or to be approximated by) a brownian process. (See [5] and [3] for details).

(Track 1) **Equation of continuity**. In brief and heuristic words, we have two things to realize. *First*, the left and right derivatives of a diffusion are adjoint, this relation commands the continuity of the diffusion. *Secondly*, Huygens' principle adapted to a diffusion commands its connection. The continuity is expressed by the closeness of a differential form ω_1. The connection compatible with Huygens'principle is expressed by the closeness of a differential form ω_2. The form ω_1 gives us the equation of continuity. The form ω_2 gives us the inertia forces and the equation of motion. The form ω_2 is the germ of the Poincaré-Cartan form in Dynamics ; the passage from a dispersive diffusion to a diffusion adapted with a wave propagation is at the root of the space-time structure.

So, let us begin with the brownian kernel G associated with the brownian motion. Between the left derivative L and the right derivative R of the brownian motion X_t, we have the basic relation

$$\int E[(f(Lg) + (Rf)g)(X_t)]dt = 0,$$

where f and g are functions with compact space-time support, expressing that L and R are adjoint with respect to the measure $\rho dq^3 dt$, with the expectation E involving the measure ρdq^3. From this basic equation, we deduce the Fokker-Planck equations $L^+\rho = 0$ and $R^+\rho = 0$ directly , where L^+ and R^+ are adjoint to L and R with respect to the space-time Lebesgue measure $dq^3 dt$. The difference between the Fokker-Planck equations gives us the continuity equation $\partial_t \rho + div(\rho v)$ where ρ, as above, is the density of the brownian kernel G with respect to the space-time Lebesgue measure, and where v is the mean velocity of the brownian paths passing by the fixed point q at the fixed time t. And we can verify that, writing up the closeness of the space-time differential form $\omega_1 = d(\log \rho D)$ explicitly, where D is the diffusion coefficient expressing the dispersion of the brownian paths, we get this equation of continuity directly, that is, this equation is obtained directly on the basis of the relation $dd=0$ where the exterior derivative is applied twice, and which is related to the relation $\partial\partial = 0$ (the boundary of a boundary is zero) in Stokes' formula [14], [12], [17], [18].

(Track 2) **Left and right derivatives.** The brownian paths arriving at the point q at time t have a mean backward velocity $v_- = q / t$ since they started from $q=0$ at time $t=0$. And the extension of these brownian paths from point q and time t, that is, the brownian paths starting from point q at time t, have the mean forward velocity $v_+ = 0$ by the very symmetry of the brownian random walk. So, taking account of the dispersion of the paths expressed by the diffusion coefficient D, we have the left (backward) and right (forward) derivatives

$$(Lf)(q) = (\partial_t f + v_-.\nabla f - \bar{D}\Delta f)(q)$$

$$(Rf)(q) = (\partial_t f + v_+.\nabla f + D\Delta f)(q)$$

here reduced to

$$Lf = \partial_t f + \frac{q}{t}.\nabla f - D\Delta f \text{ and } Rf = \partial_t f + D\Delta f$$

So, we can verify that we have the basic relation above expressing the adjointness of L and R, directly. The adjoints L^+ and R^+ with respect to the space-time Lebesgue measure are

$$L^+ g = \partial_t g + div(gv_-) + \Delta(Dg) \quad \text{and} \quad -R^+ f = \partial_t f + div(fv_+) - \Delta(Df)$$

By adding $L^+\rho = 0$ and $-R^+\rho = 0$, we get the continuity equation $\partial_t\rho + div(\rho v) = 0$ where $v = (v_+ + v_-)/2$.

(Track 1) **Germ of the Poincaré-Cartan form.** *Secondly,* Huygens' principle can be expressed, not with Huygens' spheres, but with Gauss-Huygens' spheres : By taking gaussian spherical distributions, that is, by taking the brownian kernel G instead of the usual Huygens' spherelets, and in order to get the envelope of two neighbouring Gauss-Huygens' spherelets, we here have to consider the difference between $G(q+dq)$ and $G(q)$ at time t.

The identity

$$e^{-(A+B)^2} - e^{-A^2} = e^{-A^2}(e^{-2AB-B^2} - 1)$$

gives us at fixed time t

$$G(q+dq) - G(q) = G(q)(e^{p.\theta - \frac{1}{2}\theta^2 Q} - 1)$$

where $p = q/t$, $Q = 3.2D/t$, $\theta = dq/2D$, and taking account of the Laplace torsion $dq^2 = 2Ddt$, $\theta^2 = dt/2D$. So we have

$$d(\log G) = (2D)^{-1}\left[-H_1.dq + \frac{1}{2}H_2 dt\right]$$

where we recognize the Hermite polynomials $H_1 = p$ and $H_2 = p^2 - Q$. We emphasize that we have used the Laplace torsion after a space differentiation instead of effecting an ordinary space-time differentiation, because the translation of the Poincaré-Cartan form $Mv.dq-Edt$ into $(Mv+A).dq-(E+V)dt$ resulting from Coriolis' holonomic liaisons or constraints [9] shows that the Laplace torsion gives us the right way.

And we recognize in the following differential form

$$\omega_2 = -\omega_1 = -d(\log G) = (2D)^{-1}(H_1 dq - H_2 dt)$$

the probabilistic germ of the Poincaré-Cartan form. So, writing up the closeness of the differential form ω_2 explicitly, we get the formal probabilistic germ of the inertia forces, even though such forces arise in fact from diffusions interacting on themselves, or something of the sort.

(Track 2) **The Poincaré-Cartan forms ω_1 and ω_2.** Let us now consider a diffusion. Starting from the Fokker-Planck equations $L^+\rho = 0$ and $R^+\rho = 0$, we obtain $\mathrm{grad}(\log \rho D) = D^{-1}(v_+ - v_-)/2$. So, the space Poincaré form

$$\omega_1| = d(\log\rho D) = D^{-1}\frac{v_+ - v_-}{2}.dq$$

is closed, and its extension from space to space-time is the following closed Poincaré-Cartan form

$$\omega_1 = d(\log\rho D) = D^{-1}\frac{v_+ - v_-}{2}.dq + \partial_t(\log\rho D)dt,$$

writing down $d\omega_1 = 0$ explicity gives us the equation of continuity.

Now, the following space Poincaré form

$$\omega_2| = D^{-1}\frac{v_+ + v_-}{2}.dq\ ,$$

when it is closed in space, expresses the existence of fronts resulting either from symmetry, or from some process of regularization. Then, $\omega_2|$ extended from space to space-time is the Poincaré-Cartan form

$$\omega_2 = D^{-1}\frac{v_+ + v_-}{2}.dq + (something)dt,$$

and by writing down $d\omega_2 = 0$ explicitly we get the equation of motion involving the inertia forces.

For instance, the diffusion associated with the Schrödinger equation has the Poincaré-Cartan forms $\omega_1 = d(\hbar R)$ and $\omega_2 = d(\hbar S)$ with $\psi = e^{R+iS}$ and $D = \hbar / 2M$. This diffusion takes place in the Euclidean space on which the wave function ψ is also defined, and is constructed from its right derivative according to the following diagram :

$$\begin{array}{ccc} Euclid \times Time & \rightarrow & Euclid \times Time \\ \uparrow & & \downarrow \\ R = \partial_t + v_+\nabla + D\Delta & \leftarrow & i\partial_t\psi = -D\Delta\psi \end{array}$$

The diffusion associated with the Klein-Gordon equation, and called the relativistic brownian motion [6], is such that $\omega_1 = d(\hbar R)$ and $\omega_2 = d(\hbar S)$ with $\psi = e^{R+iS}$ and $D = \hbar / 2M$. The mean space-time velocity $v = (v_+ + v_-)/2$ is such that its time component v^o = ic. The wave function ψ is defined on the Minkowski space and is a solution of the Klein-Gordon equation. This diffusion is constructed in the 4 dimensional Euclidean space according to the following diagram :

$$\begin{array}{ccc} Euclid & \rightarrow & Minkowski \\ \uparrow & & \downarrow \\ R = v_+\nabla + D\Delta & \leftarrow & \psi \text{ from K.-G. equation} \end{array}$$

The mass M is variable. In particular, in the proper frame, we have $M_o \neq m_o$ with m_o constant ; the fact that M_o is not constant expresses the deformation of the wave during its evolution, and the interaction on itself [4].

(Track 1) **Space-time periodicity and semi-group property**. Now, in order to describe the propagation of a wave on the basis of the semi-group property initiated by Huygens, Jacobi's identity, that is, the Fourier space expansion of the gaussian brownian kernel, shows us that we have to change t into it in order to extend the periodic behavior from space to space-time, as it must be since the space Poincaré form $\omega_2|$, necessarily closed for a wave, can and must be extended into the closed space-time Poincaré-Cartan form ω_2, the phase increment having to be expressed in terms of dq and dt.

The diffusion associated with this wave propagation has two distinct forms ω_1 and ω_2 Writing down $d\omega_1 = 0$ explicitly gives us the equation of continuity. Writing down $d\omega_2 = 0$ explicitly gives us the equation of motion, involving the inertia forces.

The kernel is now $K_t = G_{it}$ that is,

$$K = \frac{1}{(4\pi Dit)^{3/2}} \exp(-\frac{q^2}{4Dit}) dq^3 dt.$$

It satisfies the Schrödinger equation, of which it propagates the solutions. Or to put it in heuristic language, and observing at first that this Gauss-Huygens spherical propagator, when submitted to a Fourier expansion giving Jacobi's identity, involves and gives us the monochromatic waves, let us say that this "flat" propagator satisfies the "flat" (linear) differential equation, and propagates not only the "flat" (monochromatic) waves, but also the general solution of the "flat" equation just as well, when in fact the wave works its own propagation ("law without law", except the large numbers laws). And we ourselves are only working with a schema based on the large numbers laws (the Laplace-de Moivre theorem and the subsequent weak and strong large numbers laws), which gives us an approximation at some definite scale, and in particular equations of motion which will soon appear as conservation laws exactly.

BUT, such a description of the propagation of a wave is NOT compatible with the Galilean group, and more precisely the phase is NOT an invariant of the Galilean group. In order to recover the invariance of the phase, we must consider a propagation in space-time, in other words, the brownian kernel must be 4 dimensional, and the differential operator involved by the wave propagation appears to be the d'Alembertian, we are here at the root of the relativistic formalism.

3 - THE RELATIVISTIC FORMALISM FROM THE SEMI-GROUP PROPERTY OR FROM THE COMBINATORIAL NUMBERS

(Track 1) **The Lorentz group.** The propagator is now

$$K = \frac{1}{(4\pi DiT)^{4/2}} \exp(-\frac{q^2 - c^2t^2}{4DiT}) dq^3.cdt.$$

We have changed t into it, and we make the distinction between the coordinate time t involved in the quadratic form $q^2 - c^2t^2$ where $-c^2t^2 = (ict)^2$ and the parameter time T involved in the variance DT.

We call inertial frame any proper frame associated with a monochromatic wave, and we want to describe the propagation of a monochromatic wave seen from some inertial frame. In order to get the germs of the fundamental relativistic invariants, we approximate the propagator by a space-time binomial distribution, in other words we approximate the brownian space-time paths associated with the propagator by zigzag paths. And our time unit will be the lifetime of any zig or zag.

So, the Laplace-de Moivre theorem, heuristically written in the following form

$$\frac{\prod_{j=1,2,3,0} \binom{n}{k_j} \frac{1}{2^n}}{\frac{1}{(4\pi DiT)^{4/2}} \exp(-\frac{q^2 - c^2t^2}{4DiT}) dq^3.cdt} \to 1$$

gives us $q^2 - c^2t^2$, $dq^3.cdt$, and DT invariant. In fact, the numerator at left hand and the number 1 at right hand are invariant. Hereby, the denominator at left hand is globally invariant.

This denominator is made of three separate factors, of which we can verify the invariance separately. Let us express DT at first in the proper frame of the propagated and observed monochromatic wave, in order to have it invariant at once. For any fixed integer n, the discrete approximation of $dq^3.cdt\,/\,(4\pi DiT)^{4/2}$ is an integer invariant, and so $dq^3.cdt$ is itself invariant, and so is the exponential also, which makes the quadratic form $q^2 - c^2t^2$ invariant. And finally, we have now DT invariant when expressed in such or such inertial frame, we shall analyze this invariance just below.

Let us now start from the exponential, as it gives us the number of zigzag paths arriving at the distance $\sqrt{L^2}$ where $L^2 = q^2 - c^2t^2$, we have the quadratic form invariant directly, and so $dq^3.cdt$ is invariant also, since DT is expressed at first in the proper frame.

Putting $2D = \hbar / M$, we also have

$$\begin{aligned}\exp(-\frac{L^2}{4DiT}) &= 1 - \frac{1}{4DiT}L^2 + ...\\ &= 1 + \frac{i}{2\hbar}\frac{M}{T}L^2 + ...\end{aligned}$$

invariant, so we see that $\hbar / ML^2T^{-1}$ is invariant with L^2 invariant, and with MT^{-1} also invariant since at first expressed in the proper frame, so $\hbar$ is itself invariant. And finally, we see that DT and MT^{-1} are themselves also invariant, in particular the propagator is isotropic in any inertial frame. However, let us take notice that in $\hbar = ML^2T^{-1}$, L^2 is quadratic invariant, and that in $dq^3.cdt = L^2.LcT$, L^2 is transverse invariant with LcT invariant.

So, the phase increment *pdq-Edt* is also invariant, as it must be, and as it is shown in the following Dictionary. By Fourier transform, we shall obtain the invariance of the quadratic form $E^2 / c^2 - p^2$ also.

BUT up to now the limit velocity c is NOT invariant, this property is completely independent of the Lorentz group which is directly involved by the combinatorial numbers $\binom{n}{k}$, and we cannot hope to obtain the limit rule "c+c=c", or equivalently, the torsion rule "1+1=1", from these combinatorial numbers directly. The local invariance of c^2 is due to a brownian perspective effect, and globally or universally, c^2 is either constant, or more generally must take into account the gravifying masses.

(Track 2) **Dictionary**. From $L^2 = q^2 - c^2t^2$ invariant, we get the length contraction $L = L_0\sqrt{1-\beta^2}$ with $\beta = v / c$. Since LcT is invariant, we also have $cT = cT_0\big/\sqrt{1-\beta^2}$. We then have $cM = cM_0\big/\sqrt{1-\beta^2}$ which gives us the quadratic form $E^2/c^2 - p^2$ invariant, we shall see below that this invariance also appears whith the Fourier transform, taking account of the phase invariance (we write cT_0 for $(cT)_0$ or c_0T_0, et cetera).

The phase increment is given by

$$\begin{aligned}pdq - Edt &= (pq - E)dt\\ &= -cM(1-\beta^2).cdt\\ &= -cM_0.cdt_0 \quad \text{invariant, as it must be.}\end{aligned}$$

Now, by Fourier transform where $S_n = X_1 + ... + X_n$ with $X = \pm 1$ and $p = q = \frac{1}{2}$, from the characteristic function $E(e^{i\theta S_n}) = (\cos\theta)^n$ we get

$$P\{S_n = 2k - n\} = \frac{1}{2\pi}\int_{-\pi}^{+\pi} e^{i(2k-n)}(\cos\theta)^n d\theta$$

and so we have the Laplace-de Moivre theorem (of which, up to now, we have only used the combinatorial proof based on Stirling's formula [16])

$$\Pr ob\left\{a \le \frac{S_n}{\sqrt{n}} \le b\right\} = \sum_{a \le \frac{2k-n}{\sqrt{n}} \le b} \frac{1}{2\pi}\int_{-\pi}^{\pi} e^{i(2k-n)}(\cos\theta)^n d\theta$$

$$\to \int_a^b dx \frac{1}{2\pi}\int_{-\infty}^{\infty} e^{ixy} e^{-y^2/2} dy$$

$$= \int_a^b \frac{1}{\sqrt{2\pi}} e^{-x^2/2} dx$$

where $\frac{2k-n}{\sqrt{n}} = x$ but $\frac{1}{\sqrt{n}} = dx$, and where $\theta\sqrt{n} = y$

Since xy (which stands for $pq - Et$)) is invariant, and x^2 (which stands for $q^2 - c^2t^2$) is also invariant, y^2 (which stands for $p^2 - E^2/c^2$) is then invariant.

In brief

$q^2 - c^2t^2$ invariant $\qquad$ $E^2/c^2 - p^2$ invariant

$L = L_0\sqrt{1-\beta^2}$ $\qquad$ $cT = cT_0/\sqrt{1-\beta^2}$

$cM = cM_0/\sqrt{1-\beta^2}$

$pdq - Edt$ invariant

4 - THE INVARIANCE OF c^2 FROM THE BROWNIAN PERSPECTIVE

(Track 1) **Equation of state.** From a thermodynamical viewpoint, the wave propagation described with the 4 dimensional brownian kernel is ruled by an isothermal equation of state, when geometrically the propagation of a wave in a perfect gas is adiabatic necessarily (Laplace). Writing $2D = \hbar/M$, and expressing the constant $\hbar$ in terms of mass M, length L, and time T, we have $\hbar = ML^2T^{-1}$ invariant, with MT^{-1} invariant, and with L^2 invariant like $q^2 - c^2t^2$. The equation of state is $2DE = \hbar c^2$ (read it "volume x pressure = constant") or equivalently $E = Mc^2$, which means that all monochromatic waves are similar.

Conversely, and in heuristic language, let us associate a net with any perfect gas. In such a correspondence, we associate knots with collisions, threads segments between consecutive knots with free paths, that is, straight segments between consecutive collisions, and meshes with some local parameters such as mean free paths, mean free times, et cetera. And now, let us replace any such straight segment, or thread segment, by a brownian bridge made of (a large number of) brownian paths with primitive diffusion coefficient D_0. In such a brownian net as this, necessarily seen in space-time, and taking the space size of a mesh for L, this length is (scaling) covariant with the wave length Λ. Let us define T in order that LT be scaling invariant. And now let us change M into αM (in notation $M \to \alpha M$) where

M is proportional to the local density of knots and $\alpha \rangle 0$ a scalar. In such a scaling, we have $L \rightarrow L / \alpha$, $T \rightarrow \alpha T$, $D \rightarrow D / \alpha$, and LT, LM, M / T, DT scaling invariant, et cetera, with in particular for the wave length Λ and the period $\mathbf{T}$ the scaling variances $\Lambda \rightarrow \Lambda / \alpha$ and $\mathbf{T} \rightarrow \mathbf{T} / \alpha$, in particular the time T and the period $\mathbf{T}$ are scaling contravariant [4]. From the de Broglie-Planck relation between the phase of the matter wave and the Poincaré-Cartan form, we deduce that $E \rightarrow \alpha E$, so that E/M is scaling invariant, that is, we have $E = Mc^2$ with c^2 scaling invariant : All the monochromatic waves with a same velocity v are similar, and have the same space-time velocity, they have the same proper frame, and they are made of the same brownian thread of which the diffusion coefficient is D_0. And seen from their own proper frame, all the monochromatic waves are similar.

(Track 2) **Brownian net and Relativistic brownian motion.** The brownian net constructed just above is to be understood in particular with the general structure of a non stationary semi-group of non-negative matrices : The reduction Q* of such a semi-group Q describes the random evolution of a sample path from mesh to mesh, and this evolution tends to become a brownian motion.

But as the meshes are made of compressible brownian bridges, then waves propagate through these brownian bridges in the limit as the finite number of threads constituting the brownian bridges becomes infinite [1]. And the mean straight segments of the brownian bridges are the zigs and zags of the zigzag paths of the relativistic brownian motion constituting the diffusion defined by the wave function associated with these waves [6].

(Track 1) **Brownian perspective.** The zigzag paths of the relativistic brownian motion [6] are defined in such a way that their diffusion coefficient D is strictly positive and finite, according to the scale defined by the Laplace-de Moivre theorem. With respect to the parameter time T, this condition implies that for the approximating random walks we have $\Delta x^2 / \Delta T \rightarrow 2D$ with $0 \langle D \langle \infty$ in the limite as $DT \rightarrow 0$, Δx being the pace of the discrete coordinate x.

Let us now consider the brownian paths with primitive diffusion coefficient D_0 in space-time, the sequence of their mutual shocks defining zigzag paths of diffusion coefficient $D = D_0 T_e$ where T_e is the mean life time of such a zig or zag at the flatness situation of equilibrium, and is out of scaling (Harris' relation) [11].

In the limite as the finite number of such brownian paths (or segments) joining two distinct points in space-time becomes infinite, we recover the brownian bridges of the brownian net defined above, of which the mean straight segments in space-time are the zigs and zags of the relativistic brownian motion.

Now, since at the standard scale the zigs and zags of the relativistic brownian motion are the size of points, at such a scale we necessarily have $T_e = 0$. Since $O \langle D \langle \infty$,we then have $D_0 = \infty$ necessarily at the standard scale. In other words, the fluctuations of D_0 are not perceptible at the standard scale.

It can also be shown quite simply that we have $c^2 = 4\pi D_0 / N$ where N is the number of brownian bridges which add up to a wave length. As N is scaling invariant, c^2 appears to be, not necessarily constant in the Universe, but at least locally invariant. In the simplest case where there is no gravifying mass, c^2 is not only locally invariant, but also universally constant.

So, we now have the quadratic form $q^2 - c^2t^2$ invariant, with c^2 invariant. The invariance of $q^2 - c^2t^2$ is directly combinatorial, the invariance of c^2 is also combinatorial but through a perspective effect as it must be since we must have the limite rule "c+c=c".

5 - GRAVITATION

(Track 1) **Introduction to Gravitation.** According to these lines, a real description of Gravitation would take into account the fact that gravifying masses are dense regions in the space-time brownian net, of which the meshes are then distorted and distributed according to ergodization. But such a description is not necessary, because the static schema (Archimedes' principle and lever) is enough, and this schema is only based on the unavoidable principle according to which "the boundary of a boundary is zero" ([14] and [12][17][18]).

The material which follows is now classical, except the fact that we first deduce the Schwarzschild from this principle which, from flatness and *dG=0*, gives us *G=0* outside the gravifying masses. From the Schwarzschild we deduce the relation $c^2 = 1 - 2M/r$ entirely due to Einstein, and the Poisson equation. Then it is possible to deduce the Einstein equation $G = 8\pi T$ from the Poisson equation, but not directly as it is often tried.

From *dG=0* we have $dT = 0$, that is, the equation of continuity and the equations of motion for the gravifying masses explicitly, so these equations are linked and mixed relativistically, this remark is entirely due to Einstein, and in disagreement with a separate treatment of the equations of motion.

(Track 1) **Lagrange's equations and geodesics.** By comparing Lagrange's equations with the equation of geodesics, we are led to consider the following metric tensor

$$ds^2 = -(const. - 2V - 2T)dt^2$$

where the constant will be taken equal to one and will play the role of the velocity of light in empty space, where V is the potential (equal to *M/r* in the Schwarzschild), and where *T* is the kinetic energy. Then Langrange's equations "agree" with the equation of geodesics if d^2t/ds^2 is "negligible".

In the particular case of a static universe like the solar system (Coriolis' holonomic liaisons and Foucault's pendulum and gyroscope), that is, in the case of a metric tensor

$$ds^2 = g_{00}dt^2 + \sum_{i,j} g_{ij}dq^i dq^j,$$

where $g_{0i} = g_{i0} = 0$, and where the $g_{\alpha\beta}$ do not depend on time $(i,j = 1,2,3; \alpha,\beta = 0,1,2,3)$, then we have $d^2t/ds^2 = 0$ exactly since the Christoffel symbols $\Gamma^0{}_{\alpha\beta}$ are equal to zero whatever α and β, and so Lagrange's equations and the equation of geodesics coincide. That is of course the case when the universe is flat, with a metric tensor like in special relativity.

(Track 1) **The Schwarzschild.** Though *G* is not the most general second order tensor satisfying the relation *dG=0* [14], for brievity we shall only say that it is the simplest one. As *G=0* for a flat static universe, we shall have *G=0* outside the gravifying masses. If we consider one gravifying mass only, like in the solar system when the sun is considered as the unique gravifying mass and the planets as test masses only, we are led to solve the equation *G=0* with symmetry conditions with respect to some fixed point in space considered as the singular gravifying region. That can be done analytically [11], but that can also be seen directly from a classical property of the radius of curvature of the parabola due in fact to Apollonius of Perga, this geometric approach is due to J.-A. Wheeler [17].

So we have

$$ds^2 = smthing + \frac{1}{1-2\frac{M}{r}}dr^2 + smthing,$$

and the local relativistic LT invariance gives us the Schwarzschild

$$ds^2 = -(1-\frac{2M}{r})dt^2 + \frac{1}{1-\frac{2M}{r}}dr^2 + r^2(d\theta^2 + \sin^2\theta dq^2)$$

(Track 1) **The Newtonian potential.** The Newtonian degenerescence of the Schwarzschild is

$$ds^2 = -(1-\frac{2M}{r})dt^2 + dx^2 + dy^2 + dz^2$$

and the Newtonian potential $V = M/r$ satisfies the Poisson equation $\Delta V = -4\pi\rho$, where ρ is the density of the gravifying masses with respect to the Lebesgue measure [14][15].

(Track 1) **The Einstein equation.** From the Poisson equation we deduce that the gravitation equation is $G = 8\pi T$ where necessarily $dT = 0$ expresses the relativistic compatibility of the equation with the motion of the gravifying masses : the right hand side of the Poisson equation leads to the equation of continuity, and the continuity equation is relativistically linked and mixed with the equation of motion. In the simplest case we then have $T = \rho v \otimes v$ where $v = (v^o = ic, v^j)$is the velocity of the gravifying masses.

6 - CONCLUSION

The semi-group property, or more simply, the combinatorial numbers $\binom{n}{k}$, give us the inertia forces formally, and the equation of motion at the vicinity of the flatness equilibrium represented by a monochromatic wave which itself arises, through Jacobi's identity, from the gaussian kernel, that is, from the numbers $\binom{n}{k}$. The semi-group property, or more simply, the numbers $\binom{n}{k}$, also give us the relativistic formalism, on the only basis that, between the inertial frames, or equivalently the monochromatic waves, the description of the wave propagation must be coherent, and more precisely that the phase be invariant. The Lorentz group comes from the numbers $\binom{n}{k}$ directly, with the length contraction, the invariance of action, et cetera. The invariance of c comes from a brownian perspective effect which is induced by the numbers $\binom{n}{k}$. And all of that has come by replacing Huygens' sphere by a Gauss-Huygens' sphere.

Acknowledgements : Paul Lévy, Louis de Broglie, Gérard Petiau, John A. Beekman, I. Prigogine, Mme Lemaire-d'Agaggio, John A. Wheeler, in chronological order, and with value of dedication.

REFERENCES

[1] P.W. Anderson, More is different, Science, Vol. 177, n° 4047, pp. 393-396, (1972).

[2] V.I. Arnold, Sur la géométrie différentielle des groupes de Lie de dimension infinie et ses applications à l'hydrodynamique des fluides parfaits, Ann. Inst. Fourier, Grenoble, 16, pp. 319-361, (1966).

[3] R. Baire, Leçons sur les Fonctions Continues, and Thèse, Ann. di Mat. Vol. 3, n° 3, pp 1-32 (1899).

[4] L. de Broglie, La Thermodynamique de la Particule Isolée ou Thermodynamique cachée des Particules, Gauthier-Villars, Paris, (1966).

[5] J.-P. Caubet, Semi-groupes généralisés de matrices positives, Ann. Inst. H. Poincaré, Vol. 1, n° 3, pp. 239-310, (1965). Développement et régularisation d'un semi-groupe généralisé de matrices positives, C.R. Acad. Sc. Paris, t. 268, pp. 1348-1350, (1969).

[6] J.-P. Caubet, Le Mouvement Brownien Relativiste, Lect. Notes in Math., 559, Springer, Heidelberg, (1976). Principe de Huygens, ondes browniennes, ondes brogliennes, mécanique quantique et mécanique classique, C.R. Acad. Sc. Paris, t. 285, pp. 817-820, série A, (1977).

[7] J.-P. Caubet, The Great Fugue of the brownian theory of light, J. Stoch. Anal. Appl., 3, 2, pp. 119-151, (1985).

[8] J.-P. Caubet, Commentaires, in Colloquio sul tema "La Dimensione Scientifica dello Sviluppo Culturale", promosso dall' Unesco e dall' Accademia Europea sotto gli auspici dell' Accademia Nazionale dei Lincei, Atti dei Convegni Lincei 83, pp. 191-194, Roma, (1990).

[9] G. Coriolis, Sur le principe des forces vives dans les mouvements relatifs des machines, Journ. Ecole Polytechnique, Cahier 21, tome 13, Paris, (1832). Sur les équations du mouvement relatif des systèmes de corps, Journ. Ecole Polytechnique, Cahier 24, tome 15, Paris, (1835).

[10] Th. Frankel, Gravitational Curvature, an Introduction to Einstein's Theory, Freeman, San Francisco, (1979).

[11] T.E. Harris, Diffusion with collisions between particles, Journ. Appl. Prob., 2, pp. 323-338, (1965).

[12] A. Kheyfets and J.A. Wheeler, Boundary of a boundary principle and geometric structure of field theories, Intern. Journ. Theor. Physics, Vol. 25, n° 6, pp. 573-580, (1986).

[13] J.-L. Lagrange, Méchanique Analitique, La Veuve Desaint, Paris, (1788).

[14] C.W. Misner, K.S. Thorne and J.A. Wheeler, Gravitation, Freeman, San Francisco, (1973).

[15] H. Poincaré, Théorie du Potentiel Newtonien, Carré et Naud, Paris, (1899).

[16] H. Poincaré, Calcul des Probabilités, 2ème édition, Gauthier-Villars, Paris, (1912).

[17] J.A. Wheeler, A. Journey into Gravity and Spacetime, Scientific American Library, Freeman, New York, (1990).

[18] J.A. Wheeler, On recognizing law without law, Amer. J. Phys., 51, pp. 398-404, (1983). World as system self-synthesized by quantum networking, IBM Journ. Res. Develop., Vol. 32, n° 1, (January 1988). Information, Physics, Quantum : TheSearch for Links, Proc. 3rd Int. Symp. Foundations of Quantum Mechnics, pp. 354-363, Tokyo, (1989).

4th Int. Coll. on Differential Equations, pp. 83-93
D. Bainov and V. Covachev (Eds)

MONODROMIC UNBOUNDED POLYCYCLES

Marcello Galeotti
Istituto di Matematica e Statistica - Università di Ancona, Italy.

Abstract
A conjecture of R. Conti **[1]** is so rephrased: "A plane polynomial vector field of degree n cannot have a monodromic unbounded polycycle with more than n-1 connected components".
Here we prove that these components cannot be more than n.

Keywords: Polynomial vector fields. Poincaré compactification. Monodromic polycycles.

INTRODUCTION

In **[1]** R. Conti has posed the following problem.
Let $\mathcal{V}$ be a polynomial vector field in $\mathbf{R}^2$ of degree n and S a center of the flow generated by $\mathcal{V}$. Suppose N_S, the maximal region of closed orbits surrounding S, is unbounded and denote by ∂N_S its boundary.
How many connected components can ∂N_S have at most?
In **[2]** R. Conti has showed, with examples, that this number is at least n-1, for any n, while it cannot exceed n+1.
He has conjectured that **n-1** is in fact the required maximum.
Conti's Conjecture can be extended by introducing the concept of monodromic unbounded polycycle (of which ∂N_S, described above, is a particular case). Let $\mathcal{V}$ be a polynomial vector field on $\mathbf{R}^2$ and $\tilde{\mathcal{V}}$ its Poincaré compactification on $\mathbf{S}^2$ **[3]**, via the central projection

$\pi : S^2 - S^1 \to T$, where S^1 is the equator and T the tangent plane at the north pole of S^2.

Definition If $\tilde{\Gamma}$ is a monodromic polycycle **[4]** of $\tilde{\vartheta}$, such that $\tilde{\Gamma} \cap S^1 \neq \phi, \tilde{\Gamma} \neq S^1$, then $\Gamma = \pi(\tilde{\Gamma})$ is called a **monodromic unbounded polycycle**.

We recall that a monodromic polycycle is, roughly speaking, a finite union of singular points and open trajectories joining them, no part of which can be contracted on itself, such that a return map is defined in a lateral neighborhood of it.
A connected component of Γ will be called, briefly, an arc.
An open unbounded semi-trajectory (positive or negative) $\gamma \subset \Gamma$ will be called a semi-arc.
We define $B_{k,n}$ as the class of polynomial vector fields of R^2 of degree **n** having a monodromic unbonded polycycle with **k** arcs.

Hence we re-state as follows

Conti's Conjecture $B_{k,n} = \phi$ *if* $k > n-1$

In **§3** we will prove

Theorem 3.1 $B_{k,n} = \phi$ *if* $k > n$

1. Let

$$\vartheta = P(x,y)\frac{\partial}{\partial x} + Q(x,y)\frac{\partial}{\partial y}$$

a polynomial vector field on $\mathbb{R}^2$, with $\deg\mathcal{V} = \max(\deg P, \deg Q) = n$, P and Q having no common factor. Consider $\tilde{\mathcal{V}}$, the Poincaré compactification of $\mathcal{V}$.

If (U;u,v) is the chart of $S^2 = \{(x_1,x_2,x_3): x_1^2 + x_2^2 + x_3^2 = 1\}$

defined by $x_1 > 0$, $u = \frac{x_2}{x_1}$, $v = \frac{x_3}{x_1}$, then

$$\tilde{\mathcal{V}}_{|U} = A(u,v)\frac{\partial}{\partial u} + B(u,v)\frac{\partial}{\partial v}$$

is given [3] by

$$(2) \quad \begin{aligned} A(u,v) &= -u^{n-1} P\left(\frac{1}{u},\frac{v}{u}\right) \\ B(u,v) &= u^{n}\left[Q\left(\frac{1}{u},\frac{v}{u}\right) - v P\left(\frac{1}{u},\frac{v}{u}\right)\right] \end{aligned}$$

In particular

$$(3) \quad \tilde{\mathcal{V}}_{|U\cap S^1} = F(v) = Q_n(1,v) - v P_n(1,v),$$

where Q_n, P_n denote the homogeneous components of degree n of Q and P.

In [5] M. Villarini has proved the following results:

Theorem 1.1 If $\tilde{\mathcal{V}}_{|S^1} \equiv 0$, then **Conti's Conjecture** is true.

Lemma 1.2 (Multiplicity Lemma) Suppose $p = (0,\bar{v})$ is a singular point of $\tilde{\mathcal{V}}_{|U\cap S^1}\left(\tilde{\mathcal{V}}_{|U\cap S^1} \not\equiv 0\right)$, and let k= multiplicity

of $\bar{v}$ as a root of F(v) (k will be called the tangential multiplicity of p).
Then in the half-plane u>o (u<0) of the chart U there are at most k alternately oriented characteristic trajectories at p (that is, k trajectories having p alternately as α and ω limit set).

Corollary 1.3 $B_{k,n} = \phi$ if **k > n+1**.

Because of **Theorem 1.1**, we will assume, in the following, $\tilde{\mathcal{V}}_{|S^1} \not\equiv 0$ and, in particular, $P_n(0,1) \neq 0$ (that is, deg P_n (1,v)= n).

2. In this section we prove preliminary results to

Theorem 3.1 $B_{n+1,n} = \phi$

Throughout we assume, by absurd

$$(H)\, \mathcal{V} \in B_{n+1,n}, \qquad \tilde{\mathcal{V}}_{|S^1} \not\equiv 0$$

Lemma 2.1 If **(H)** holds, then

(i) n is odd;
(ii) if p is a singular point at infinity of $\mathcal{V}$ (i.e. a singular point of $\tilde{\mathcal{V}}_{|S^1}$), its tangential multiplicity is even.

Sketch of the Proof Let $\Gamma = \pi(\tilde{\Gamma})$ a monodromic unbounded polycycle of $\mathcal{V}$ with n+1 arcs, $\tilde{\Gamma} \subset \bar{S}^2_+$ anti-clockwise oriented, $P_n(0,1) \neq 0$.

Let $p_1, p_2, \ldots, p_s$ be the distinct singular points of $\tilde{\mathcal{V}}_{|U\cap S^1}$, ordered according to the anti-clockwise orientation of S^1 $(s \geq 1)$.
If γ_1 is the first semi-arc of Γ in the half-plane $x>0$, according to the anti-clockwise orientation, and $\pi(\tilde{\gamma}_1)=\gamma_1$, it is easily seen that p_1 cannot be the α limit set of $\tilde{\gamma}_1$, otherwise the isocline $P=0$ would have $n+1$ intersection with the projective line $\mathbf{RP}^1$.
Hence p_1 is the ω limit set of $\tilde{\gamma}_1$ and, by the same argument, p_s is the α limit set of $\tilde{\gamma}_{n+1}$.
Therefore $n+1$ is even and n is odd, proving (i).
By a linear change of coordinates in $\mathbf{R}^2$, not reverting the orientation, we can make p_1 be the last singular point of $\tilde{\mathcal{V}}_{|U\cap S^1}$.
Since, by what was said above, the first semi-arc of $\tilde{\Gamma}$ at p_1 has p_1 as ω limit set and the last one has p_1 as α limit set, it follows from the **Multiplicity Lemma** (**Lemma 1.2**) that the tangential multiplicity of p_1 is even.

Any other singular point of $\tilde{\mathcal{V}}_{|S^1}$ can be made to play the role of p_1, so (ii) is proved too ■

Lemma 2.2 If **(H)** holds,

(i) the closure of a semi-arc of $\tilde{\Gamma}$ is not tangent to S^1;

(ii) no isocline is tangent to $\mathbf{RP}^1$.

The proof utilizes elementary facts of algebraic geometry.

Corollary 2.3 If **(H)** holds,

(i) $Q_n - \lambda P_n \not\equiv 0$ for any $\lambda \in \mathbf{R}$;

(ii) if **p=(0,0)** is a singular point of $\tilde{\mathcal{V}}_{|U\cap S^1}$ of tangential multiplicity 2k, then $(1,0,0)\in RP^1$ is a point of multiplicity 2k for $Q=0$ and of multiplicity 2k-1 for $P=0$.

With the assumptions of **Corollary 2.3**, then

$$Q(x,y)=x^{n-2k}\varphi(y)+\text{terms of lower degree in } x$$
$$P(x,y)=x^{n-2k+1}\psi(y)+\text{terms of lower degree in } x$$

By **Lemma 2.2**, it follows, furtherly, that if $\tilde{\gamma}$ is a semi-arc of $\tilde{\Gamma}$ "tending" to $p=(0,0)\in S^1\cap U$ (for t>0 or t<0), $\gamma=\pi(\tilde{\gamma})$ is asymptotically tangent to a line r) $y=y_0$, y_0 being a root of $\varphi(y)$.

Definition The line r, described above, is called a characteristic line at infinity of $\mathcal{V}$, and its multiplicity is defined as the multiplicity of the root y_0 of $\varphi(y)$.

Utilizing the method of σ-process, or directional blow-up (cfr. **[6]**), we are then able to prove the following Lemmas, assuming **(H)**.

Lemma 2.4 Every characteristic line at infinity of $\mathcal{V}$ has even multiplicity.

Lemma 2.5 If p=(0,0) is a singular point of $\tilde{\mathcal{V}}_{|U\cap S^1}$ of multiplicity 2k and y=0 is a characteristic line at infinity of $\mathcal{V}$ of multiplicity 2h, the line v=0 has at least 2k+2h contacts with $\tilde{\mathcal{V}}_{|U}$ in (0,0).

Lemma 2.6 If r_1 and r_2 are two consecutive characteristic lines at infinity of $\mathcal{V}$, there is a region of R^2

filled by open trajectories asymptotically tangent to both r_1 and r_2.

Lemma 2.7 There exists a characteristic line at infinity of $\mathcal{V}$, which we can identify with y=0, such that

(i) there are two semi-arcs, γ' and γ'', of Γ in the half-plane x>0 (x<0) such that, when $|x|$ is large enough,

$$\gamma')\, y = \xi_1(x) = \lambda x^{-q} + o(x^{-q})$$
$$\gamma'')\, y = \xi_2(x) = \lambda x^{-q} + o(x^{-q}),$$

with $\lambda \in R - \{0\}$, q a positive integer, $|\xi_1(x)| < |\xi_2(x)|$;

(ii) if $y = \xi_i(x)$ is the equation of another semi-arc of Γ asymptotically tangent to y=0 in the half-plane x>0 (x<0), then

$$|\xi_i(x)| \leq |\xi_2(x)|$$

when $|x|$ is sufficiently large

3. The results of **§2** allow us to prove

Theorem 3.1 $B_{n+1,n} = \phi$

First of all, we state one more Lemma, whose proof is a matter of straightforward computation.

Lemma 3.2 Let $\tilde{\mathcal{V}}_{|v}$ be given by **(2)** and C) f(u,v)=0 be an algebraic curve of degree r, no branch of which is invariant for $\tilde{\mathcal{V}}_{|v}$.

Then the contacts of C with $\tilde{\mathcal{V}}_{|v}$ are at most **(n+r-1)r**.

Proof of Theorem 3.1 Assuming **(H)** to hold, we construct a cubic **C**, in the plane (u,v), having too many contacts with $\tilde{\mathcal{V}}_{|U}$.

Firstly, we choose coordinates (x,y) in $\mathbf{R}^2$, where $\mathcal{V}$ is defined, and the direction of time t such that:

(i) Γ is anti-clockwise oriented;

(ii) all the singular points of $\tilde{\mathcal{V}}_{|U\cap S^1}$ have non-negative ordinates;

(iii) **y=0** is a characteristic line at infinity of $\mathcal{V}$ satisfying the requirements of **Lemma 2.7** (let 2h the multiplicity of y=0 and 2k≥2h the tangential multiplicity of $p=(0,0)\in S^1\cap U$);

(iv) λ**>0**;

(v) the lines **x=0** and **x=1** cross all the semi-arcs of Γ respectively in the half-planes x≤0 and x>0.

(vi) in the region $D=\{x\leq 0\}\cup\{x\geq 1\}$ $\mathcal{V}$ has no singular point nor any semi-arc of Γ intersects y=0.

In the chart (U;u,v) we call $\gamma_1^-,\cdots,\tilde{\gamma}_{n+1}^-$; $\gamma_1^+,\cdots,\tilde{\gamma}_{n+1}^+$ the semi-arcs of $\tilde{\Gamma}$ (the pre-image of Γ) respectively in the u<0 and u>0 half-plane, ordered by their quotas. Then, by the above assumptions and recalling the proof of **Lemma 2.1,** $\tilde{\gamma}_1^-$ and $\tilde{\gamma}_1^+$ have (0,0) as ω limit set.

Besides, it follows from (v) that each semi-arc $\tilde{\gamma}_i^-$ is unbounded, tending to infinity with direction $\frac{v}{u}=y_i$ (where $(0,y_i)=\gamma_i^-\cap\{x=0\}$), while **u=1** crosses $\gamma_1^+,\cdots,\tilde{\gamma}_{n+1}^+$.

It also follows from (v),(vi) and **Lemma 2.6** that the open region bounded by $\tilde{\gamma}_1^-$ and the negative v half-axis is filled by open unbounded trajectories having (0,0) as ω limit set and tending to infinity with a finite slope $\left(\lim_{u\to\infty}\frac{v}{u}\ \text{finite}\right)$.

The cubic **C** we consider, in the plane (u,v), is given by equation

$$(4)\quad a_1u^3+a_2u^2+\left(b_1u^2+b_2u-1\right)v+\left(c_1u+c_2\right)v^2=0.$$

In fact a certain number of cases have to be examined, but here we bound ourselves to discuss the simplest one, i.e. when 2h=2k=2 (the other ones are treated in a fairly similar way).

Taking into account (i)-(vi) above, we impose on **C** the further conditions:

(c_1) C passes through $P_1=(u_1,0)$, $u_1<0$;

(c_2) $a_2<0$;

(c_3) C passes through $P_2=(0,v_2)$, $v_2<0$;

(c_4) C passes through $P_3=(u_3,v_3)$, $0<u_3<1$, $v_3>0$, P_3 lying above $\tilde{\gamma}^+_{n+1}$;

(c_5) C passes through $P_4=(u_4,v_4)$, $u_3<u_4\leq 1$, $v_4>0$, P_4 lying below $\tilde{\gamma}^+_1$ ($\tilde{\gamma}^+_1$ has equation $v=\lambda u^{q+1}+o\left(u^{q+1}\right)$, $\lambda>0$, and does not cross v=0 when $0<u\leq 1$);

(c_6) C is tangent to $u=u_4$ at P_4.

From **(4)** and **(c_1)-(c_6)**, it follows, through elementary geometric considerations, that **C** consists of two connected components: a Jordan curve J_1 through P_3 and P_4, and an unbounded component J_2 through P_1, O, P_2, tangent to a line $u=\bar{u}$, $\bar{u}<0$, when $v\to\pm\infty$ (**Fig. 1**).

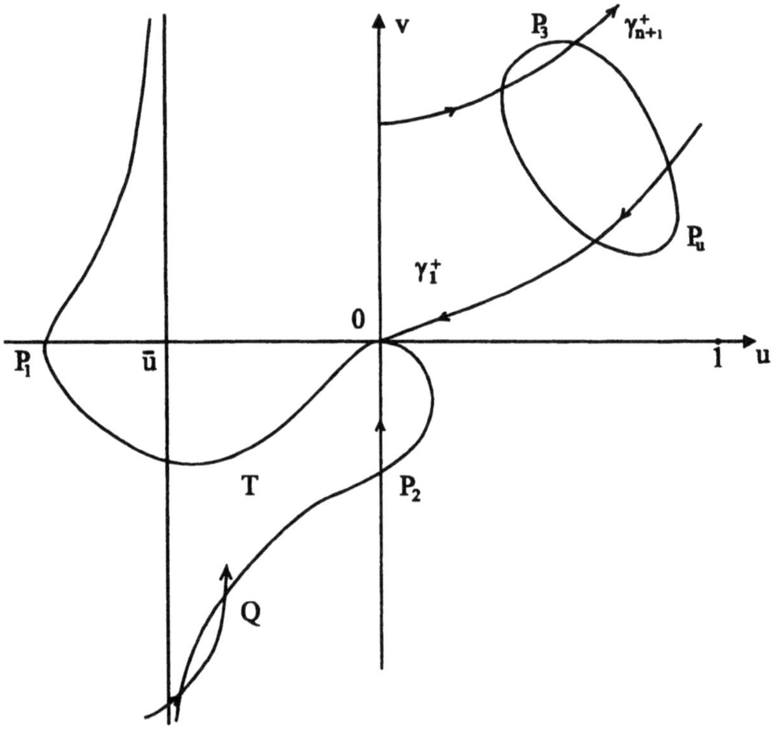

Fig. 1

Let us count the contacts of **C** with $\tilde{\mathcal{V}}_\mu$.

The contacts of J_1 are immediately seen to be at least 2n+2.

As to J_2, it follows from Lemma 2.7 that it has at least 2k+2h (in our case 4) contacts with $\tilde{\mathcal{V}}_\mu$ in (0,0).

Consider the two open arcs $J_2', J_2'' \subset J_2$, with vertex in O, being respectively J_2' the one through P_1 and J_2'' the one through P_2.

Let T be the region bounded by $u=\bar{u}, J_2', u=0, J_2''$ and Q a point of $J_2'' \cap T$, $Q \neq P_2$.

Then, if $\tilde{\mathcal{V}}_\mu$ points inside T at Q, the negative semi-orbit of Q must cross J_2'' again, as it follows from the discussion of assumptions (i)-(vi).

Otherwise, if $\tilde{\vartheta}_{\psi}$ points outside T at Q, there is a contact of J_2'' with $\tilde{\vartheta}_{\psi}$ between Q and P_2.

In any case, J_2'' has at least one contact with $\tilde{\vartheta}_{\psi}$.

Therefore the contacts of the cubic **C** with $\tilde{\vartheta}_{\psi}$ are at least

$$2n+2+4+n+1 = 3n+7$$

whereas, by **Lemma 3.2**, they can be at most

$$(n+3-1)3 = 3n+6 .$$

This contradiction proves the **Theorem** ■

REFERENCES

[1] **R. Conti** - *Centers of Polynomial Systems* - Quaderni Ist. Mat. U. Dini, Univ. di Firenze (1987-88).

[2] **R. Conti** - *On Centers of Type B of Polynomial Systems* - Archivum Mathematicum (Brno), Vol. 26, n.2 (1990), pp. 93-100.

[3] **E. Gonzales Velasco** - *Generic Properties of Polynomial Vector Fields at Infinity* - Trans. Am. Math. Soc., 143 (1969), pp. 201-222.

[4] **Yu.S. Ilyashenko** - *Dulac's Memoir "On Limit Cycles" and Related Problems of the Local Theory of Differential Equations* - Russiam Math. Surveys 40:6 (1985), pp. 1-49.

[5] **M. Villarini** - *Punti Singolari di Tipo Centro per Sistemi Polinomiali nel Piano* - Tesi di Dottorato di Ricerca in Matematica, Firenze (1990).

[6] **V.I. Arnold, Yu.S. Ilyashenko** - *Ordinary Differential Equations*-in Dynamical Systems I, V.I. Arnold and D.V.Anosov eds., Springer - Verlag (1988).

4th Int. Coll. on Differential Equations, pp. 95-103
D. Bainov and V. Covachev (Eds)

UNIFORM SCHEMES FOR SINGULARLY PERTURBED SCALAR INITIAL VALUE PROBLEMS

González-Vera P., González-Pinto S. and Casasús L.
Dpto. Análisis Matemático, Univ. de La Laguna
38271, Tenerife, Canary Islands, Spain.

Abstract. *In this paper certain difference schemes to integrate singularly perturbed scalar initial value problems of the form (1.1) are given.*

Keywords: uniform convergence, difference schemes, singularly perturbed initial value problems.

1. Introduction

Let the singularly perturbed scalar problem be considered,

$$\varepsilon y'(x) = f(x,y),\ y(0) = y_0,\ 0 \le x \le X,\ 0 < \varepsilon << 1, \tag{1.1}$$

where prime (') denotes, as usual, derivative with respect to x.

It is well known that under adequate restrictions on f and the initial value (see hypotheses (H1)-(H4) given later on) this problem has an initial layer with width of order $O(\varepsilon|\log(\varepsilon)|)$, (see, e.g. [1]).

In this paper we are interested in difference schemes with uniform convergence on the small parameter "ε" to the solution $y_\varepsilon(x)$, of (1.1). Schemes of this type have been considered in [1], [2], [3], making use of special non-uniform meshes.

On the other hand, there exist some schemes with uniform convergence on ε, on uniform meshes, namely $\{x_k = kh,\ k = 0,1,\ldots,N=X/h\}$, which are only valid for particular problems. As for the linear case see [1], [4], [5] and for Riccati-type, [6], [7] and [8].

Here, for the general problem (1.1) we give an scheme of these characteristics (uniform convergence on ε on uniform meshes). Such this scheme requires the exact calculation of

$$F(c,y) = \int dy/f(c,y), \text{ where c is a constant.} \tag{1.2}$$

2. The Continuous Problem

In order to guarantee the existence and uniqueness of the solution $y_\varepsilon(x)$, for the problem (1.1) and the uniform convergence of the scheme (3.2)-(3.4) (given later on) we assume the following hypotheses (sufficient but not necessary):

(H1) $f \in C^{(1)}(\Omega)$, *with* $\Omega = [0,X]\times[Y_0,Y_1]$, $Y_0 < Y_1$ *being two real numbers.*

(H2) *There exist two real numbers* $A > 0$ *and* B, *such that*

$$f_y(x,y) \leq -A + Bf(x,y), \text{ for all } (x,y) \text{ in } \Omega.$$

(H3) $f(x,Y_0) > 0$ *and* $f(x,Y_1) < 0$, *for all* x *in* $[0,X]$.

(H4) $Y_0 \leq y_0 \leq Y_1$.

Under these assumptions we have the following result,

Theorem 1. *a) There exists a unique function* $w(x)$, $w \in C^{(1)}([0,X])$ *such that*

$$Y_0 < w(x) < Y_1 \text{ and } f(x,w(x)) = 0, \text{ for all } x \text{ in } [0,X]. \tag{2.1}$$

b) For all x *in* $[0,X]$ *we have*

$$\left.\begin{array}{l} f(x,t) > 0, \text{ if } Y_0 \leq t < w(x) \\ f(x,t) < 0, \text{ if } w(x) < t \leq Y_1 \end{array}\right\} \tag{2.2}$$

c) The problem (1.1) has a unique solution satisfaying

$$m_0(x) \leq y_\varepsilon(x) \leq m_1(x), \text{ for all } \varepsilon \text{ and all } x \tag{2.3}$$

where $m_0(x) = \min\{y_0, w(t) : t \in I_x = [0,x]\}$ *and* $m_1(x) = \max\{y_0, w(t) : t \in I_x\}$.

d) There exist three positive constants α, β *and* γ *independent of* x *and* ε *such that*

$$|y_\varepsilon(x) - w(x)| \leq \beta\varepsilon, \text{ if } x \geq \alpha\varepsilon|\log(\varepsilon)| \tag{2.4}$$

$$|dy_\varepsilon(x)/dx| \leq \gamma(1 + \varepsilon^{-1}\exp(-\varepsilon^{-1}\alpha^{-1}x)). \tag{2.5}$$

Proof. By using (H1)-(H3), it is not a difficult matter to prove a) and b). In order to show c) let us assume that $y(t) > m_1(t)$ for some t in $(0,X)$ ($y(t)$ denotes the exact solution for a fixed ε). Let ζ be in $[0,t]$ such that $y(x)$ takes the absolute maximum in $[0,t]$), then $\zeta \in (0,t)$ because $y'(t) < 0$ (from b)) and $y(t) > y_0$. Therefore, $0 = f(\zeta, y(\zeta))$ and consequently, $y(t) \leq y(\zeta) = w(\zeta) \leq m_1(t)$ (absurd). The case $y(t) < m_0(t)$ is also impossible.

Let us now check (2.4). For this purpose, we assume for simplicity that $B = 0$, then from a)

$$\varepsilon w'(x) = f(x,w(x)) + \varepsilon w'(x), \quad w(0) = w_0. \tag{2.6}$$

From here, by making $d(x) = y(x) - w(x)$ and subtracting (2.6) of (1.1) we get

$$\varepsilon d'(x) = f_y(x,z(x))\, d(x) - \varepsilon w'(x), \quad d(0) = y_0 - w_0, \tag{2.7}$$

where $Y_0 < z(x) < Y_1$ and $f_y(x,z(x))$ is a continuous function on $[0,X]$. Now, integration of (2.7) yields (2.4) (after using (H2)).

In order to show (2.5), derivation with respect to x in (1.1) gives,

$$\varepsilon v'(x) = f_x(x,y) + f_y(x,y)v, \text{ with } v(x) = y'(x). \tag{2.8}$$

Now, by a strightforward integration and (H2), (2.5) is achieved ■

3. The numerical scheme

We next propose an algorithm based on the exact integration of the following autonomous problem (k=0,1,2,...,N=X/h)

$$\varepsilon u'(x) = f(\sigma_k,u),\ u(x_k) = u_k\ (u_0 = y_0),\ x \in [x_k,x_{k+1}], \tag{3.1}$$

where $x_0 = 0$, $x_{k+1} = x_k+h$ and σ_k is any constant fixed in $[x_k,x_{k+1}]$.
Therefore, if F(x,y) is a function such that

$$F_y(x,y) = \frac{1}{f(x,y)} \tag{3.2}$$

with the partial derivatives continuous on $\Omega^* = \Omega - \{(x,w(x)) : x \in [0,X]\}$, then one has the difference scheme (either explicit or implicit),

$$F(\sigma_k,u_{k+1}) - F(\sigma_k,u_k) = \rho,\ \text{if } f_k \neq 0,\ \text{for } k = 0,1,\dots,N-1, \tag{3.3}$$

$$u_{k+1} = u_k,\ \text{if } f_k = 0 \text{ and } u_0 = y_0,$$

where $\rho = h/\varepsilon$, $f_k = f(\sigma_k,u_k)$.

Observe that $\{\sigma_k\}_{k=0}^N$ can be arbitrarily chosen, provided that $\sigma_k \in [x_k,x_{k+1}]$, and a perturbation on the initial value, say $u_0 = y_0 + O(h)$ is also allowed. Furthermore, a global approximation u(x) in the whole interval [0,X] and not only at the mesh-points can be obtained, by putting,

$$F(\sigma_k,u(x)) - F(\sigma_k,u_k) = (x-x_k)/\varepsilon,\ \text{if } x \in I_k = [x_k,x_{k+1}],\ k=0,1,\dots \tag{3.4}$$

On the other hand, in Theorem 2, we shall prove that (3.4) exhibits uniform convergence of order h to the exact solution of (1.1).

Now, we give a second order method (i.e. the order of convergence is h^2) by using the solution of (3.2)-(3.4) as $\sigma_k = x_k$, $k = 0,1,\dots,N-1$:

$$z_{k+1} = z_k f_k^{-1} f(x_k,u_{k+1}) + \varepsilon f(x_k,u_{k+1}) \times$$
$$\{G(x_k,u_{k+1},u_k) - G(x_k,u_k,u_k)\},\ \text{if } f_k \neq 0, \text{ and} \tag{3.5.a}$$

$$z_{k+1} = z_k t_k + hA_k + B_k(1-t_k),\ \text{if } f_k = 0, \tag{3.5.b}$$

where $z_0 = y_0 - u_0$, $f_k = f(x_k,u_k)$, $f_x^k = f_x(x_k,u_k)$, $f_y^k = f_y(x_k,u_k)$, $t_k = \exp(\rho f_y^k)$, $A_k = -f_x^k/f_y^k$, $B_k = \varepsilon A_k/(f_y^k)$ and the function G verifies

$$G_y(x,y,z) = [F(x,y)-F(x,z)]f_x(x,y)/[f(x,y)]^2. \tag{3.6}$$

A global function on [0,X] can be also obtained from (3.5) when replacing z_{k+1} by z(x), u_{k+1} by u(x) and x varying in $[x_k,x_{k+1}]$.

We shall prove in Theorem 3 that $|y(x)-u(x)-z(x)| \leq \lambda h^2$, with λ a constant independent of ε, x and h.

We shall illustrate the application of these schemes for two simple cases, namely:

3.1. The Linear case

Set: $f(x,y) = a(x)y + b(x).$ (3.7)

In this case (3.2)-(3.4) become (for $k=0,1,\dots,N=X/h$; $x_k=kh$)

$$u(x) = (u_k+b_k/a_k)\exp(a_k(x-x_k)/\varepsilon) - b_k/a_k, \quad \text{if } x \in I_k = [x_k,x_{k+1}] \qquad (3.8)$$

$$u_{k+1} = u(x_{k+1}),$$

where $a_k=a(\sigma_k)$ and $b_k=b(\sigma_k)$ with σ_k arbitrarily fixed in I_k, and $u_0 = y_0$.

Now, from (3.5) and (3.8) the estimation z_{k+1} takes the form:

$$z_{k+1} = z_k e_k + h(2a_k)^{-1}\{\beta_k s_k g_k e_k + 2\alpha_k[(e_k-1)/g_k-1]\}, \qquad (3.9)$$

where $z_0 = y_0-u_0$, $\sigma_k = x_k$, and

$s_k = a_k u_k+b_k$, $g_k = \rho a_k$, $e_k = \exp(g_k)$, $\beta_k = a'_k/a_k$, $\alpha_k = (a_k b'_k-a'_k b_k)/a_k$, with

$\rho = h/\varepsilon$, $a'_k = a'(x_k)$, $b'_k = b'(x_k)$.

As a consequence of Theorem 2 we can obtain the uniform convergence (of order h) for u(x) given by (3.8) and if, in adition, we assume that a''(x) and b''(x) are continuous, then $v_k = u_k + z_k$ presents uniform convergence of order h^2 (see Theorem 3, given later on).

3.2. The Riccati Equation

Set: $f(x,y) = c(x)(y-a(x))(y-b(x))$ (3.10)

In this case the scheme (3.4) becomes ($k=0,1,2,\dots,N=X/h$)

$$u(x) = (a_k-b_k t_k(x))/(1-t_k(x)), \text{ if } x \in I_k = [x_k,x_{k+1}], \qquad (3.11)$$

$$u_{k+1} = u(x_{k+1}),$$

where $t_k(x) = (u_k-a_k)(u_k-b_k)^{-1}\exp\{\varepsilon^{-1}c_k(a_k-b_k)(x-x_k)\}$ and a_k, b_k, c_k are obtained by evaluation of the three respective functions a(x), b(x) and c(x) on $x = \sigma_k$, with σ_k arbitrarily fixed in I_k.

4. Convergence

In this section we justify the convergence of the algorithms given by (3.2)-(3.6). We shall denote by δ, λ and τ three constants independent of ε,h.

Theorem 2. *Assume (H1)-(H4) and $f_{yy} \in C(\Omega)$. Then there exists h_0 such that, for all $0 < h \le h_0$, all $0 < \varepsilon \le \varepsilon_0$ and all x in [0,X],*

$$|y_\varepsilon(x)-u_\varepsilon(x)| \le \tau h, \text{ provided that } |y_0-u_0| \le \delta h. \qquad (4.1)$$

Here, $u_\varepsilon(x)$ denotes any continuous function given by (3.2)-(3.4).

Proof. We consider any uniform mesh $\{x_k = kh,\ k = 0, 1,..., N = X/h\}$ and any fixed ε. Now, we rewrite the problem (3.1) in the form

$$\varepsilon u'(x) = f_h(x,u),\ u(0) = u_0 \tag{4.2}$$

where $f_h(x,u) = f(\sigma_k,u)$, if $x \in [x_k,x_{k+1})$, $k = 0,1,...,N-1$, and σ_k is a constant arbitrarily chosen in I_k. It follows, as a consequence of Theorem 1, (consider $\Omega_k = I_k\times[Y_0,Y_1]$, $k = 0,1,..,N$ and apply this Theorem for each k) that (4.2) has a unique continuous solution u(x) (observe that f_h can be discontinuous at the mesh-points $\{x_k\}_{k=1}^{N-1}$). Now, by making $d(x) = y(x)-u(x)$ and subtracting (4.2) from (1.1), we can write

$$\varepsilon d'(x) = f_y(x,y(x))d(x) + r(x),\ d(0) = d_0 = y_0-u_0 \tag{4.3}$$

where $r(x) = [f(x,u(x)-f_h(x,u(x))] - \frac{1}{2}f_{yy}(x,z(x))d^2(x)$, with z(x) belonging to the interval whose end points are u(x) and y(x). Moreover, r(x) is a continuous and bounded function on $[0,X]-\{x_k\}_{k=1}^{N-1}$.
Hence, integration of (4.3) gives

$$d(x) = d_0\exp(\varepsilon^{-1}\int_0^x f_y(s,y(s))ds) + \varepsilon^{-1}\int_0^x r(t)\exp(\varepsilon^{-1}\int_t^x f_y(s,y(s))ds)dt.$$

Now, from (H2) and,

$|f(x,u(x))-f_h(x,u(x))| \le Lh$ and $|f_{yy}(x,z(x))| \le F$, where
$L = \max\{|f_x(x,y)| : (x,y)\in\Omega\}$ and $F = \max\{|f_{yy}(x,y)| : (x,y)\in\Omega\}$,
we get

$$|d(x)| \le Sh + (2\varepsilon)^{-1}RF\int_0^x d^2(t)\exp(\varepsilon^{-1}A(t-x))dt, \tag{4.4}$$

where $R = \exp(|B|(Y_1-Y_0))$ and $S = R(\delta+L/A)$.
If $F = 0$, the proof is trivial, and if $F > 0$, then by defining

$$m(x) = \max\{h^{-1}|d(t)| : t \in [0,x]\},$$

we can write from (4.4) that,

$$2m(x) < 2S + hQm^2(x), \text{ for all } x \text{ in } [0,X], \tag{4.5}$$

where $Q = RF/A$. The proof is easily completed by using the continuity of the function m(x) and the inequality: $m(0) \le \delta < S$ ▪

Theorem 3. *Assume (H1)-(H4) and $f \in C^{(2)}(\Omega)$. Then there exists h_0 such that for all $0 < h \le h_0$, all $0 < \varepsilon \le \varepsilon_0$ and all x in [0,X], we have*

$$|y_\varepsilon(x)-u_\varepsilon(x)-z_\varepsilon(x)| \le \lambda h^2 \tag{4.6}$$

where $u_\varepsilon(x)$ is given by (3.2)-(3.4) as $\sigma_k=x_k$, $k = 0,1,...,N=X/h$, and $z_\varepsilon(x)$ is obtained from (3.5)-(3.6) (note that $z_\varepsilon(x)$ depends on $u_\varepsilon(x)$).

Proof. In this case we set $\sigma_k=x_k=kh$, $k = 0,1,..,N=X/h$. Moreover u(x) satisfies the differential equation given by (3.1). Then, by making

$$d(x) = y(x)-u(x),$$

it follows, by subtracting (3.1) from (1.1) and by using Taylor's Formula, that

$$\varepsilon d'(x) = f_x(x_k,u(x))(x-x_k) + f_y(x,u(x))d(x) + \phi_k(x)h^2, \text{ if } x \in I_k, \quad (4.7)$$

where $I_k = [x_k,x_{k+1}]$, $k = 0,1,\ldots,N-1$ and $\phi_k(x)$ being a continuous function in (x_k,x_{k+1}) which is bounded in the form $|\phi_k(x)| \le 2^{-1}F(1+\tau)^2$ (F and τ are two constants given in Theorem 1).

Now, integration of (4.7) yields, for all x in [0,X],

$$d(x) = d_0 \exp\left(\varepsilon^{-1}\int_0^x g(s)ds\right) + \varepsilon^{-1}\int_0^x r(t)\exp\left(\varepsilon^{-1}\int_t^x g(s)ds\right)dt \quad (4.8)$$

where $d_0 = y_0 - u_0$, $r(x) = p(x) + q(x)h^2$ and

$p(x) = f_x(x_k,u(x))(x-x_k)$, $q(x) = \phi_k(x)$, $g(x) = f_y(x_k,u(x))$, if $x \in (x_k,x_{k+1})$.

Therefore

$$d(x) = z(x) + v(x)h^2, \quad (4.9)$$

where

$$z(x) = d_0 \exp\left(\varepsilon^{-1}\int_0^x g(s)ds\right) + \varepsilon^{-1}\int_0^x p(t)\exp\left(\varepsilon^{-1}\int_t^x g(s)ds\right)dt, \text{ and} \quad (4.10a)$$

$$v(x) = \varepsilon^{-1}\int_0^x q(t)\exp\left(\varepsilon^{-1}\int_t^x g(s)ds\right)dt. \quad (4.10b)$$

Next, the two following statements will be proved,

1) $|v(x)| \le \nu$ (ν is a constant independent of ε and h),

2) z(x) coincides with the function obtained from (3.5)-(3.6).

Assertion 1) follows from

$$\exp\left(\varepsilon^{-1}\int_t^x g(s)ds\right) \le R\exp(\varepsilon^{-1}A(t-x)), \text{ for all } 0 \le t \le x,$$

where $R = \exp(|B|(Y_1-Y_0))$ (use (H2) and (3.1)).

In order to show 2), we rewrite z(x) in the following way,

$$z(x) = z_k \exp\left(\varepsilon^{-1}\int_{x_k}^x f_y(x_k,u(s))ds\right) +$$

$$\varepsilon^{-1}\int_{x_k}^x f_x(x_k,u(t))(t-x_k)\exp\left(\varepsilon^{-1}\int_t^x f_y(x_k,u(s))ds\right)dt, \ \forall x \in I_k, \quad (4.11)$$

with $z_{k+1} = z(x_{k+1})$, $k = 0,1,,\ldots,N-1$, and $z_0 = y_0 - u_0$. Two cases should be now considered:

a) $f(x_k,u_k) = 0$. Then, $u(x) = u_k$ for all x in I_k, because u satisfies

$$\varepsilon u'(x) = f(x_k,u(x)), \ u(x_k) = u_k. \quad (4.12)$$

Therefore, $f_x(x_k,u(x)) = f_x(x_k,u_k)$ and $f_y(x_k,u(x)) = f_y(x_k,u_k)$. From here, (3.5.b) follows.

b) $f(x_k,u_k) \ne 0$. This implies that $f(x_k,u(x)) \ne 0$ for all x in I_k. Moreover, by using (4.12) and (3.2) we have

$$F(x_k,u(x)) - F(x_k,u_k) = \int_{x_k}^{x} u'(t)/f(x_k,u(t))dt = (x-x_k)/\varepsilon,\ x \in I_k. \qquad (4.13)$$

On the other hand, derivation with respect to x in (4.12) gives,

$$\exp(\varepsilon^{-1}\int_t^x f_y(x_k,u(s))ds) = f(x_k,u(x))/f(x_k,u(t)),\ \text{if } x_k \le t \le x \le x_{k+1}. \qquad (4.14)$$

Now, by substituting (4.13) and (4.14) into (4.11) and bearing in mind (4.12), we obtain

$$z(x) = z_k f(x_k,u(x))/f_k + \varepsilon f(x_k,u(x)) \int_{x_k}^{x} [F(x_k,u(t))-F(x_k,u_k)]\times$$
$$f_x(x_k,u(t))[f(x_k,u(t))]^{-2}u'(t)dt, \qquad (4.15)$$

where $f_k = f(x_k,u_k)$ ∎

5. Numerical Examples

In this section we test our schemes on two scalar problems treated in the literature and whose exact solution is known in advance. Comparison with some classical schemes and with other ones more specific for singularly perturbed problems will be made. In order to integrate these problems we consider throughout uniform meshes with mesh-step h.

Example 1 [5].

$$y' = -200\,(y-g(x)) + g'(x),\ y(0) = 10$$
$$g(x) = 10 - (10+x)\exp(-x),\ 0 \le x \le 10.$$

The Table 1 show the exact solution at some points.

Table 1

x	0	2^{-4}	2^{-3}	2^{-2}	2^{-1}	1	2	2^2	2^3	10
y	10	0.55	1.06	2.02	3.63	5.95	8.38	9.74	9.99	10.00

We have used the following methods to integrate this problem:

T-R: Trapezoidal Rule, B-E: Backward Euler Formula,

FDS: Exponentially fitted, uniformly second order accurate, method [5],

D(2,2): Scheme DIRK(2,2) method [5], [9],

D(2,3): Scheme DIRK (2,3) method [5], [9],

EFDIRK: Exponentially fitted 2 stage DIRK scheme [5],

A(i): Scheme (3.8), where i = 1, 2 or 3 according to σ_k= kh, (k+1/2)h or (k+1)h respectively.

B(1): Scheme (3.8)-(3.9) with σ_k= kh.

In Table 2 we display the absolute error

$$E_h = \max\{\,|y(kh) - y_k| : 0 \le k \le N = 10/h\}$$

where y(x) denotes the exact solution and y_k the approximated solution on the knots $x_k = kh$, obtained by the scheme under consideration.

Table 2

SCHEME	$h = 2^{-3}$	$h = 2^{-4}$	$h = 2^{-5}$	$h = 2^{-6}$
T-R	8.52 E+0	7.24 E+0	5.17 E+0	2.63 E+0
B-E	3.83 E-1	7.40 E-1	1.36 E+0	1.98 E+0
FDS	2.14 E-3	1.01 E-3	4.18 E-4	1.38 E-4
D(2,2)	1.35 E+0	1.92 E+0	2.00 E+0	1.24 E+0
D(2,3)	6.26 E+0	5.33 E+0	3.77 E+0	1.81 E+0
EFDIRK	1.00 E-3	4.90 E-4	2.22 E-4	4.05 E-5
A(1)	1.02 E+0	5.02 E-1	2.32 E-1	9.95 E-2
A(2)	4.75 E-1	2.26 E-1	9.37 E-2	3.08 E-2
A(3)	4.03 E-2	4.26 E-2	4.31 E-2	3.74 E-2
B(1)	5.55 E-2	1.30 E-2	2.82 E-3	5.55 E-4

These data show that classic schemes (with A-stability) do not provide good results at this problem for moderate step-size h (because of the initial layer). Our schemes comfirm their convergence order, h for A(i) and h^2 for B(1), since the absolute error is approximately divided by 2 or 4 when the step size changes from h to h/2, and can compete with FDS and EFDIRK.

Example 2 ([4], [7], [8])

$$\varepsilon y' = (1-x)y - y^2, \quad y(0) = 0.5, \quad 0 \le x \le 1. \tag{5.1}$$

This is the Dahlquist's knee problem. Its roots present a crossover at the point x = 1. Thus the root y = 1-x, which is stable if x < 1, loses stability when $x \to 1^-$ and then the root y = 0 gains stability there. Therefore, we cannot guarantee the uniform convergence in this case because the hypothesis (H2) is not satisfied at the point x = 1, y = 0. However, the numerical results given at Table 3 allow to conjecture the uniform convergence.

Here, we compute the maximum difference

$$E_h^{\varepsilon} = \max \{(u_k - v_k),\ k = 0,1,\ldots,N = 1/h\}, \tag{5.2}$$

where u_k and v_k are obtained from (3.11) when we take $\sigma_k = kh$ and $\sigma_k = (k+1)h$ respectively (in both cases we consider the initial value, $u_0 = v_0 = 0.5$).

Observe that in this case a(x) = 1-x and b(x) = 0. Moreover, we shall prove that the exact solution is bounded in this way,

$$v_k \le y(x_k) \le u_k, \text{ where } x_k = kh,\ k = 0,1,\ldots,N = 1/h, \tag{5.3}$$

hence E_k^{ε} is an upper bound of the absolute error.

In order to show (5.3) we first suppose that

$$v_k \le y(x_k) \le u_k,$$

and then we shall prove,

$$v_{k+1} \le y(x_{k+1}) \le u_{k+1}.$$

Observe that $u_{k+1} = u(x_{k+1})$, where the function u(x) satisfies

$$\varepsilon u' = a_k u - u^2, \; u(x_k) = u_k, \text{ with } a_k = a(x_k). \tag{5.4}$$

On the other hand the exact solution of (5.1) verifies

$$\varepsilon y' = a(x)y - y^2, \; y(x_k) = y_k, \tag{5.5}$$

then by making d(x) = u(x) - y(x) and s(x) = u(x) + y(x), it follows

$$\varepsilon d' = (a_k - a(x))u(x) + (a(x)-s(x))d, \; d(x_k) \geq 0,$$

and from here $d(x_{k+1}) \geq 0$, since $(a_k - a(x))u(x) \geq 0$ for all $x \geq x_k$. The lower bound can be also proved in a similar way.
In Table 3 we display the values of E_h^ε .

Table 3

E_h^ε	$\varepsilon = 10^{-1}$	$\varepsilon = 10^{-2}$	$\varepsilon = 10^{-3}$	$\varepsilon = 10^{-4}$	$\varepsilon = 10^{-5}$	$\varepsilon = 10^{-6}$
h = 5.E-2	4.52E-2	5.06E-2	5.00E-2	5.00E-2	5.00E-2	5.00E-2
h = 1.E-2	9.02E-3	1.01E-2	1.00E-2	1.00E-2	1.00E-2	1.00E-2
h = 5.E-3	4.51E-3	5.04E-3	5.07E-3	5.00E-3	5.00E-3	5.00E-3
h = 1.E-3	9.02E-4	1.01E-3	1.02E-3	1.00E-3	1.00E-3	1.00E-3

Observe that although Tables 2 and 3 allow us to guess that an extrapolation procces could be done for our schemes, the existence of an asymptotic error formula in powers of h, has not been proved, so that we can not assure, in general, such that statement.

References

1. *I.P. Boglaev*, Comput. Maths. Math. Phys., 25, no. 4, 30-39, Great Britain (1985).
2. *I.P. Boglaev*, Comput. Maths. Math. Phys., 27, no. 1, 39-48, Great Britain (1987).
3. *S. González-Pinto, L.Casasús and P. González-Vera*, J. Comp. Appl. Math. 35, 217-225, North Holland (1991).
4. *E.P. Doolan, J.J.H. Miller and W.H.A. Schilders*, Uniform numerical methods for problems with initial and boundary layers, Boole Press, Dublin (1980).
5. *J.Carroll*, BAIL III, 167-172, Boole Press Conf. Ser., 6, Boole, Dún Loaghaire Dublin (1984).
6. *J. Carroll*, J. Comp. Appl. Math. 16, no. 1, 9-25 (1986).
7. *M.J. O'Reilly*, Numer. Math. 50, no. 4, 483-501 (1987).
8. *S. González-Pinto, L. Casasús and P. González-Vera*, Computers Math. Applic., Vol 23 no. 1, 75-85, Great Britain (1992).
9. *R. Alexander*, Siam J. Numer. Anal., 16, 1006-1021 (1977).

4th Int. Coll. on Differential Equations, pp. 105-114
D. Bainov and V. Covachev (Eds)

The Operator Theoretical Treatment of Partial Differential Equations: Recent Developments with Applications

MARIÉ GROBBELAAR
Department of Mathematics, Applied Mathematics and Astronomy
University of South Africa, P.O. Box 392, Pretoria 0001, South Africa

ABSTRACT We discuss recently developed abstract theories which generalize the classical theories of semigroups and fractional powers of a closed operator. Applications to linear and nonlinear partial differential equations are presented.

Keywords: B-evolution, fractional powers of a closed pair, empathy

1. INTRODUCTION

Although the concept "semigroup" was named in 1904, the theory started developing rapidly only in the 1940's with the realization that semigroup theory has wide applicability to boundary-value problems for partial differential equations by considering them from the point of view of an abstract evolution equation of the form $u' = Au$, $u(0) = u_0$. In the 1950's the theory of fractional powers of a closed operator in Banach space was developed by among others Balakrishnan [1], Kato [15] and Yosida [27] with the aim of studying nonlinear evolutions equations of the form $u' = A(t,u)u$, $u(0) = u_0$ or $u' = Au + F(u)$, $u(0) = u_0$. During the 1970's several mathematicians studied "two operator" abstract evolution equations by once more using semigroup theory. So for instance Showalter [22] studied the evolution problem $Mu' + Lu = 0$, $u(0) = u_0$, which serves as model for Sobolev equations, by showing that $M^{-1}L$ generates a semigroup. However, it was shown later that the conditions under which such an approach is possible, are not generally met (cf. [26] where Sobolev equations are treated more generally within the framework of the more recently developed theory of B-evolutions - see Section 2). A second related evolution problem is encountered in [3], where H. Brill shows that the evolution equation $\frac{d}{dt}(Bu) = Au$, $(Bu)(0) = y$ where A and B are operators from a Banach space X to a Banach space Y, has unique solution $u(t) = \exp(tAB^{-1})y$, provided that A and B are both closed and B surjective. Since there are cases where boundary-value problems may be reformulated in this form, but where these conditions are not all satisfied (see [18]), there was indeed a strong need to develop a theory which would provide a framework within which "two operator" problems could be studied under weaker conditions.

2. ABSTRACT THEORIES DURING THE EIGHTIES

We introduce two abstract theories developed during the eighties, viz. the theories of B-evolutions [18] and the theory of fractional powers of the generating pair of a holomorphic B-evolution [8].

2.1 The theory of B-evolutions (1982)

Definition 2.1.1 Let X and Y be (complex) Banach spaces and B a linear operator with domain $\mathcal{D}(B)$ in X and range $\mathcal{R}(B)$ in Y. A family $\{S(t): t > 0\}$ of bounded linear operators is called a B-evolution if

$$S(t)Y \subset \mathcal{D}(B) \text{ for all } t > 0,$$

and

$$S(t+s) = S(s)BS(t) \text{ for all } t,s > 0.$$

Theorem 2.1.2 Suppose $<A,B>: X \to Y \times Y$ is a jointly closed pair[1] of linear constant operators with $\mathcal{R}(B)$ dense in Y. If additionally B is closable, then $<A,B>$ is the generating pair of an L-type B-ev $S(t)$ (L-type denotes that $S(t)$ is strongly continuous and that B commutes with the Laplace transform of $S(t)$) iff

$B^{-1}: \mathcal{R}(B) \to X$ is bounded,

AB^{-1} generates a uniformly bounded C_0 semigr $E(t) = BS(t)$, the so-called associate semigroup.

Moreover, if $<A,B>$ is the generating pair of such a B-evolution, then the evolution problem

$$\frac{d}{dt}(Bu) = Au, \; u(t) \in \mathcal{D} \subset X$$

$$(Bu)(0) = y \in Y \qquad (Pr\ A)$$

[1]This is a weaker condition than the closedness of A and B.

has unique solution $u(t) = S(t)y \ \forall \ y \in \mathscr{R}(B)$.

Remark 2.1.3 The closability of B can be lifted in the holomorphic case, i.e. when the associate semigroup is holomorphic, in which case we call the B-evolution $S(t)$ holomorphic. In this case $S(t)y$ solves $(Pr\ A)$ uniquely for all $y \in Y$. If, however, B is non-closable and $E(t)$ non-holomorphic, an additional condition, which is hard to verify in applications, is required [18,p.296].

The reader is referred to [25,10] for the construction of $B(t)$-evolutions with generating pair the time-dependent pair $<A(t),B(t)>$ by Grobbelaar.

In order to treat nonlinear "two space" evolution problems, a second abstract theory was developed, viz.

2.2 Fractional Powers of the generating pair of a B-evolution (1986)

The concept of fractional powers of the generating pair of a B-evolution was introduced by Grobbelaar [8] with the aim of studying nonlinear parabolic partial differential equations with dynamic boundary conditions in the form of nonlinear evolution equations of the form $\frac{d}{dt}(Bu) = A(t,Bu)u$ [9] or also $\frac{d}{dt}(Bu) = Au + F(u)$ [11,12]. A heuristic approach was used: by recalling that whenever $<-A,B>$ is the generating pair of a B-evolution, then $-AB^{-1}$ is the infinitesimal generator of a semigroup, we constructed fractional powers of A and B separately such that the relation $B^{[\alpha]}$[2]$A^{-[\alpha]} = (AB^{-1})^{-\alpha}$, where the fractional powers, $(AB^{-1})^{\alpha}$, $0 < |\alpha| < 1$, are the fractional powers in the classical sense [27], are satisfied. We first constructed fractional powers $B^{[\alpha]}$, $0 < |\alpha| \leq 1$, of the operator B, by generalizing a representation for B^{-1} which holds in B-evolution theory [18]. Fractional powers $A^{[\alpha]}$ of A were then constructed by making use of the newly constructed fractional powers of B and the fractional powers in the classical sense of AB^{-1}.

Definition 2.2.1 Let the jointly closed pair $<-A,B>$: $\mathscr{D} \subset X \to Y \times Y$ be the generating pair of a holomorphic[3] B-evolution. Then the fractional powers $<A,B>^{[\alpha]}$ of the pair $<A,B>$ are defined by $<A,B>^{[\alpha]} = <A^{[\alpha]},B^{[\alpha]}>$ where $A^{[\alpha]}$

---

[2]The notation $B^{[\alpha]}$ is used to emphasize the fact that the fractional powers thus defined are not fractional powers in the usual arithmetical sense. This property is only recovered by the composition $B^{[\alpha]}A^{-[\alpha]}$.

[3]In our seminal paper [8] we restricted ourselves to the holomorphic case in order to avoid the additional condition [18,p.296] which has to be imposed in cases where B is non-closable.

and $B^{[\alpha]}$ are respectively defined by

$$(B^{[-\alpha]})y = \frac{1}{\Gamma(\alpha)} \lim_{n\to\infty} \int_0^\infty ne^{-ns} S(s)(ns)^{\alpha-1} y\,ds \;\forall\; y \in \mathcal{R}(B) = \mathcal{D}(AB^{-1});$$

$$A^{[-\alpha]}y = (B^{[-\alpha]}) \frac{1}{\Gamma(\alpha)} \int_0^\infty s^{\alpha-1} e^{-sAB^{-1}} y\,ds \;\forall\; y \in Y;$$

$$B^{[\alpha]} = (B^{[-\alpha]})^{-1}, \quad A^{[\alpha]} = (A^{[-\alpha]})^{-1}.$$

The reader is referred to [8] for the properties of the generalized fractional powers of the operator pair $<A,B>$, e.g.

For $0 < \alpha < 1$, $0 < \beta < 1$, $\alpha + \beta < 1$

$$B^{-[\alpha+\beta]}y = B^{-[\alpha]}BB^{-[\beta]}y = B^{-[\beta]}BB^{-[\alpha]}y \;\forall\; y \in \begin{cases} Y \text{ if } B \text{ is closed} \\ \mathcal{R}(B) \text{ otherwise} \end{cases},$$

$$A^{-[\alpha+\beta]}y = B^{-[\beta]}BA^{-[\alpha]}B^{[\beta]}A^{-[\beta]}y = B^{-[\alpha]}BA^{-[\beta]}B^{[\alpha]}A^{-[\alpha]}y \;\forall\; y \in Y.$$

3. FURTHER ABSTRACTIONS DURING THE NINETIES

3.1 The theory of empathies (1988-1993)

During the late eighties and early nineties Sauer developed the concept of empathies of operators [19,20,21]. This is a far reaching generalization of the theory of B-evolutions with among others the advantage that in applications to evolution problems of the form $\frac{d}{dt}(Bu) = Au$, $u(t) \in \mathcal{D} \subset X$, $(Bu)(0) = y \in Y$, the closability of the operator B is not required, even in the non-holomorphic case. We will restrict ourselves to (C_0) empathies:

Definition 3.1.1 Let X and Y be Banach spaces with $\{S(t):\ t > 0\}$ a family of bounded linear operators from Y into X and $\{E(t):\ t > 0\}$ a semigroup in Y. The family $\{<S(t),E(t)>:\ t > 0\}$ is called an empathy of class (C_0) if

$S(t+s) = S(t)E(s) = S(s)E(t)$; $\quad s, t > 0$,
$E(t)$ is of class (C_0).

Theorem 3.1.2 Suppose the pair $<A,B>$: $\mathscr{D} \subset X \to Y \times Y$ is jointly closed and that B has a bounded inverse B^{-1}: $\mathscr{R}(B) \to X$. If AB^{-1} generates a semigroup $\{E(t)$: $t > 0\}$ of class (C_0) in Y, then $<A,B>$ is the generating pair of an empathy $\{<S(t),E(t)>$: $t > 0\}$ of class (C_0).

Moreover, if $<A,B>$ is the generating pair of an empathy $\{<S(t),E(t)>$: $t > 0\}$ of class (C_0), then $v(t) = S(t)y$ solves the evolution problem $\frac{d}{dt}(Bv) = Av$, $(Bv)(0) = y$ for all y in the domain of AB^{-1}, the infinitesimal generator of the semigroup $\{E(t)$: $t > 0\}$.

Remark 3.1.3 If $<A,B>$ is the generating pair of an empathy $\{<S(t),E(t)>$: $t > 0\}$ of class (C_0), then $S(t)$ is almost a B-evolution, i.e. we have $S(t+s)y = S(t)E(s)y = S(t)BS(s)y = S(s)BS(t)y$ for all $y \in \mathscr{D}(AB^{-1})$, s, $t > 0$. Also, if the conditions of Theorem 3.1.2. are met, then B^{-1} has representation in terms of the Laplace transform of $S(t)$, i.e. if $P(\lambda)$ denotes the Laplace transform $\int_0^\infty e^{-\lambda t}S(t)dt$ of $S(t)$, then $B^{-1}y = \lim_{\lambda \to \infty} \lambda P(\lambda)y$ for all $y \in \mathscr{D}(AB^{-1})$ and all λ such that $\mathrm{Re}\lambda > 0$.

From the above it is clear that the concept of an empathy of operators contains that of a B-evolution in the sense that it is easier to show that a pair $<A,B>$ generates an empathy. Moreover, the representation of the inverse of the operator B, which was used in the construction of fractional powers of the generating pair $<A,B>$, of a holomorphic B-evolution, is still at our disposal. Thus the theory of empathies of operators enables us to construct fractional powers of closed pairs of operators under weaker conditions than in [8].

3.2 Fractional powers related to an empathy of operators

Definition 3.2.1 Let the jointly closed pair $<-A,B>$: $\mathscr{D} \subset X \to Y \times Y$ be the generating pair of a (C_0) empathy $\{<S(t),E(t)>$: $t > 0\}$. Then the fractional powers $<A,B>^{[\alpha]}$ of the pair $<A,B>$ are defined by $<A,B>^{[\alpha]} = <A^{[\alpha]},B^{[\alpha]}>$ where $A^{[\alpha]}$ and $B^{[\alpha]}$ are respectively defined by

$$(B^{[-\alpha]})y = \frac{1}{\Gamma(\alpha)} \lim_{n \to \infty} \int_0^\infty ne^{-ns}S(s)(ns)^{\alpha-1}y ds \;\forall\; y \in \mathscr{R}(B) = \mathscr{D}(AB^{-1});$$

$$A^{[-\alpha]}y = (B^{[-\alpha]})\frac{\sin\pi\alpha}{\alpha}\int_0^\infty \lambda^{-\alpha}B(\lambda B + A)^{-1}y\,d\lambda \ \forall\ y \in \mathcal{R}(B) = \mathcal{D}(AB^{-1});$$

$$B^{[\alpha]} = (B^{[-\alpha]})^{-1}, \quad A^{[\alpha]} = (A^{[-\alpha]})^{-1}.$$

Exactly as before we have $B^{[\alpha]}\ A^{-[\alpha]} = (AB^{-1})^{-\alpha}$ where $(AB^{-1})^{-\alpha}$ is the fractional power in the sense of Balakrishnan [1] of AB^{-1}, the infinitesimal generator of the (C_0) semigroup in the empathy. Similarly as in the theory of fractional powers of the generating pair of a holomorphic B-evolution, we can also verify the properties:

For $0 < \alpha < 1$, $0 < \beta < 1$, $\alpha + \beta < 1$

$$B^{-[\alpha+\beta]}y = B^{-[\alpha]}BB^{-[\beta]}y = B^{-[\beta]}BB^{-[\alpha]}y,$$

$$A^{-[\alpha+\beta]}y = B^{-[\beta]}BA^{-[\alpha]}B^{[\beta]}A^{-[\beta]}y = B^{-[\alpha]}BA^{-[\beta]}B^{[\alpha]}A^{-[\alpha]}y \ \forall\ y \in \mathcal{R}(B).$$

The reader is referred to [20], where it is pointed out that the theory of C-semigroups [23] can under special assumptions be brought within the scope of the theory of empathies of operators and that problems treated within the framework of integrated semigroups [16] can be treated with the aid of the theory of empathies by allowing solution and initial states (effect and cause) to be in different spaces.

4. APPLICATIONS

We apply the abstract theories discussed above to two problems with dynamic boundary conditions, viz a Navier-Stokes problem and a boundary-value problem for the beam equation. Finally we point out the role played by the basic idea in B-evolution theory, i.e. "cause and effect in different spaces", in the study of the equations of motion for second grade fluids.

4.1 A Navier-Stokes problem with dynamic boundary conditions (NSPDBC)

When a rigid body performs a rotation in a viscous incompressible fluid, the mathematical model is the following Navier-Stokes problem:

$$\frac{\partial v}{\partial t} - \nabla \cdot \mathbf{T}(v) + (v \cdot \nabla)v = 0 \qquad in\ \Omega \times (0,T)$$

$$\nabla \cdot v = 0 \qquad in\ \Omega \times (0,T)$$

$$\frac{d}{dt}[\mathbf{J}\omega(t)] - \int_{\Sigma}(\mathbf{T}(v)\cdot n)\Lambda x\, ds = 0 \qquad on\ \Sigma \times (0,T)$$

$$v = \omega(t)\Lambda x \qquad on\ \Sigma \times (0,T)$$

$$v = 0 \qquad on\ \partial\Omega_1 \times (0,T)$$

$$v(x,0) = v_0(x),\ \omega(0) = \omega_0.$$

Here v is the velocity field for the fluid, $\mathbf{T} = -p + 2\nu\mathbf{D}$ is the stress tensor, $\mathbf{J}$ denotes the moment of inertia tensor of the rigid body, ω its angular velocity, while $\Omega = \Omega_1 \setminus \mathscr{Cl}(\Omega_2)$, with Ω_1 and Ω_2 respectively the interiors of the three-dimensional vessel containing the fluid and the rigid body, denotes the region filled with the fluid. The boundaries of Ω_1 and Ω_2 are denoted by $\partial\Omega_1$ and Σ respectively.

We briefly discuss our Lebesgue space results for this problem [12] (see [11] for our Hilbert space results). We followed the approach of a non-homogeneous evolution problem of the form $\frac{d}{dt}(B_r v) + A_r v = F(v)$, $(B_r u)(0) = y$ where A_r and B_r are operators with common domain $\mathscr{D}$ in $L^{r,3}(\Omega)$ and ranges in $L^{r,3}(\Omega) \times \mathbb{R}^3$, which act on v only. This entailed the construction of a projection in $L^{r,3}(\Omega) \times \mathbb{R}^3$ in order to eliminate the pressure terms occurring in both the Navier-Stokes equation and the dynamic boundary condition. By showing that the pair $<-A_r, B_r>$ generates a holomorphic B_r-evolution which solves the linear homogeneous evolution problem, and then using the method of iterated approximations to treat the non-homogeneous evolution problem, we were able to establish the existence of a unique solution, local in time, for (NSPDBC), for all pairs $<v_0(x), \omega(0)>$ in $Y_r = P_r[L^{r,3}(\Omega) \times \mathbb{R}^3]$, $r > 3$. A crucial result in the treatment of the nonlinear inhomogeneous evolution equation, is the estimate $\|P_r<(v\cdot\nabla)v,0>\|_{Y_r} \le M\|A_r^{\frac{1}{2}}B_r^{-\frac{1}{2}}B_r v\|_{Y_r}^2$, $v \in \mathscr{D}(<A_r^{\frac{1}{2}}B_r^{-\frac{1}{2}}B_r, B_r>)$ for the nonlinear term $F(v) = -P_r<(v\cdot\nabla)v,0>$. This was achieved by among others making use of an interesting relationship between the projection P_r and the classical projection constructed by Fujiwara and Morimoto [6].

4.2 A BVP with dynamic boundary conditions for the elastic beam equation

When a load is attached to an elastic beam of length l which has its end $x = 0$ fixed, while to the end $x = l$ a load is attached and if damping of structural type is incorporated, the following boundary-value problem can be derived by using energy methods [13]:

$$u_{tt} - 2\rho u_{xxt} + u_{xxxx} = 0, \qquad 0 < x < l, \quad 0 < t,$$

$$u(0,t) = u_t(0,t) = u_x(0,t) = u_{xt}(0,t) = 0,$$

$$[u_x + u_{xx}](l,t) = 0,$$

$$[u_{tt} + 2\rho u_{xt} - u_{xxx}](l,t) = 0,$$

$$u(x,0) = u_0(x), \quad u_t(x,0) = v_0(x),$$

$$u(l,0) = \eta, \quad u_t(l,0) = \mu.$$

In this problem the theory of empathies of operators and their related fractional powers play an important part, since the underlying operator pair $<-L,M>$, with L the beam operator pair given by $Lu = <u'''',-u'''(l)>$ for u in a domain $\mathscr{D}$ in $H^4(0,l)$, does not generate a holomorphic M-evolution, but only a (C_0) empathy [14]. In view of the boundary conditions, it turns out that LM^{-1} is self-adjoint and that the damping pair $<-u'',u'(l)>$ is precisely the fractional power $L^{\frac{1}{2}}M^{-\frac{1}{2}}Mu$ in which $L^{\frac{1}{2}}M^{-\frac{1}{2}}$ is self-adjoint. By adding the boundary condition $[u_{xx} - \alpha_1 u_{xxx}](l,t) = 0$, $\alpha_1 \geq 0$ (see [13]) and reformulating the problem as an abstract evolution problem $\frac{d}{dt}(\mathscr{M}U) + \mathscr{L}U = 0$, $(\mathscr{M}U)(0) = F$ in which $\mathscr{L}$ and $\mathscr{M}$ are matrices of operators, the problem could be treated with the aid of the holomorphic $\mathscr{M}$-evolution generated by $<-\mathscr{L},\mathscr{M}>$ and the fractional powers related to the (C_0) empathy generated by $<-L,M>$. Indeed, we obtained a unique solution which exhibits exponential decay with time for any initial displacement in $\mathscr{D}(<L^{\frac{1}{2}}M^{-\frac{1}{2}}M,M>) = \{u = u(x)\colon u \in H^2(0,l),\ u(0) = u'(0) = 0\}$ and any initial velocity in $L^2(0,l)$. This shows that the incorporation of the structural damping pair has a stabilizing effect on the system.

4.3 The equations of motion for second grade fluids

To describe the behaviour of second grade fluids the following set of equations have been formulated (see e.g. [24]):

$$\frac{\partial}{\partial t}(v-\alpha_1\Delta v) + v\cdot\nabla(v-\alpha_1\Delta v) - \mu\Delta v + \nabla p = \sum_{i=1} N_i(v) \qquad \text{in } \Omega,$$

$$\nabla\cdot v = 0 \qquad \text{in } \Omega,$$

$$v = 0 \qquad \text{on } \partial\Omega.$$

This problem, or special cases thereof, was addressed by several authors, among others [17], [4], all of whom succeeded only in showing the existence of unique solutions in the distributional sense. Very recently the question of the existence of a unique classical solution was investigated by G.P. Galdi, N. Sauer and M. Grobbelaar-Van Dalsen [7]. It turned out to be of crucial importance to append not an initial condition of the form $v\big|_{t=0} = v_0(x)$, but of the form $Bv\big|_{t=0} = w_0(x)$ (the use of such a substitution is underlying to the B-evolution approach), where in this case $Bv = P(v-\alpha_1\Delta v)$ with P the classical projection operator in $L^2(\Omega)$. By now using the technique of subdivision of the problem into a "tentatively" linear problem and a Stokes problem and employing Schauder's fixed point theorem[4] [5,p.171], the existence of a unique solution, local in time, was shown, provided that the domain Ω and the data have a certain degree of regularity, while if the data are sufficiently small and regular enough and α_1, the constant of elasticity, is large enough, global existence of a classical solution was attained.

References

1. A.V. Balakrishnan. Fractional powers of closed operators and the semigroups generated by them, *Pac. J. Math.* **10**, (1960), 419-437.
2. H. Brill. A semilinear Sobolev equation in a Banach space. *J. Differential Equations* **24** (1977), 412-425.
3. H. Beirão Da Veiga. On an Euler type equation in hydrodynamics, *Ann. Mat. Pura App.* **125** (1980), 279-294.
4. D. Cioranescu & O. El Hacène. Existence and uniqueness for fluids of second grade, In: *Research Notes in Mathematics*, Pitman, Boston (1984), 178-197.
5. A. Friedman. *Partial Differential Equations*, Holt, New York, 1969.

[4]The technique is derived from Beirão Da Veiga who used it in a different context [2].

6. D. Fujiwara & H. Morimoto. An L_r theorem for the Helmholtz decomposition of vector fields, *J. Fac. Sci. Univ. Tokyo Sect* 1A *Math.* **24**, (1977), 685-700.
7. G.P. Galdi, M. Grobbelaar-Van Dalsen & N. Sauer. Existence and uniqueness of classical solutions of the equations of motion for second grade fluids, *Arch. Rational Mech. Anal.*, to appear.
8. M. Grobbelaar-Van Dalsen. Fractional powers of a closed pair of operators, *Proc. Roy. Soc. Edinburgh* **105 A** (1986), 49-58.
9. M. Grobbelaar-Van Dalsen. Semilinear evolution equations and fractional powers of a closed pair of operators, *Proc. Roy. Soc. Edinburgh* **105 A**, (1987), 101-115.
10 M. Grobbelaar-Van Dalsen. On *B*-evolutions and dynamic boundary conditions on a portion of the boundary, *Appl. Anal.* **40** (1991), 151-172.
11. M. Grobbelaar-Van Dalsen & N. Sauer. Dynamic boundary conditions for the Navier-Stokes equations, *Proc. Roy. Soc. Edinburgh* **113 A**.(1989), 1-11.
12. M. Grobbelaar-Van Dalsen & N. Sauer. Solutions in Lebesgue spaces for the Navier-Stokes equations with dynamic boundary conditions, *Proc. Roy. Soc. Edinburgh*, to appear.
13. M. Grobbelaar-Van Dalsen. On the solvability of the boundary-value problem for the elastic beam with attached load, *Math. Models Methods Appl. Sci.*, to appear.
14. M. Grobbelaar-Van Dalsen. On fractional powers related to an empathy of operators and mathematical models for the stabilization of the elastic beam with attached load, Research Report **164/93(25)** Dept. Of Maths. Appl. Maths. and Astronomy, University of South Africa.
15. T. Kato. Fractional powers of dissipative operators, *J. Math. Soc. Japan* **13** (1961), 246-274.15.
16. F. Neubrander. Integrated semigroups and their application to the abstract Cauchy problem, *Pac. J. Math.*, **135**, 111-155.
17. A.P. Oskolkov. On some model non-stationary systems in the theory of non-Newtonian fluids, *Proc. Steklov Inst. Math.* **127**, (1975), 37-66.
18. N. Sauer. Linear evolutions equations in two Banach spaces, *Proc. Roy. Soc. Edinburgh*, **91 A**, (1982), 287-303.
19. N. Sauer & J. Singleton. Evolution operators in empathy with a semigroup, *Semigroup Forum*, **39** (1989), 85-94.
20. N. Sauer. Empathy, C-semigroups and integrated semigroups, *Proceedings: International Conference on evolution equations, Baton Rouge*, 1993, to appear.
21. N. Sauer. *Dynamic boundary conditions and related evolution processes*, Elsevier Sc. Publ., Amsterdam, to appear.
22. R. E. Showalter. The Sobolev Equation, *Appl. Anal.* **5** (1975),15-22, 81-99.
23. I. Tanaka. On exponentially bounded C-semigroups, *Tokyo J. Math.* **10** (1987), 107-117.
24. C. Truesdell & W. Noll. *The nonlinear field theories of Mechanics*, Handbuch der Physik **III/3**, Springer Verlag, Heidelberg, 1965.
25. M. Van Dalsen. Evolution equations involving non-stationary evolutions between two Banach spaces. *Quaestiones Math.* **8** (1985), 97-129, 199-229.
26. A. van der Merwe. *B*-evolutions and Sobolev equations, *Appl. Anal.* 29, (1988), 91-105.
27. K. Yosida. Fractional powers of infinitesimal generators and analyticity of the semigroup generated by them. *Proc. Japan Acad.* **36**, (1960), 86-89.

4th Int. Coll. on Differential Equations, pp. 115-124
D. Bainov and V. Covachev (Eds)

Hilbert-Arnold problem for cubic Hamiltonians and limit cycles

E. HOROZOV, I. D. ILIEV
University of Sofia, Faculty of Mathematics and Informatics, 1126 *Sofia, BULGARIA*
Institute of Mathematics of the Bulgarian Academy of Sciences, 1090 *Sofia, BULGARIA*

Abstract. In this survey paper we consider the present state of the infinitesimal version of the 16-th Hilbert problem and its relation to the existence and positions of the limit cycles of quadratic vector fields in the plane.

Keywords and phrases: limit cycles, bifurcations, Abelian integrals, quadratic systems.

0. INTRODUCTION

V. Arnold [1] formulated the infinitesimal version of the celebrated 16-th Hilbert problem (second part) as follows. Let $f(x,y)$, $g(x,y)$ be polynomials of degree less or equal to n and $H(x,y)$ be a polynomial of degree $n+1$. Let $\delta(h)$, $h \in \Delta$, be a continuous family of ovals of the algebraic curves $\Gamma_h : \{H(x,y) = h\}$. Then the Hilbert-Arnold problem is:

(HA) *Find an upper bound $Z(n)$ for the number of the zeros of the integrals*

$$I(h) = \int_{\delta(h)} -f\,dy + g\,dx, \quad h \in \Delta, \tag{0.1}$$

which depends only on n but not on the coefficients of the polynomials f, g, H.

The infinitesimal Hilbert problem is also known as the *weakened 16-th Hilbert problem.* It naturally arises in attempts to determine the number of limit cycles in small perturbations (given by f and g) of the Hamiltonian system with Hamiltonian H. The exact formulation of the latter problem is:

(LC) *Find an upper bound $c(n)$ for the number of the limit cycles of the system*

$$\begin{aligned} \dot{x} &= H_y + \varepsilon f \\ \dot{y} &= -H_x + \varepsilon g \end{aligned} \tag{0.2}$$

for small enough ε.

The connection between the problems (HA) and (LC) becomes apparent after expanding in ε the Poincaré map $P_\varepsilon(h)$ (or more precisely – the displacement function d_ε) of the perturbed system (0.2):

$$d_\varepsilon \equiv P_\varepsilon(h) - h = \varepsilon I(h) + \varepsilon^2 J(h) + \varepsilon^3 K(h) + \dots, \tag{0.3}$$

in which $I(h)$ is given by (0.1). As it is well-known, the isolated zeros of $I(h)$ in Δ correspond to limit cycles of (0.2) which tend to periodic orbits of the corresponding Hamiltonian system as $\varepsilon \to 0$.

In the row of results toward solving these problems we first schould point out that the *existence* of the upper bound $Z(n)$ in (HA) has been proved by A. Varchenko [2] and A. Khovanski [3] in 1984. Their proofs do not make it possible to find *explicitly* any bound for $Z(n)$ even for $n = 2$.

The present paper concerns the above stated problems (HA) and (LC) for the simplest nontrivial case $n = 2$, i. e. when the system (0.2) is quadratic. Our goal is to make a review of recent developments (some of them not published elsewhere) of these problems and their correlation. We consider results towards obtaining sharp bounds for $Z(2)$ and $c(2)$. A natural formulation for the first one, introduced in Petrov [4], [5], uses the notion of Chebyshev space.

Definition. A d dimensional $\mathbb{R}$-linear space of functions is said to be *Chebyshev* if any nonzero function from this space has at most $d-1$ zeros.

General conjecture. *The space of Abelian integrals* (0.1) *is Chebyshev.*

We have to point out that even if this result were established it will not yield straightforward the bound for $c(2)$. Moreover, it is well-known [6] that $c(2) \geq 3$. We conjecture this is the exact bound: $c(2) = 3$ (see also Żoładek [7]).

In the next section we are going to pose detailed conjectures for different classes of qubic Hamiltonians both for the zeros of the Abelian integrals (0.1) and for the number of the limit cycles of (0.2). For this we need to make a classification suitable for our purposes.

1. CLASSIFICATION OF THE CUBIC HAMILTONIANS HAVING A CENTRE AND PRECISE VERSIONS OF THE STATEMENTS

Consider an arbitrary quadratic system:

$$
\begin{aligned}
\dot{x} &= F(x,y) = \sum_{0\leq i+j\leq 2} A_{ij}x^iy^j \\
\dot{y} &= G(x,y) = \sum_{0\leq i+j\leq 2} B_{ij}x^iy^j.
\end{aligned}
\tag{1.1}
$$

In the corresponding 12-dimensional parameter space the Hamiltonian systems form a co-dimension three space, given by the restriction

$$A_{10} + B_{01} = 0,\ 2A_{20} + B_{11} = 0,\ A_{11} + 2B_{02} = 0.$$

Under these conditions the Hamiltonian function H can be written as a cubic polynomial

$$
\begin{aligned}
H = E - B_{00}x + A_{00}y - \frac{B_{10}}{2}x^2 + A_{10}xy + \frac{A_{01}}{2}y^2 \\
-\frac{B_{20}}{3}x^3 - \frac{B_{11}}{2}x^2y + \frac{A_{11}}{2}xy^2 + \frac{A_{02}}{3}y^3.
\end{aligned}
$$

We shall consider Hamiltonians, for which the corresponding Hamiltonian systems

$$
\begin{aligned}
\dot{x} &= H_y \\
\dot{y} &= -H_x
\end{aligned}
\tag{1.2}
$$

have a centre. Quadratic systems (1.1) with a centre are called *integrable*. The terminology reflects the fact that each quadratic system with a centre has also a first integral (a constant of motion). There exist four classes of integrable systems, depending on the form of the first integral [7], [8]: Q_2, Q_3^H, Q_3^{NH} and Q_4. The subscripts show the co-dimension of the corresponding class within the space of all quadratic systems. Q_3^H is the Hamiltonian case. Our nearby goal is to classify the cubic Hamiltonians with a centre. Topological classifications of the possible phase portraits of systems (1.2), made by several authors [9], [10], disregard some details important for our problem – for example the presence of an invariant line or of an axial symmetry in the system. In this connection often the Hamiltonians and the corresponding systems, having a certain specific property, are called "non-generic". There are several definitions of the notion "generic" even for quadratic Hamiltonian vector fields. Throughout this paper we say the cubic H is *generic* if the corresponding Hamiltonian system (1.2) has a centre and does not belong simultaneously to any of other integrable classes of quadratic systems. The latter condition has a simple geometrical interpretation: non-generic are exactly those Hamiltonians for which the level curves in suitable coordinates have an axis of symmetry. For such Hamiltonians we will simply say that they have an axis of symmetry.

To make a simple and explicit classification, which is useful also in the proofs, we introduce normal forms for the Hamiltonians. In any generic quadratic Hamiltonian vector field a separatrix contour consists of a saddle point and a homoclinic trajectory, which contains a unique centre in its interior. The separatrix cycle and the centre inside are the bounds of a continuous set of periodic trajectories (the periodic annulus). If we place the centre $C = (x_c, y_c)$ in the origin of the coordinate system and the saddle point $S = (x_s, y_s)$ – in $(1,0)$, then the Hamiltonian will become

$$H = \frac{x^2+y^2}{2} - \frac{x^3}{3} + axy^2 + \frac{b}{3}y^3. \tag{1.3}$$

It is enough to consider Hamiltonians H, corresponding to values of the parameters (a,b) from the set

$$\Omega = \left\{(a,b): \ -\frac{1}{2} \le a \le 1, \ \ 0 \le b \le (1-a)\sqrt{1+2a}\right\}.$$

The remaining Hamiltonians can be obtained from these by affine transformations. The subset of non-generic Hamiltonians (see the bifurcation diagram in Figure 1) consists of the curve $\Gamma_{ng} = \partial\Omega$,

$$\Gamma_{ng} = \{b=0\} \cup \{b^2 = (1-a)^2(1+2a)\}.$$

The curve Γ_∞,

$$\Gamma_\infty = \left\{b^2 - 4a^3 = 0\right\}$$

presents on the bifurcation diagram Hamiltonians with a critical point at infinity and the curve Γ_2,

$$\Gamma_2 = \left\{b^2 + 8a^2 + 4a = 0\right\}$$

presents those with a cuspidal point. These statements can be proved by simple computations.

The set $\Omega \setminus \{\Gamma_{ng} \cup \Gamma_\infty \cup \Gamma_2\}$ consists of three open components Ω_1^0, Ω_2^0, Ω_3^0:

1) Ω_1^0 – corresponding to generic Hamiltonians with one centre and one saddle point;
2) Ω_2^0 – corresponding to generic Hamiltonians with two centres and two saddle points;
3) Ω_3^0 – corresponding to generic Hamiltonians with one centre and three saddle points.

In fact it is easy to verify that for the Hamiltonians with an axis of symmetry (i. e. represented by points of Γ_{ng}), the corresponding system belongs to $Q_3^H \cap Q_3^{NH}$.

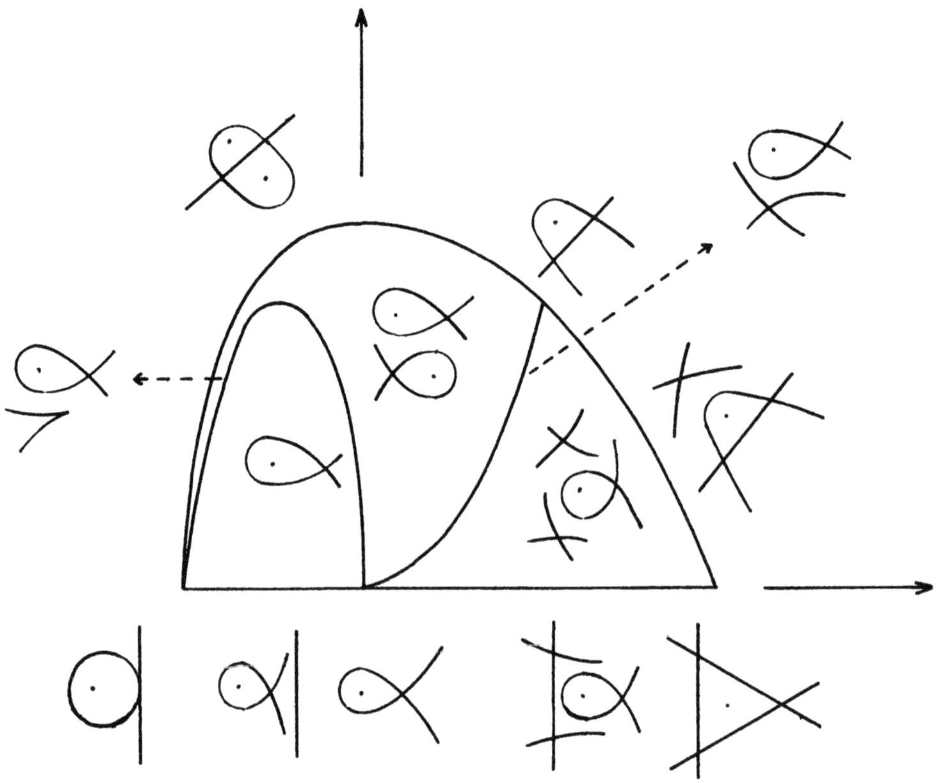

Figure 1.

The Hamiltonian, presented by the point $(a, b) = (1, 0)$ in Fig. 1:

$$H = \frac{x^2 + y^2}{2} - \frac{x^3}{3} + xy^2,$$

has three axes of symmetry. In this case the corresponding system belongs simultaneously to the classes Q_3^H, Q_3^{NH} and Q_2 (this is the so called "Hamiltonian triangle").

Note also that for Hamiltonians (1.3) the set Δ where $H = h$ has compact closed components is the interval $(0, \frac{1}{6})$ if H has a unique centre and the union $(0, \frac{1}{6}) \cup (\frac{x_s^2+y_s^2}{6}, \frac{x_c^2+y_c^2}{6})$ if H has two centres. Here (x_c, y_c) and (x_s, y_s) are the coordinates of the second centre and the attached saddle; one has $x_c^2 + y_c^2 > x_s^2 + y_s^2 \geq 1$.

It is an easy task to specify the dimension d in (0.1) for each cubic H.

Lemma 1.1. *The dimension of the space of functions* (0.1) *is as follows:*
(i) $d = 3$ *in the generic case;*
(ii) $d = 2$ *for Hamiltonians with one axis of symmetry;*
(iii) $d = 1$ *for Hamiltonians with three axes of symmetry.*

Proof. Writting the polynomials f, g in the form

$$f(x, y) = \sum_{0 \leq i+j \leq 2} a_{ij} x^i y^j, \quad g(x, y) = \sum_{0 \leq i+j \leq 2} b_{ij} x^i y^j,$$

we have $f_x + g_y = -(\alpha x + \beta y + \gamma)$, where

$$\alpha = -(2a_{20} + b_{11}),\ \beta = -(a_{11} + 2b_{02}),\ \gamma = -(a_{10} + b_{01}).$$

Then applying in (0.1) the Stokes formula we obtain

$$I(h) = \iint_{Int\,\delta(h)} (\alpha x + \beta y + \gamma)\, dx\, dy = \alpha X(h) + \beta Y(h) + \gamma M(h), \tag{1.4}$$

where $Int\,\delta(h)$ denotes the interior of the oval $\delta(h)$. In the generic case, $I(h) \equiv 0$ implies $\alpha = \beta = \gamma = 0$ (see for example [11]), which yields (i). To prove (ii), let first suppose that H is symmetric with respect to the axis $y = 0$. Then $Y(h) \equiv 0$. In the general case we can apply a linear change to obtain that a certain linear combination of the functions X, Y, M is identically zero. Hence $d \leq 2$. If $d = 1$, then $\xi = X/M$ and $\eta = Y/M$ are constants, which is impossible (for example because this contradicts the regularity of the centroid curve, see section 2 below). Finally, to prove (iii) we see as above that two linear combinations of the integrals X, Y, M are identically zero, which yields that X and Y are proportional to $M > 0$. The lemma is proved.

Now the general conjecture can be precised:

Conjecture 1. *For any cubic Hamiltonian with a centre, the space of Abelian integrals* (0.1) *forms a Chebyshev system. Namely, the non-zero functions in* (0.1) *have:*
(i) *at most two zeros if H is generic;*
(ii) *at most one zero if H has one axis of symmetry;*
(iii) *no zeros if H has three axes of symmetry.*

The above estimates are exact. Concerning (LC), we state

Conjecture 2. *The exact estimate for the number of limit cycles for quadratic systems of the form* (0.2) *is $c(2) = 3$. More precisely,*
(i) *for generic Hamiltonians and Hamiltonians with one axis of symmetry the system* (0.2) *has no more than two limit cycles;*
(ii) *for Hamiltonians with three axes of symmetry* (*a Hamiltonian triangle*) *the system* (0.2) *has no more than three limit cycles.*

These estimates are exact too, as they are equal to the cyclicity of the centres in (1.2).

In the rest of this paper we are going to consider to what extent these conjectures are reliable. We are going to review the results obtained in this direction, to point out some of the difficulties one faces in attempts to prove them. At the end we speculate over eventual constructions which may lead to proofs of the remaining parts.

2. DEFINITION AND PROPERTIES OF THE CENTROID CURVE

We begin by pointing out that the integrals $X(h)$, $Y(h)$, $M(h)$ defined in section 1 have well-known mechanical meaning. The first two, X and Y, are the momenta with respect to the axes Ox and Oy of $Int\,\delta(h)$. The integral $M(h)$ is the area of $Int\,\delta(h)$. Then the ratios

$$\xi(h) = \frac{X(h)}{M(h)},\quad \eta(h) = \frac{Y(h)}{M(h)},$$

define the coordinates of the centre of gravity or the centroid point [12] of $Int\,\delta(h)$. The fact that the integral $I(h)$ from (1.4) has a zero at $h = h_0$ means geometrically that the line $\ell:\ \alpha x + \beta y + \gamma = 0$ passes through the centroid point $(\xi(h_0), \eta(h_0))$ of $Int\,\delta(h_0)$. Now following [13], we define the *centroid curve* as

$$L = \{(\xi(h), \eta(h)) :\ h \in [h_s, h_c]\},$$

where h_s and h_c are the values of the Hamiltonian H at the saddle S and the centre C.

The importance of the concept of the centroid curve lies in the fact that its geometry contains the complete information about the zeros of each integral $I(h)$ (i. e. for all values α, β, γ and all perturbations f, g), although the definition of L depends only on H. We mention also the following obvious facts: a) L is affine invariant; b) the endpoints of L are C and the centroid of the loop area Z; c) L is a line segment for Hamiltonians with one axis of symmetry; d) L is a single point for Hamiltonians with three axes of symmetry. The main part of this survey is devoted to finding these properties of L which would lead to the proof of the following conjectures:

Conjecture 3. *The centroid curve L is regular for all Hamiltonians except for the Hamiltonian triangle. Moreover a certain linear combination of $\xi(h)$ and $\eta(h)$ is strictly monotone.*

Conjecture 4. *For each generic Hamiltonian, the centroid curve L is strictly convex.*

We will show in this paper that both Conjectures 3 and 4 imply Conjecture 1. For this, first we make precise the connection between the intersections (with multiplicities) of the line ℓ and the centroid curve L on one hand and the zeros of the Abelian integral (1.4) on the other hand. Although the proofs of the theorems formulated below are quite simple (see [13]), their statements constitute the dictionary between the analytic and the geometric language. We suppose that H is an arbitrary cubic Hamiltonian and that $I(h) \not\equiv 0$ unless the opposite is claimed. We begin with the following obvious but basic result:

Theorem 2.1. [13] (i) *For any $h_0 \in (h_s, h_c)$ the equality $I(h_0) = 0$ holds if and only if the line ℓ passes through the point $(\xi(h_0), \eta(h_0))$.*

(ii) *If L is regular at $h_0 \in (h_s, h_c)$, then the equalities $I(h_0) = I'(h_0) = 0$ hold if and only if ℓ is tangent to the centroid curve L at the point $(\xi(h_0), \eta(h_0))$.*

(iii) *If L is regular at h_0 and $I(h_0) = I'(h_0) = 0$, then $I''(h_0) = 0$ holds if and only if the curvature of L at $(\xi(h_0), \eta(h_0))$ is zero.*

In general we can prove by induction:

Corollary 2.2. *If L is regular at $h_0 \in (h_s, h_c)$ and $I(h_0) = I'(h_0) = 0$, then $I(h)$ has at h_0 a zero of order $k \geq 3$ if and only if the curvature of L at $(\xi(h_0), \eta(h_0))$ has a zero of order $k-2$.*

Along the same line of investigation we study how the properties of L are connected to the asymptotics of $I(h)$ at the ends of the interval $[h_s, h_c]$. Let us consider first the case $h = h_c$ corresponding to a centre of (1.2). In this case as is well-known the function $I(h)$ has the expansion

$$I(h) = f_1(h - h_c) + f_2(h - h_c)^2 + f_3(h - h_c)^3 + \dots$$

Theorem 2.3. [13] (i) *The coefficient f_1 is zero if and only if ℓ passes through the centre (x_c, y_c).*

(ii) *Both coefficients f_1 and f_2 are zero if and only if ℓ is tangent to L at (x_c, y_c).*

(iii) (Bautin [6]) $f_1 = f_2 = f_3 = 0$ *if and only if* $I(h) \equiv 0$.

For the next theorem we are going to use the expansion of $I(h)$ near the value $h = h_s$, corresponding to a saddle-loop:

$$I(h) = c_1 + c_2(h - h_s)\ln(h - h_s) + c_3(h - h_s) + \dots$$

Theorem 2.4. [13] (i) *The coefficient c_1 is zero if and only if ℓ passes through $Z = (\xi(h_s), \eta(h_s))$.*

(ii) *The condition $c_1 = c_2 = 0$ is equivalent to any of the following:* 1) *the line ℓ passes through Z and (x_s, y_s);* 2) *the line ℓ is tangent to L at the point Z.*

(iii) [11] *The condition $c_1 = c_2 = c_3 = 0$ is equivalent to $I(h) \equiv 0$.*

We also need some global properties of the centroid curve. A simple consequence of Conjecture 3 is the following theorem:

Theorem 2.5. *Running L, the tangent vector rotates within an angle less than π.*

Proof. Given $h \in \Delta$, denote by $\overline{\ell}$ the tangent vector at the corresponding point of L: $\overline{\ell} = (\xi'(h), \eta'(h))$. From Conjecture 3 it follows the existence of a vector $\overline{p} = (r, s)$ with the property that the inner product of $\overline{p}$ and $\overline{\ell}$ is non-zero for all h. Therefore $\overline{\ell}$ deviates with respect to $\overline{p}$ on angle less than $\pi/2$ which proves the proposition.

3. ZEROS OF ABELIAN INTEGRALS: RESULTS, DIFFICULTIES AND SPECULATIONS

In this section we describe several results giving some evidence for the Conjectures 1, 3 and 4. For non-generic Hamiltonians (except for the Hamiltonian triangle), Conjecture 3 is in fact a simple consequence of the results in [14], [15]. Indeed, it is easily seen that each non-generic Hamiltonian has also a straight line as a level curve (note that in the case of standard elliptic Hamiltonians presented by the point $(a, b) = (0, 0)$ on the bifurcational diagram, this line is gone to infinity). Next we can suppose that in (0.2) the perturbation is chosen so that (0.2) has an invariant line. Using [15] this gives the monotonicity of $I(h)$ for all α, β, γ except for the case when $L \subset \ell$. Then $I'(h) \neq 0$ implies $\xi'^2(h) + \eta'^2(h) \neq 0$. Formulating these facts in the context of the present survey, we obtain

Theorem 3.1. *For each non-generic Hamiltonian except for the Hamiltonian triangle, the centroid curve (each of two centroid curves) is a regular line-segment curve.*

In [13], [16] we have verified Conjecture 3 for Hamiltonians from $\Omega_2^0 \cup \Omega_3^0$ thus making in these cases Theorems 2.1 and 2.5 unconditional.

Theorem 3.2. *Conjecture 3 holds for Hamiltonians (1.3) with $(a, b) \in \Omega_2^0 \cup \Omega_3^0$.*

Next we emphasize on the result of [13], where we have proved the following theorem:

Theorem 3.3. *For Hamiltonians (1.3) with $(a, b) \in \Omega_3^0$ the centroid curve is strictly convex.*

Using Theorems 2.1 and 2.5 one gets the following corollary:

Corollary 3.4. *For generic Hamiltonians with three saddles and one centre the integral $I(h)$ has no more than two zeros.*

We consider Corollary 3.4 to be a quite strong evidence for Conjecture 1 as it gives a sharp estimate for the number of the zeros of $I(h)$ for entire component Ω_3^0 of the set of generic Hamiltonians. One may argue that in the case of Ω_2^0, when two centroid curves are present, both of them can produce zeros of $I(h)$. The next theorem shows that the total number cannot be greater than in the case of only one centroid curve.

Theorem 3.5. *If both centroid curves are strictly convex then any line can intersect their union at most in two points (counting the multiplicities).*

Proof. Denote by L_1 and L_2 the two centroid curves corresponding to two periodic annuli. We wish to prove that each line can intersect $L = L_1 \cup L_2$ at most in two points.

We let ℓ_s^1 and ℓ_c^1 denote the tangent lines to L_1 at its endpoints Z_1 and C_1 respectively. Recall that C_1 is the centre inside the loop, Z_1 is the centroid point of the loop area and that the tangent ℓ_s^1 passes also through the saddle S_1. Using Theorem 2.5 we see that ℓ_s^1 and ℓ_c^1 do intersect at a point O_1 placed from the convex side of L_1. Thus the plane is divided by ℓ_s^1 and ℓ_c^1 into four sectors. Denote by R_1 the sector where L_1 lies and by R_1^{op} the opposite sector. The same construction applies to the second centroid curve L_2. Our goal will be achieved if we prove that $L_1 \subset R_2^{op}$ and $L_2 \subset R_1^{op}$. For this we need the equations of the tangents at the endpoints of L_1, L_2. Unfortunately it is extremely difficult to obtain the equations of ℓ_s^1 and ℓ_s^2. For this reason we are unable to perform a direct proof using explicit formulas. Instead of that, we can proceed as follows. First recall the well-known fact that the four critical points form a convex quadrilateral with two saddles as a pair of opposite vertices and two centres as another pair. Denote by m_s the line passing through the saddles S_1 and S_2 and respectively by m_c the line through the centres C_1 and C_2. Next, let O be the intersection point of m_s and m_c and Δ_1 (resp. Δ_2) be the interior of the triangle with vertices C_1, S_1, O (resp. C_2, S_2, O). Then the needed result will follow if we check that $O_1 \in \Delta_1$ and $O_2 \in \Delta_2$. The latter is a consequence of the inclusions $L_1 \subset \Delta_1$ and $L_2 \subset \Delta_2$, which we establish next. It suffices to prove the first of them.

Fix the coordinate system in which the Hamiltonian H has a normal form (1.3); hence $S_1 = (1,0)$ and $C_1 = (0,0)$. Moreover we have to concentrate on the case $(a,b) \in \Omega_2^0$ (see section 1). If $a = 0$, an easy computation yields that S_2 and C_2 lie in the half-plane $y < 0$. As in a quadratic system three singular points are never collinear, then S_2 and C_2 lie in the same half-plane for each $(a,b) \in \Omega_2^0$. Further, it is easily seen that in the case $a = 0$, the curve $L_1 \setminus C_1$ lies on the same side of m_c where the saddle S_1 does. To conclude that this remains true in the general case, we can use Bautin's result: in a generic quadratic Hamiltonian system the total cyclicity of the centres is at most two. Geometrically this implies (see Theorem 2.3)) that the lines ℓ_c^1 and m_c cannot coincide in the generic case. Conjecturing L_1 is convex, this yields that for each $(a,b) \in \Omega_2^0$ the curve L_1 is placed entirely in the half-plane with respect to m_c where S_1 lies. The next step is to prove that $L_1 \setminus C_1$ is contained in the lower half-plane $y < 0$. But this is a immediate consequence of Theorem 2.4 and Corollary 3.3 in [13]. The last step is to prove that L_1 lies entirely in the same half-plane with respect to m_s, where C_1 lies. For this we apply the result from [11] concerning the position of the saddle loops. Namely using as coordinate axes the tangents to the loop at the saddle S_1, then the loop through S_2 is contained entirely in an adjacent quadrant. Hence the loop through S_1 surrounding C_1, can never intersect the line m_s in another point except S_1. Therefore the curve L_1 being inside the loop is contained in the same half-plane determined by m_s, where the centre C_1 lies. Thus we proved that $L_1 \subset \Delta_1$, which implies $O_1 \in \Delta_1$. Similarly, $O_2 \in \Delta_2$.

The hypothesis of Theorem 3.5 has been verified in [16] for generic Hamiltonians possessing central symmetry: $H(x - x_0, y - y_0) = -H(x_0 - x, y_0 - y)$. Centrosymmetric Hamiltonians correspond to the points (a,b) with $a = 0$, $b \in (0,1]$.

Theorem 3.6. *For generic centrosymmetric Hamiltonians both centroid curves are strictly convex.*

Corollary 3.7. *For generic centrosymmetric Hamiltonians the integral $I(h)$ has no more than two zeros in the union Δ of the intervals where $H = h$ contains an oval.*

As the curvature of L in the non-generic case is identically zero, a priori one can expect any irregularity of L for generic Hamiltonians close to the non-generic ones. But via small perturbations of non-generic H, one can prove this is not the case:

Theorem 3.8. *The centroid curve of any generic Hamiltonian presented by a point* (a,b) *close to* Γ_{ng} *is regular and strictly convex.*

We consider this theorem to be another strong evidence in favour of Conjectures 3 and 4. For example, if Conjecture 4 fails then the bifurcation diagram of the curvature of L considered as a function on Ω should contain a closed contour, disconnected from the rest of the diagram. A phenomenon of this type seems highly inconsistent to the intuition built up by singularity theory [17]. We have two different proofs of the fact stated above. One of them uses the techniques of Petrov in the spirit of [18] and the other rests on careful study of Picard-Fuchs equations. The shortage of space compels us to publish the proofs elsewhere.

In view of Theorem 3.8, Corollary 2.2 with $k = 4$ and the fact that in $\Omega \setminus \Gamma_{ng}$ the curvature of L near its endpoints is nowhere zero and is of the same sign, the proof of Conjecture 4 reduces to the proof of the following one:

Conjecture 5. *The integral* $I(h)$ *has no zero of multiplicity four in* Δ.

In fact the proof of Theorem 3.6 is done by proving Conjecture 5 for symmetric Hamiltonians.

4. BOUNDS FOR THE NUMBER OF THE LIMIT CYCLES

The results of the previous section yield in some cases the exact estimate for the number of the limit cycles in (0.2). More precisely for generic Hamiltonians the main result of [11] (based on Roussarie [19]) shows that the study of the first term $I(h)$ of the displacement function (0.3) is enough to judge for the number of the limit cycles near to separatrices. First of all, a simple computation [11] shows:

Theorem 4.1. *For any generic* H, *if* $I(h) \equiv 0$ *then system* (0.2) *is Hamiltonian.*

Then combining the results from [6], [11], [19] and that of Theorems 2.1 – 2.5, Corollaries 3.4 and 3.7, we obtain:

Theorem 4.2. *For generic Hamiltonians* H *with three saddles and one centre no more than two limit cycles can appear in* (0.2) *for small enough* ε.

Theorem 4.3. *For generic symmetric Hamiltonians* H *no more than two limit cycles can appear in* (0.2) *for small enough* ε.

Provided Conjecture 1 (i) holds, the estimate in Conjecture 2 for generic Hamiltonians out of the bifurcation diagram (i. e. for $(a,b) \in \Omega_1^0 \cup \Omega_2^0 \cup \Omega_3^0$) also follows from Theorem 4.1 repeating the proof from [13].

Estimates for the number of the zeros of $I(h)$ do not lead ultimately to the corresponding estimates for limit cycles. We point out that apart from the cases listed above the estimates of the limit cycles for all Hamiltonians from the bifurcation diagram faces quite serious extra difficulties. We start with the principal one.

For Hamiltonians from the boundary of Ω the identity $I(h) \equiv 0$ does not yield necessarily that system (0.2) is integrable [11]. Obviously in order to count the number of the limit cycles we need the second term $J(h)$ of the displacement function (see [20], [21]).

A second difficulty is the study of the saddle-connection bifurcations, where the results of our paper [11] do not apply. Hopefully the second term of the displacement function again will help (work in progress) in the case of a saddle-loop. But understanding the heteroclinic connection is far from satisfactory up to now. The reason is that the analogues of the Roussarie's result [19] for polycycles of that type which have been recently proved [22] do not apply to perturbations of Hamiltonian systems.

At the end we have to describe the bifurcations of the degenerate singularities. The cusp points appearing in the Hamiltonian systems from Γ_2 are of infinite co-dimension, which makes it impossible to apply the results of Dumortier et al [23] as Dumortier and Roussarie kindly explained to us. The curve Γ_∞ corresponds to Hamiltonians with a degenerate singularity at infinity. After suitable but small perturbation this singularity gives birth to a focus around which there can exist (and certainly there exist in some directions of the perturbation) limit cycles. A recent paper [24], devoted to this case, studies the limit cycles in the finite part of the plane, but the ones born from infinity still wait an adequate approach.

Acknowledgements. This work is partially supported by Grants MM-1/91 and MM-48/91 of the NFSR of Bulgaria.

REFERENCES

1. V. I. Arnol'd, *Geometrical methods in the theory of ordinary differential equations*, Springer-Verlag, Berlin, Heidelberg, New York (1988).
2. A. N. Varchenko, *Funct. Anal. Appl.* **18** (1984), 98 - 108.
3. A. G. Khovansky, *Funct. Anal. Appl.* **18** (1984), 119 - 128.
4. S. G. Petrov, *Funct. Anal. Appl.* **20** (1986), 37 - 40.
5. S. G. Petrov, *Funct. Anal. Appl.* **22** (1988), 72 - 73.
6. N. N. Bautin, *Amer. Math. Soc. Transl.* **100** (1954), 1 - 19.
7. H. Żoładek, Quadratic systems with center and their perturbations, *J. Differential Equations* (to appear).
8. D. Schlomiuk, Algebraic particular integrals, integrability and the problem of the center, *Transactions of the AMS* (to appear).
9. N. I. Vulpe, *Diff. Equations* **19** (1983), 273 - 280.
10. J. C. Artés and J. Llibre, Quadratic Hamiltonian vector fields, Preprint (1991).
11. E. Horozov and I. D. Iliev, On saddle-loop bifurcations of limit cycles in perturbations of quadratic Hamiltonian systems, *J. Differential Equations* (to appear).
12. W. F. Freiberger (Ed.), *The International Dictionary of Applied Mathematics*, D. van Nostrand Company Inc., Princeton, New Jersey (1960).
13. E. Horozov and I. D. Iliev, On the number of limit cycles in perturbations of quadratic Hamiltonian systems, *Proc. Lond. Math. Soc.* (to appear).
14. G. S. Rychkov, *Diff. Equat.* **8** (1972), 1748 - 1750.
15. W. A. Coppel, *Dynamics Reported* **2** (1989), 61 - 88.
16. E. Horozov and I. D. Iliev, Perturbations of quadratic Hamiltonian systems with symmetry, Preprint (1993).
17. V. I. Arnol'd, A. N. Varchenko and S. M. Gusein-Zade, *Singularities of differentiable maps* I, Birkhäuser Verlag, Basel (1988).
18. G. S. Petrov, *Funct. Anal. Appl.* **24** (1990), 205 - 210.
19. R. Roussarie, *Bol. Soc. Bras. Math.* **17** (1986), 67 - 101.
20. C. Chicone and M. Jacobs, *J. Diff. Equations* **91** (1991), 268 - 326.
21. Zhang Zhi-Fen and Li Bao-Yi, *Ann. Mat. Pura ed Appl.* **CLXI** (1992), 181 - 212.
22. A. Mourtada, These de doctorat, Univ. de Bourgogne (1990).
23. F. Dumortier et al, *Lect. Notes in Math.* **1480** (1991).
24. Zhi-fen Zhang and Chengzhi Li, On the number of limit cycles of a class of quadratic Hamiltonian systems under quadratic perturbations, *Res. Rept Peking Univ. No* 33 (1993).

4th Int. Coll. on Differential Equations, pp. 125-135
D. Bainov and V. Covachev (Eds)

Periodic solutions of neutral impulsive systems with a small delay

L. JODAR[1], R. J. VILLANUEVA[1] and V. CHR. COVACHEV[2]

[1] Departamento de Matematica Aplicada, Universidad Politecnica de Valencia, P.O. Box 22 012, Valencia, SPAIN

[2] Institute of Mathematics, Bulgarian Academy of Science, Sofia, BULGARIA

Abstract: For a neutral impulsive system with delay under certain assumptions it is proved that if the corresponding system without delay has an isolated ω-periodic solution, then in any neighbourhood of this orbit the system considered also has an ω-periodic solution, if the delay is small enough.

Keywords: periodic; neutral; impulsive.

1. INTRODUCTION

In the mathematical simulation of the evolution of real processes in physics, chemistry, population dynamics, radio engineering, etc. which are subject to perturbation of duration negligible with respect to the total duration of the process it is often convenient to assume that the disturbances are "momentary", in the form of impulses. This leads to the investigation of differential equations and systems with discontinuous trajectories, or with impulse effect, called for the sake of brevity impulsive differential equations and systems.

In many applications it is assumed that the system considered is subject to the law of causality, i.e. the future state of the system does not depend on the past states and is determined by the present only. But in the more detailed investigation it often becomes obvious that the law of causality is just a first approximation to the real situation, and a more realistic model should involve some of the past states of the system. Moreover, many problems lose their sense if the dependence on the past is not considered. All this leads to differential equations with delay of the argument [1].

The existence of periodic solutions of differential equations is one of the classical problems of their qualitative theory. One of the traditional approaches is to investigate the system in variations with respect to a periodic solution of the unperturbed system satisfying certain nondegeneracy assumptions.

In the present paper for a neutral impulsive system it is proved that if the corresponding system without delay has an isolated ω-periodic solution, then in a neighbourhood of this orbit the system considered also has an ω-periodic solution for small values of the delay. Thus we generalize an analogous result obtained for impulsive systems with delay [2].

2. STATEMENT OF THE PROBLEM

Consider the system with impulses at fixed moments

$$\dot{x}(t) = D(t)\dot{x}(t-h) + f(t, x(t), x(t-h)),\ t \neq t_i + kh,\ i \in \mathbf{Z},\ k \in \mathbf{N} \cup \{0\},$$

$$\Delta x(t_i) = I_i(x(t_i), x(t_i - h)),\ i \in \mathbf{Z}, \tag{1}$$

$$\Delta x(t_i + kh) = 0 \text{ if } t_i + kh \neq t_l \quad \forall l \in \mathbf{Z},$$

where $x \in \Omega \subset \mathbf{R}^n$, D: $\mathbf{R} \longrightarrow \mathbf{R}^{n\times n}$, f: $\mathbf{R} \times \Omega \times \Omega \longrightarrow \mathbf{R}^n$, I_i: $\Omega \times \Omega \longrightarrow \mathbf{R}^n$ $(i \in \mathbf{Z})$, $\{t_i\}_{i\in\mathbf{Z}}$ is a strictly increasing sequence of fixed points such that $\lim\limits_{i\to\pm\infty} t_i = \pm\infty$, Ω is a domain in $\mathbf{R}^n$, $\Delta x(t_i) = x(t_i + 0) - x(t_i - 0)$, $h \geq 0$ is the delay.

As usual in the theory of the impulsive differential equations, at the points of discontinuity t_i of the solution $x(t)$ we assume that $x(t_i) \equiv x(t_i - 0)$. It is clear that, in general, the derivatives $\dot{x}(t_i + kh)$ do not exist. However, there exist the limits $\dot{x}(t_i + kh \pm 0)$. According to the above convention, we assume

$$\dot{x}(t_i + kh) \equiv x(t_i + kh - 0).$$

For the sake of convenience we shall also use the notation:

$$\overline{x}(t) = x(t-h),\ x_i = x(t_i).$$

For a vector $x \in \mathbf{R}^n$ we denote by $|x|$ its Euclidean norm, and for an $n \times n$-matrix A we define the associated norm

$$|A| = \sup\ \{|Ax|/|x|;\ x \in \mathbf{R}^n \setminus \{0\}\}.$$

Introduce the following conditions:

H1. The function $f(t, x, \overline{x})$ is continuous, ω-periodic and continuously differentiable with respect to t, twice continuously differentiable with respect to x, $\overline{x}$, and its second derivatives are locally Lipschitz continuous with respect to x, $\overline{x}$.

H2. The functions $I_i(x, \overline{x}) \in C^2(\Omega \times \Omega, \mathbf{R}^n)$, $i \in \mathbf{Z}$, and their second derivatives are locally Lipschitz continuous.

H3. There exists a positive integer m such that $t_{i+m} = t_i + \omega$, $I_{i+m}(x,\overline{x}) = I_i(x,\overline{x})$, $i \in \mathbf{Z}$, $x,\overline{x} \in \Omega$.

H4. The matrix $D(t) \in C^1(\mathbf{R}, \mathbf{R}^{n\times n})$, is ω -periodic and $\sup_{t\in[0,\omega]}|D(t)| = \mu_0 \langle 1$.

Together with (1) we consider the so called *generating* system (E is the unit matrix)

$$\overset{\bullet}{x}(t) = (E - D(t))^{-1} f(t, x(t), x(t)), \ t \neq t_i,$$
$$\Delta x(t_i) = I_i(x_i, x_i), \ i \in \mathbf{Z}, \tag{2}$$

obtained from (1) for $h = 0$, and suppose that

H5. The generating system (2) has an ω -periodic solution $\psi(t)$ such that $\psi(t) \in \Omega$ for all $t \in \mathbf{R}$.

We may note that the invertibility of the matrix $E - D(t)$ follows from the inequality $\mu_0 \langle 1$ (condition H4). Moreover, $\sup_{t\in[0,\omega]}\left|(E - D(t))^{-1}\right| \leq (1 - \mu_0)^{-1}$.

Next define the system in variations with respect to $\psi(t)$:

$$\overset{\bullet}{y} = (E - D(t))^{-1} A(t) y, \ t \neq t_i, \tag{3}$$
$$\Delta y(t_i) = B_i y_i, \quad i \in \mathbf{Z},$$

where

$$A(t) = \frac{\partial}{\partial x} f(t, x, x)\Big|_{x=\psi(t)}, \ B_i = \frac{\partial}{\partial x} I_i(x, x)\Big|_{x=\psi_i}.$$

Let $X(t)$ be the fundamental solution of (3) (i.e. $X(0) = E$). Now we make the following additional assumptions:

H6. The matrix $E - X(\omega)$ is nonsingular.

H7. The matrices $E + B_i$, $i \in \mathbf{Z}$, are nonsingular.

The last two conditions allow us to define Green's function [3]

$$G(t, \tau) = \begin{cases} X(t)(E - X(\omega))^{-1} X^{-1}(\tau), & 0 \leq \tau < t \leq \omega, \\ X(t+\omega)(E - X(\omega))^{-1} X^{-1}(\tau), & 0 \leq t \leq \tau \leq \omega \end{cases}$$

and extend it as ω -periodic with respect to t and τ.

Denote also

$$M = \sup \left\{ |G(t, \tau)| : \ t, \tau \in [0, \omega] \right\},$$

$$\beta = \sum_{i=1}^{m} |B_i|,$$

$$\mu_1 = \sup_{t\in[0,\,\omega]} |\dot{D}(t)|,$$

$$\eta(h) = \sup_{t\in[0,\,\omega]} |\dot{D}(t) - \dot{D}(t-h)|$$

is the module of continuity of the derivative $\dot{D}(t)$.

3. MAIN RESULT

Theorem 1. *Let conditions H1 - H7 hold. If*

$$\mu_0(\beta M + 5) < 1,$$

then there exists a neighbourhood $\tilde{\Omega} \subset \Omega$ *of the orbit* $x = \psi(t)$ *and a number* $h_1 > 0$ *such that for* $h \le h_1$ *system* (1) *has in* Ω *a unique* ω*-periodic solution* $x(t,\, h)$ *such that* $x(t,0) \equiv \psi(t)$.

Proof. There exists a constant $\delta_0 > 0$ such that Ω contains a closed δ_0-neighbourhood Ω_1 of the periodic orbit $x = \psi(t)$. All the supremums henceforth are taken for $x,\, \bar{x} \in \Omega_1$.

In system (1) we change the variables according to the formula

$$x = y + \psi(t), \tag{5}$$

and obtain the system

$$\begin{gathered} \dot{y} = (E - D(t))^{-1}\{A(t)y + Q(t,y) + \Delta f(t, y + \psi(t),\, \bar{y} + \bar{\psi}(t)) \\ + D(t)[\dot{\bar{y}} - \dot{y} + \dot{\bar{\psi}}(t) - \dot{\psi}(t)]\}, \quad t \ne t_i + kh, \\ \Delta y(t_i) = B_i y_i + J_i(y_i) + \Delta I_i(y_i + \psi_i,\, \bar{y}_i + \bar{\psi}_i),\ i \in \mathbf{Z}, \\ \Delta y(t_i + kh) = 0 \quad \text{for} \quad t_i + kh \ne t_l, \end{gathered} \tag{6}$$

where by definition

$$\Delta f(t,x,\bar{x}) = f(t,x,\bar{x}) - f(t,x,x),$$
$$\Delta I_i(x_i,\, \bar{x}_i) = I_i(x_i,\, \bar{x}_i) - I_i(x_i,\, x_i),$$

while

$$Q(t,y) \equiv f(t,\, y + \psi(t),\, y + \psi(t)) - f(t,\, \psi(t),\, \psi(t)) - A(t)y$$

$$= \left< \int_0^1 (1-s)\frac{\partial^2}{\partial x^2} f(t,\, x,\, x)\Big|_{x=\psi(t)+sy} ds . y,\, y \right>,$$

$$J_i(y) \equiv I_i(y + \psi_i,\, y + \psi_i) - I_i(\psi_i,\, \psi_i) - B_i y$$

$$= \left< \int_0^1 (1-s)\frac{\partial^2}{\partial x^2} I_i(x,x)\Big|_{x=\psi_i+sy} ds . y,\, y \right>.$$

Thus we have $|Q(t,y)| \le C_1|y|^2$, $t \in [0,\omega]$, $|J_i(y)| \le C_2|y|^2$, $i = \overline{1,m}$, for some positive constants C_1, C_2.

To find an ω-periodic solution of system (1) we shall find an ω-periodic solution y(t) of system (6) satisfying $|y| \le \delta \le \delta_0$ using the contraction mapping principle. Define the operator $\mathcal{G}_h$, $\mathcal{G}_h y = z$, as follows: $z(t)$ is the unique ω-periodic solution of the linear system

$$\begin{gathered}
\dot{z} = (E - D(t))^{-1} A(t) z + (E - D(t))^{-1} (Q(t, y) + \Delta f(t, y + \psi, \bar{y} + \bar{\psi})) \\
+ (E - D(t))^{-1} D(t) [(\dot{\bar{y}}(t) - \dot{y}(t)) + (\dot{\bar{\psi}}(t) - \dot{\psi}(t))],\ t \ne t_i + kh, \\
\Delta z(t_i) = B_i z_i + J_i(y_i) + \Delta I_i(y_i + \psi_i, \bar{y}_i + \bar{\psi}_i),\ i \in \mathbf{Z}, \\
\Delta z(t_i + kh) = 0 \ \text{ for } \ t_i + kh \ne t_i.
\end{gathered} \tag{7}$$

This solution [3] is given by the formula

$$\mathcal{G}_h y(t) = z(t) = \tilde{\mathcal{U}}_h y(t) + \mathcal{V}_h (y + \psi)(t), \tag{8}$$

where we have denoted

$$\tilde{\mathcal{U}}_h y(t) = \int_0^\omega G(t, \tau)(E - D(\tau))^{-1} [Q(\tau, y(\tau)) + \Delta f(\tau, y(\tau) + \psi(\tau), \bar{y}(\tau) + \bar{\psi}(\tau))] d\tau$$

$$+ \sum_{0 < t_i < \omega} G(t, t_i + 0)[J_i(y_i) + \Delta I_i(y_i + \psi_i, \bar{y}_i + \bar{\psi}_i)],$$

$$\mathcal{V}_h y(t) = \int_0^\omega G(t, \tau)(E - D(\tau))^{-1} D(\tau)(\dot{\bar{y}}(\tau) - \dot{y}(\tau)) d\tau.$$

Obviously, if $D(t) \equiv 0$, then $\mathcal{V}_h(y + \psi)(t) \equiv 0$, while $\tilde{\mathcal{U}}_h$ is reduced to the operator $\mathcal{U}_h$ considered in detail in the paper [2], whose unique fixed point gives the periodic solution for the case $D(t) \equiv 0$. The extra factor $(E - D(t))^{-1}$ under the integral in $\tilde{\mathcal{U}}_h y(t)$ causes only slight modifications in the estimates for $\mathcal{U}_h y(t)$ in [2]. So we shall just reproduce these estimates in Lemmas 2 and 4 omitting the details. On the other hand, the estimates of $\mathcal{V}_h(y + \psi)(t)$ are new and some details will be given.

Suppose, for the sake of definiteness, that

$$0 < t_1 < t_2 < \ldots < t_m < \omega.$$

Let $h_0 > 0$ be so small that for any $h \in [0, h_0]$ we should have

$$t_i + h < t_{i+1},\ i = \overline{1, m-1},\ t_m + h < \omega. \tag{9}$$

Denote by $\tilde{C}_{\omega,n}$ the space of all ω-periodic, piecewise continuous functions $w\colon \mathbf{R} \to \mathbf{R}^n$ with discontinuities of the first kind at t_i and $w(t_i - 0) = w(t_i)$, $i \in \mathbf{Z}$, supplied with the norm

$$\|w\| = \sup_{t \in \mathbf{R}} |w(t)|.$$

By $\tilde{C}^{(1)}_{\omega,n}$ we denote the space of all functions $w \in \tilde{C}_{\omega,n}$ having continuous derivatives on each interval $[0,t_1)$, (t_i,t_{i+1}), $i=\overline{1,m-1}$, and $(t_m,\omega]$, with the possible exception of the points t_i+kh, $i=\overline{1,m}$, $k\in\mathbf{N}$, and such that there exist the limits $\dot{w}(t_i+kh\pm 0)$.

Define the sets $\Delta_h=\bigcup_{i=1}^{m}(t_i,\ t_i+h)$, $I_h=[0,\omega]\setminus\Delta_h$. Further on we assume that $h\in[0,h_0]$, $\delta\in(0,\ \delta_0]$. For $\varepsilon_1>0$, $\varepsilon_2>0$ define the set of functions

$$T_{\delta,\varepsilon_1,\varepsilon_2}=\left\{y\in\tilde{C}^{(1)}_{\omega,n}\colon \|y\|\le\delta,\ \sup_{t\in[0,\omega]}|\dot{y}(t)|\le\varepsilon_1,\ \max_{i=\overline{1,m}}|\Delta y(t_i)|\le\varepsilon_2\right\}.$$

We find a dependence between $h,\ \delta,\ \varepsilon_1,\ \varepsilon_2$ so that the operator $\mathcal{J}_h$ defined by (8) maps $T_{\delta,\varepsilon_1,\varepsilon_2}$ into itself as a contraction. In all the estimates henceforth $y\in T_{\delta,\varepsilon_1,\varepsilon_2}$.

We begin with estimating the difference $\overline{y}(t)-y(t)$. If $t\in I_h$, then y is continuous in $[t-h,\ t]$, and $\dot{y}(t)$ exists inside this segment, with the exception of finitely many points. Thus we have

$$|\overline{y}(t)-y(t)|=|y(t-h)-y(t)|\le h\varepsilon_1 \tag{10}$$

and, analogously, since ψ is a solution of (2),

$$|\overline{\psi}(t)-\psi(t)|\le h\ \sup|f|.$$

These estimates imply

$$\begin{aligned}|\Delta I_i(y_i+\psi_i,\overline{y}_i+\overline{\psi}_i)| &= |I_i(y_i+\psi_i,\overline{y}_i+\overline{\psi}_i)-I_i(y_i+\psi_i,y_i+\psi_i)|\\ &\le \sup|\partial_{\overline{x}}I_i|.(|\overline{y}_i-y_i|+|\overline{\psi}_i-\psi_i|)\le h\sup|\partial_{\overline{x}}I_i|(\sup|f|+\varepsilon_1)\end{aligned} \tag{11}$$

and (again for $t\in I_h$)

$$|\Delta f(t,\ y+\psi,\ \overline{y}+\overline{\psi})|\le h\sup|\partial_{\overline{x}}f|(\sup|f|+\varepsilon_1). \tag{12}$$

Now we shall obtain analogous estimates for $t\in\Delta_h$. Then the interval $(t-h,\ t)$ contains just one point of discontinuity t_i of $y(t)$, thus

$$|\overline{y}(t)-y(t)|\le|y(t-h)-y_i|+|\Delta y(t_i)|+|y(t_i+0)-y(t)|\le h\varepsilon_1+\varepsilon_2 \tag{13}$$

and, analogously,

$$|\overline{\psi}(t)-\psi(t)|\le h\sup|f|+\sup|I_i|.$$

Further on, we obtain

$$|\Delta f(t,y+\psi,\overline{y}+\overline{\psi})|\le\sup|\partial_{\overline{x}}f|[h(\sup|f|+\varepsilon_1)+\sup|I_i|+\varepsilon_2]. \tag{14}$$

We see that the estimates obtained for $t\in\Delta_h$ are rougher than those for $t\in I_h$ (and thus valid for this case as well).

We want to show that $z \in T_{\delta,\varepsilon_1\varepsilon_2}$. Suppose that the inequality $\|z\| \le \delta$ has been already proved. Then, making use only of the rougher estimates (obtained for $t \in \Delta_h$), we can fix $\varepsilon_1, \varepsilon_2, \widetilde{h} \in (0, h_0]$ and $\widetilde{\delta} \in (0, \delta_0]$ so that for $h \in [0, \widetilde{h}]$ and $\delta \in (0, \widetilde{\delta}]$ we have $|\dot{z}(t)| \le \varepsilon_1$ and $|\Delta z(t_i)| \le \varepsilon_2$. To prove the inequality $\|z\| \le \delta$ requires the precise estimates obtained for the cases $t \in I_h$ and $t \in \Delta_h$, and makes use of the fact that the measure of the "bad" set Δ_h where $|\overline{y}(t) - y(t)|$ is not small is mh, i.e. it is small for h small. Namely, we shall repeatedly use the following lemma or arguments of its proof.

Lemma 1. *Let* $y(t) \in \widetilde{C}_{\omega,n}$ *be such that in each of the intervals* $(0, t_1), (t_1, t_2), \dots, (t_m, \omega)$ $\dot{y}(t)$ *exists, except for a finite number of points, and is bounded. Then there exists a constant* $C > 0$ *such that for any function* $\varphi(t)$ *integrable by Riemann in* $[0, \omega]$ *and any* $h > 0$ *we have*

$$\left|\int_0^\omega \varphi(t)(y(t) - y(t-h))dt\right| \le Ch \sup_{t \in [0,\omega]} |\varphi(t)|.$$

Proof. It suffices to prove the assertion for h small enough, say for $h \in (0, h_0]$. Then inequalities (9) hold and we define the sets Δ_h and I_h as above.

If $t \in I_h$, analogously to inequality (10) we obtain

$$|y(t) - y(t-h)| \le h \sup_{\tau \in [0,\omega]} |\dot{y}(\tau)|.$$

If $t \in \Delta_h$, then analogously to (13) we have

$$|y(t) - y(t-h)| \le h \sup_{\tau \in [0,\omega]} |\dot{y}(\tau)| + \sup_{i=\overline{1,m}} |\Delta y(t_i)|.$$

Then, since the measures of the sets I_h and Δ_h are respectively $\omega - mh$ and mh,

$$\left|\int_0^\omega \varphi(t)(y(t) - y(t-h))dt\right|$$

$$\le \sup_{t \in [0,\omega]} |\varphi(t)| \left\{(\omega - mh)h \sup_{t \in [0,\omega]} |\dot{y}(t)| + mh\left[h \sup_{t \in [0,\omega]} |\dot{y}(t)| + \sup_{i=\overline{1,m}} |\Delta y(t_i)|\right]\right\}$$

$$= h \sup_{t \in [0,\omega]} |\varphi(t)| (\omega \sup_{t \in [0,\omega]} |\dot{y}(t)| + m \sup_{i=\overline{1,m}} |\Delta y(t_i)|),$$

thus it suffices to choose $C = \omega \sup_{t \in [0,\omega]} |\dot{y}(t)| + m \sup_{i=\overline{1,m}} |\Delta y(t_i)|$. □

Now in order to prove that the operator $\mathcal{G}_h$ maps the set $T_{\delta,\varepsilon_1,\varepsilon_2}$ into itself for h and δ small enough it remains to show that for $y \in T_{\delta,\varepsilon_1,\varepsilon_2}$ the function $z = \mathcal{G}_h y$ defined by equality (8) satisfies $\|z\| \le \delta$. This follows from the subsequent two lemmas.

Lemma 2. *Let* $y \in T_{\delta,\varepsilon_1,\varepsilon_2}$. *Then*

$$|\tilde{\mathcal{U}}_h y(t)| \le C\delta^2 + \tilde{C}h,$$

where C *and* $\tilde{C}$ *are positive constants.*

Proof. We have

$$|\tilde{\mathcal{U}}_h y(t)| \le \int_0^\omega |G(t,\tau)| \|(E - D(\tau))^{-1}\| |Q(\tau, y(\tau))| d\tau$$

$$+\int_0^\omega |G(t,\tau)| \|(E - D(\tau))^{-1}\| |\Delta f(\tau, y(\tau) + \psi(\tau), \bar{y}(\tau) + \bar{\psi}(\tau))| d\tau$$

$$+\sum_{i=1}^m |G(t, t_i + 0)| |J_i(y_i)| + \sum_{i=1}^m |G(t, t_i + 0)| |I_i(y_i + \psi_i, \bar{y}_i + \bar{\psi}_i)|.$$

The first and third term on the right -hand side admit estimates of the form const.δ^2. The fourth term by (11) is estimated by const.h. Finally, the second term by estimates (12), (14) and arguments similar to those in the proof of Lemma 1 is also estimated by const.h.□

Lemma 3. *For* $y \in C^{(1)}_{\omega,n}$ *we have*

$$|\mathcal{V}_h y(t)| \le \text{const.}h.$$

Proof. By a change of variables using the ω−periodicity of all functions with respect to τ we have

$$\mathcal{V}_h y(t) = \int_0^\omega \{G(t,\tau)(E - D(\tau))^{-1} D(\tau) - G(t, \tau - h)(E - D(\tau - h))^{-1} D(\tau - h)\} \dot{\bar{y}}(\tau) d\tau.$$

To the last integral we apply Lemma 1 with $\dot{\bar{y}}(\tau)$ instead of $\varphi(t)$ and $G(t,\tau)(E - D(\tau))^{-1}D(\tau) - G(t,\tau - h)(E - D(\tau - h))^{-1}D(\tau - h)$ considered as a function of τ for any fixed t instead of $y(t)$. This yields the required inequality, moreover, it is easily seen that the constant can be chosen independent of t.□

We can also apply Lemma 3 to ψ instead of y and obtain

$$|\mathcal{V}_h \psi(t)| \le \text{const.}h.$$

From Lemmas 2 and 3 we have

$$|z(t)| \le C\delta^2 + \widetilde{C}h,$$

where C and $\widetilde{C}$ are some positive constants.

Choose $\widetilde{\delta}_0 = \min\{\widetilde{\delta}, (2C)^{-1}\}$. Then for any $\delta \in (0, \widetilde{\delta}_0]$ we have $C\delta^2 \le \delta/2$. Thus $|z(t)| \le \delta/2 + \widetilde{C}h$. Denote $\bar{h}(\delta) = \min\{\widetilde{h}, \delta/(2\widetilde{C})\}$. Then for any $\delta \in (0, \widetilde{\delta}_0]$ and $h \in [0, \bar{h}(\delta)]$ the operator $\mathcal{J}_h$ maps the set $T_{\delta,\varepsilon_1,\varepsilon_2}$ into itself. It remains to show that this operator is a contraction, possibly decreasing once more $\widetilde{\delta}_0$ and $\bar{h}(\delta)$.

Let $y, \widetilde{y} \in T_{\delta,\varepsilon_1,\varepsilon_2}$, $\delta \in (0, \widetilde{\delta}_0]$, and $z = \mathcal{J}_h y, \widetilde{z} = \mathcal{J}_h \widetilde{y}$. Then

$$z - \widetilde{z} = (\widetilde{\mathcal{U}}_h y - \widetilde{\mathcal{U}}_h \widetilde{y}) + \mathcal{V}_h(y - \widetilde{y}).$$

It is sufficient to investigate the contraction properties of the nonlinear operator $\widetilde{\mathcal{U}}_h$ and the linear operator $\mathcal{V}_h$.

Lemma 4. *The following estimate is valid*

$$\|\widetilde{\mathcal{U}}_h y - \widetilde{\mathcal{U}}_h \widetilde{y}\| \le \|y - \widetilde{y}\|(\delta K_1(\delta) + hK_2(h, \delta)),$$

where K_1 and K_2 are polynomials with positive coefficients.

The proof of Lemma 4 is very similar to the proof of the contraction property of the operator $\mathcal{U}_h$ in [2], so we omit it.

Lemma 5. *Let $\zeta \in C^{(1)}_{\omega,n}$. Then*

$$|\mathcal{V}_h \zeta(t)| \le \left\{ const.h + \frac{M\omega}{1-\mu_0}\left[1 + \frac{2\mu_1}{1-\mu_0}\right]\eta(h) + \mu_0 \frac{4+\beta M}{1-\mu_0} \right\} \|\zeta\|.$$

Proof. We have

$$\mathcal{V}_h \zeta(t) = \int_0^{\omega} \{G(t,\tau)(E - D(\tau))^{-1} D(\tau) - G(t, \tau - h)(E - \bar{D}(\tau))^{-1} \bar{D}(\tau)\} \dot{\bar{\zeta}}(\tau) d\tau.$$

Let $t \ne 0, t_1, \ldots, t_m, \omega$. We integrate by parts, taking into account that the function $G(t, \tau)$ is discontinuous at $\tau = t_1, \ldots, t_m$ and $\tau = t$; ζ is discontinuous at $t_1, \ldots, t_m$.

Thus we obtain

$$\mathcal{V}_h \zeta(t) = (E - D(t))^{-1} D(t)(\bar{\zeta}(t) - \zeta(t))$$

$$-\sum_{i=1}^{m} \{G(t, t_i + h)(E - D(t_i + h))^{-1} D(t_i + h) - G(t, t_i + 0)(E - D(t_i))^{-1} D(t_i)\} \times$$

$$\times (\zeta(t_i + 0) - \zeta(t_i))$$

$$+\sum_{i=1}^{m} G(t,\ t_i+0)B_i(E-D(t_i))^{-1}D(t_i)\overline{\zeta}(t_i)$$

$$+\int_0^{\omega}\{G(t,\ \tau)(E-D(\tau))^{-1}A(\tau)(E-D(\tau))^{-1}D(\tau)$$

$$-G(t,\ \tau-h)(E-\overline{D}(\tau))^{-1}\overline{A}(\tau)(E-\overline{D}(\tau))^{-1}\overline{D}(\tau)\}\ \overline{\zeta}(\tau)d\tau$$

$$-\int_0^{\omega}\{G(t,\ \tau)[(E-D(\tau))^{-1}\dot{D}(\tau)]^2-G(t,\ \tau-h)[(E-\overline{D}(\tau))^{-1}\dot{\overline{D}}(\tau)]^2\}\overline{\zeta}(\tau)d\tau$$

$$-\int_0^{\omega}\{G(t,\ \tau)(E-D(\tau))^{-1}\dot{D}(\tau)-G(t,\ \tau-h)(E-\overline{D}(\tau))^{-1}\dot{\overline{D}}(\tau)\}\ \overline{\zeta}(\tau)d\tau.$$

Making use of Lemma 1 and of the module of continuity $\eta(h)$ of $\dot{D}(\tau)$, we derive the assertion of the lemma. □

From Lemmas 4 and 5 (with $\zeta=y-\tilde{y}$) we obtain

$$|z(t)-\tilde{z}(t)| \le \|y-\tilde{y}\|\ \{\delta K_1(\delta)+hK_2(h,\ \delta)$$

$$+\frac{M\omega}{1-\mu_0}\left[1+\frac{2\mu_1}{1-\mu_0}\right]\eta(h)\ +\ \mu_0\frac{4+\beta M}{1-\mu_0}\},$$

where K_1 and K_2 are polynomials with positive coefficients. By condition (4) we have

$$\mu_0(\beta M+5)<1$$

which implies that

$$\tilde{\mu}\equiv\mu_0\frac{4+\beta M}{1-\mu_0}<1.$$

Choose a number $q\in(\tilde{\mu},\ 1)$ and denote $r=q-\tilde{\mu}$. Then there exists $\delta_1\in(0,\ \tilde{\delta}_0]$ such that for any $\delta\in(0,\ \delta_1]$ we have $\delta K_1(\delta)\le r/3$. Afterwards we can choose $h_1\in(0,\ \overline{h}(\delta_1)]$ such that for any $h\in[0,\ h_1]$ the following inequality is valid

$$\max\left\{hK_2(h,\ \delta_1),\ \frac{M\omega}{1-\mu_0}\left[1+\frac{2\mu_1}{1-\mu_0}\right]\eta(h)\right\}\le r/3.$$

Thus for any $\delta\in(0,\ \delta_1]$ and $h\in[0,\ h_1]$ we have

$$\|\mathcal{G}_h y-\mathcal{G}_h\tilde{y}\| \le \|y-\tilde{y}\|\left(\frac{r}{3}+\frac{r}{3}+\frac{r}{3}+\tilde{\mu}\right)=q\|y-\tilde{y}\|,\ q<1,$$

for any $y, \tilde{y} \in T_{\delta,\varepsilon_1,\varepsilon_2}$. So the operator $\mathcal{G}_h$ has a unique fixed point in $T_{\delta,\varepsilon_1,\varepsilon_2}$, which is an ω-periodic solution $y_h(t)$ of system (6). Since $y(t) \equiv 0$ is the solution of system (7) for $h = 0$, then $y_0(t) \equiv 0$. Thus $x_h(t) = y_h(t) + \psi(t)$ is the unique ω-periodic solution of system (1) in a neighbourhood of $\psi(t)$ for $h \geq 0$ small enough and $x_0(t) \equiv \psi(t)$. This completes the proof of Theorem 1.□

ACKNOWLEDGEMENTS

This work has been supported by a Generalitat Valenciana grant and by the Bulgarian Ministry of Education and Science under Grant MM-8. The third author acknowledges support by the Commission of the European Communities for his stay in Valencia, project No. ERB3510PL90295, as well as to C.N.R.S. for his stay in Pau (France) where a part of the work was realised. He expresses his sincere gratitude to Prof. O. Arino for his interest in the work and helpful discussions.

References

1. J. K. Hale, *Theory of Functional Differential Equations,* Springer - Verlag, New York - Heidelberg - Berlin, 1977.

2. D. D. Bainov and V. Chr. Covachev, *Periodic solutions of impulsive systems with delay viewed as a small parameter* (submitted).

3. A. M. Samoilenko and N. A. Perestyuk, *Differential Equations with Impulse Effect,* Viščа Škola, Kiev (1987) (in Russian).

4th Int. Coll. on Differential Equations, pp. 137-146
D. Bainov and V. Covachev (Eds)

Analytic-numerical solutions of variable coefficient second order delay differential equations

L. JODAR[1], R. J. VILLANUEVA[1], V. CHR. COVACHEV[2] and J. A. MARTIN[3]

[1] Departamento de Matematica Aplicada, Universidad Politecnica de Valencia, P.O. Box 22 012, Valencia, SPAIN

[2] Institute of Mathematics, Bulgarian Academy of Science, Sofia, BULGARIA

[3] Departamento de Analisis Matematico y Matematica Aplicada, Universidad de Alicante, Alicante, SPAIN

Abstract: The exact solution of the initial value problem for scalar second order delay differential equations with variable coefficients is constructed by means of the method of steps and the method of Frobenius. Then, in a bounded domain, a finite analytic solution with error bounds is provided.

Keywords: delay equations, variable coefficients.

1. INTRODUCTION

In many fields of the contemporary science and technology systems with delaying links are often met and the dynamical processes in these are described by systems of delay differential equations [1], [2], [3], [4]. The delay appears in complicated systems with logical and computing devices, where certain time for information processing is needed.

The theory of linear delay differential equations has been developed in the fundamental monographs [1], [2], [5], [6]. Analytic solutions of some linear systems of delay differential equations have been investigated in [7], [8], [9], [10].

In this paper we consider initial value problems for second order delay differential equations of the form

$$\begin{gathered} X''(t)+A(t)X'(t)+B(t)X(t)+B_1(t)X(t-\omega)=F(t),\ 0<t<a, \\ X(t)=G(t),\ t\in[-\omega,0],\ \omega>0, \end{gathered} \tag{1.1}$$

where $A(t)$ and $B(t)$ are analytic complex-valued functions in $|t|<a,\ \omega<a\le+\infty$, $B_1(t)$ is a complex-valued continuous function, the unknown $X(t)$ as well as $F(t)$ and $G(t)$ are complex-valued functions, with $F(t)$ continuous in $0<t<a$ and $G(t)$ is a continuously differentiable function in $[-\omega,0]$.

Problem (1.1) can be transformed into the equivalent first order system

$$Z'(t)+\begin{pmatrix} 0 & -1 \\ B(t) & A(t) \end{pmatrix} Z(t) + \begin{pmatrix} 0 & 0 \\ B_1(t) & 0 \end{pmatrix} Z(t-\omega) = \begin{pmatrix} 0 \\ F(t) \end{pmatrix}, \quad 0<t<a,$$

$$Z(t)=\begin{pmatrix} G(t) \\ G'(t) \end{pmatrix}, \quad -\omega \le t \le 0, \tag{1.2}$$

but this approach has some drawbacks, such as the increase of the computational cost and the lack of explicitness due to the relationship

$$X(t)=(1,0)Z(t).$$

The aim of this paper is twofold. First of all we construct a series solution of problem (1.1) by means of a method of Frobenius and the method of steps, but dealing directly with (1.1). Secondly, we truncate the series solution and provide error bounds for the continuous finite approximate solution when $t \in [n\omega, (n+1)\omega]$ and n is a positive integer. For the constant coefficient case, systems of second order delay differential equations have been recently studied in [8], [9] avoiding the transformation of the problem into an equivalent extended first order system.

This paper is organized as follows. In sec. 2. we construct a series solution of problem (1.1). Error analysis of the finite truncated series in terms of the data, for a given interval $[n\omega, (n+1)\omega] \subset (0,a)$ is studied in sec. 3.

2. A SERIES SOLUTION OF THE PROBLEM

We begin this section considering the homogeneous differential equation

$$X''(t)+A(t)X'(t)+B(t)X(t)=0, \tag{2.1}$$

where $A(t)$, $B(t)$ are complex-valued analytic functions in $|t|<a$ and

$$A(t)=\sum_{\nu=0}^{\infty} A_\nu t^\nu, \; B(t)=\sum_{\nu=0}^{\infty} B_\nu t^\nu, \; |t|<a. \tag{2.2}$$

The coefficients A_ν, B_ν satisfy the Cauchy inequalities

$$|A_\nu|\rho^\nu \le L, \; |B_\nu|\rho^\nu \le L, \; 0<\rho<a, \; \nu \in \mathbf{N} \cup \{0\}, \tag{2.3}$$

for some positive constant L. Let us look for solutions of (2.1) of the form

$X(t)=\sum_{\nu=0}^{\infty} C_\nu t^\nu$, where the coefficients C_ν are to be determined. We get the recurrent equations

$$(\nu+2)(\nu+1)C_{\nu+2} = -\sum_{j=0}^{\nu}\left((j+1)A_{\nu-j}C_{j+1}+B_{\nu-j}C_j\right), \; \nu \ge 0, \tag{2.4}$$

where C_0, C_1 are arbitrary complex numbers. Making use of (2.3), we can show that for any pair of complex numbers C_0 and C_1 the series $X(t)=\sum_{\nu=0}^{\infty} C_\nu t^\nu$ with coefficients C_ν defined by (2.4) is absolutely convergent in $|t|<a$. Let us denote by $X_1(t)$ the solution of (2.1) constructed by the above procedure with $C_0=1$, $C_1=0$, and let $X_2(t)$ be defined in the same way with $C_0=0$, $C_1=1$. Then the pair $\{X_1, X_2\}$ is a fundamental system of solutions of (2.1) in $|t|<a$, and the set of all complex solutions of (2.1) in $|t|<a$ is given by

$$X(t)=X(0)X_1(t)+X'(0)X_2(t). \tag{2.5}$$

Now let $H(t)$ be a continuous complex-valued function in $|t|<a$, and let us consider the nonhomogeneous equation

$$X''(t)+A(t)X'(t)+B(t)X(t)=H(t),\ |t|<a. \tag{2.6}$$

The matrix

$$W(t)=\begin{pmatrix} X_1(t) & X_2(t) \\ X_1'(t) & X_2'(t) \end{pmatrix} \tag{2.7}$$

is invertible in $|t|<a$, and its inverse matrix is

$$W^{-1}(t)=exp(\int_0^t A(s)ds)\begin{pmatrix} X_2'(t) & -X_2(t) \\ -X_1'(t) & X_1(t) \end{pmatrix}. \tag{2.8}$$

By the variation of parameters method, the general solution of the linear equation (2.6) is given by

$$X(t)=X(0)X_1(t)+X'(0)X_2(0)+\int_0^t exp(\int_0^\tau A(s)ds).(X_1(\tau)X_2(t)-X_1(t)X_2(\tau))H(\tau)d\tau.$$

We introduce the integral operator $\mathbf{U}$ defined by

$$(\mathbf{U}H)(t)=\int_0^t K(t,\tau)H(\tau)d\tau,\ 0\le t<a, \tag{2.9}$$

where the kernel $K(t, \tau)$ is given by

$$K(t,\tau)=\exp(\int_0^\tau A(s)ds).(X_1(\tau)X_2(t)-X_1(t)X_2(\tau)). \tag{2.10}$$

It is clear that $K(t, \tau)=0$ and that the operator $\mathbf{U}$ is linear, i.e.

$$\mathbf{U}(\alpha H+\beta J)=\alpha\mathbf{U}H+\beta\mathbf{U}J.$$

Here and below, when possible, we will drop the dependence on *t* for the sake of brevity.

Now we can state the result obtained as follows.

Theorem 1. *Consider equation* (2.6) *where* $A(t), B(t)$ *are analytic complex-valued functions with power expansions* (2.2), *and let* $H(t)$ *be a continuous function. Let* $\{X_1(t), X_2(t)\}$ *be the pair of analytic solutions of* (2.1) *satisfying* $X_1(0)=1, X_1'(0)=0, X_2(0)=0, X_2'(0)=1$. *Then the unique solution* $X(t)$ *of* (2.6) *satisfying the initial conditions* $X(0)=G_1, X'(0)=G_2$ *is given by*

$$X = G_1X_1 + G_2X_2 + \mathbf{U}H, \tag{2.11}$$

where the operator $\mathbf{U}$ *is defined by* (2.9).

Let us consider the delay differential equation (1.1) written in the form

$$X''(t)+A(t)X'(t)+B(t)X(t)=F(t)-B_1(t)X(t-\omega),\ t>0,$$
$$X(t)=G(t),\ t\in[-\omega,0]. \tag{2.12}$$

where $A(t), B(t)$ are analytic complex-valued functions, and note that for $t\in[0,\ \omega]$ the right-hand side of (2.12) is a known continuous function. Then by Theorem 1 the solution of (2.12) can be written in the form

$$X=\mathfrak{X}+\mathbf{U}(F-B_1X_{-\omega})=\mathfrak{X}+\mathbf{U}F-\mathbf{U}(B_1X_{-\omega}), \tag{2.13}$$

where by definition

$$X_{-\omega}(\tau)=X(\tau-\omega)$$

and

$$\mathfrak{X}(t)=G(0)X_1(t)+G'(0)X_2(t),\ 0\le t<a. \tag{2.14}$$

Note that (2.13) is a feedback expression that gives the solution of (1.1) in an interval of length ω in terms of the solution in the previous intervals of length ω. In order to find a closed form expression for the solution of (1.1) in any interval $[n\omega,\ (n+1)\omega]\subset[0,a)$, we introduce a recurrent sequence of integral operators. If H is a continuously differentiable function in $[-\omega,a)$, we define

$$\mathbf{U}_1(B_1H)=\mathbf{U}(B_1H), \tag{2.15}$$

and for an integer $k>1$

$$\mathbf{U}_k(B_1H)=\mathbf{U}(B_1(\mathbf{U}_{k-1}(B_1H))_{-\omega}). \tag{2.16}$$

Here, as above, $(\mathbf{U}_{k-1}(B_1H))_{-\omega}(t)=(\mathbf{U}_{k-1}(B_1H))(t-\omega)$. Now by induction on n we prove that the solution of (1.1) can be written in the compact form

$$X(t)=G(t),\ t\in[-\omega,0], \tag{2.17}$$

and for $t\in[n\omega,\ (n+1)\omega]$

$$X=\mathfrak{X}+\mathbf{U}F+\sum_{k=1}^{n}(-1)^k\mathbf{U}_k(B_1(\mathbf{U}F)_{-\omega})+\sum_{k=1}^{n}(-1)^k\mathbf{U}_k(B_1X_{-\omega})+(-1)^{n+1}\mathbf{U}_{n+1}(B_1G_{-\omega}). \tag{2.18}$$

This proves the following result.

Theorem 2. *Let us consider problem* (1.1) *under the hypothesis of Theorem 1. Let* $\{\mathbf{U}_k\}_{k\ge1}$ *be the sequence of operators defined by* (2.15) *and* (2.16), *where* $B_1(t)$ *is a complex-valued continuous function and let* $K(t,\tau)$ *be defined by* (2.10). *If* $\mathfrak{X}(t)$ *is defined by* (2.14), *then the exact solution of* (1.1) *in the interval* $[n\omega,\ (n+1)\omega]\subset[0,a)$ *for* $n\ge0$ *is given by* (2.18).

Remark. It is easy to show that the integral operators $\mathbf{U}_k$ defined by (2.15), (2.16) can be written in terms of the data in the form

$$(\mathbf{U}_k(B_1H))(t)=\int_0^t\int_0^{t_n-\omega}\cdots\int_0^{t_{p+1}-\omega}[K(t,t_n)B_1(t_n)K(t_n-\omega,t_{n-1})\ldots$$

$$\ldots B_1(t_{p+1})K(t_{p+1}-\omega,t_p)B_1(t_p)H(t_p)]\,dt_p dt_{p+1}\ldots dt_n, \tag{2.19}$$

where $p=n-k+1$.

From a computational point of view, the solution provided by Theorem 2 has the drawback that the expression of $K(t,\tau)$ and $\mathfrak{X}(t)$ is given in terms of infinite series involving $X_1(t)$, $X_2(t)$ and $e^{A(t)}$ defined by Theorem 1. In the sequel we construct finite approximate solutions of (1.1) obtained by truncation of the quoted infinite series.

3. FINITE ANALYTIC APPROXIMATE SOLUTIONS AND ERROR BOUNDS

Let

$$X_1(t)=\sum_{\nu=0}^{\infty}C_\nu t^\nu,\ X_2(t)=\sum_{\nu=0}^{\infty}D_\nu t^\nu,\ \ |t|<a, \tag{3.1}$$

be the pair of analytic series solutions of (2.1) given by Theorem 1. For any positive integer i let us define the truncated series

$$X_1^i(t)=\sum_{\nu=0}^{i}C_\nu t^\nu,\ \ X_2^i(t)=\sum_{\nu=0}^{i}D_\nu t^\nu, \tag{3.2}$$

$$X^i(t)=G(0)X_1^i(t)+G(0)X_2^i(t) \tag{3.3}$$

and if $(w(\tau))_i$ is the McLaurin expansion of an analytic function $w(\tau)$ up to power i, set

$$K_i(t,\tau)=(\exp(\int_0^\tau A(s)ds))_i(X_1^i(\tau)X_2^i(t)-X_1^i(t)X_2^i(\tau)). \tag{3.4}$$

In order to evaluate the error due to this truncation on some fixed interval $[n\omega,\ (n+1)\omega]\subset[0,a)$, we shall need some estimates. First choose $\theta>0$ so that $(n+1)\omega(\theta+1)<a$. Then for a fixed nonnegative integer $q\le n$ we denote

$$r=(q+1)\omega(\theta+1), \tag{3.5}$$

$$S_q=\max\left\{\sup_{|t|\le r}|X_1(t)|,\ \sup_{|t|\le r}|X_2(t)|\right\}, \tag{3.6}$$

$$R_q=\sup_{|t|\le r}\left|\exp(\int_0^t A(s)ds)\right|. \tag{3.7}$$

In fact, S_q can be expressed in terms of $A(t)$ and $B(t)$, but the dependence is too cumbersome and we omit it.

In the notation (3.6), (3.7) for $0 \le t \le (q+1)\omega$ we have

$$|X(t)| \le (|G(0)| + |G'(0)|)S_q, \tag{3.8}$$

$$|K(t,\tau)| \le 2S_q^2 R_q. \tag{3.9}$$

From the Cauchy inequalities it follows that

$$|C_\nu| \le S_q r^{-\nu}, \quad |D_\nu| \le S_q r^{-\nu}, \tag{3.10}$$

where r was defined by (3.5). Then for $0 \le t \le (q+1)\omega$ we have

$$|X_1(t) - X_1^i(t)| \le \sum_{\nu=i+1}^{\infty} |C_\nu| |t|^\nu \le S_q \sum_{\nu=i+1}^{\infty} \left(\frac{|t|}{r}\right)^\nu \le S_q \sum_{\nu=i+1}^{\infty} (1+\theta)^{-\nu} = \frac{S_q}{\theta(1+\theta)^i} \equiv S_q E_i \tag{3.11}$$

and, analogously,

$$|X_2(t) - X_2^i(t)| \le S_q E_i. \tag{3.12}$$

Obviously,

$$\lim_{i\to\infty} E_i = 0. \tag{3.13}$$

Then we deduce

$$|X_j^i(t)| \le |X_j(t) - X_j^i(t)| + |X_j(t)| \le S_q(1+E_i), \; j = 1,2, \tag{3.14}$$

$$|X(t) - X^i(t)| \le (|G(0)| + |G'(0)|)S_q E_i, \tag{3.15}$$

$$|X^i(t)| \le (|G(0)| + |G'(0)|)S_q(1+E_i), \tag{3.16}$$

$$\left|(\exp(\int_0^t A(s)ds))_i\right| \le R_q(1+E_i), \tag{3.17}$$

$$|K_i(t,\tau)| \le 2S_q^2 R_q (1+E_i)^3, \tag{3.18}$$

$$|K(t,\tau) - K_i(t,\tau)| \le 2R_q S_q^2[(1+E_i)^3 - 1] \equiv 2R_q S_q^2 \gamma_i, \tag{3.19}$$

where

$$\lim_{i\to\infty} \gamma_i = 0. \tag{3.20}$$

Now for $i \in \mathbf{N}$, $0 \le t < a$ we define the operators

$$(\mathbf{U}^i H)(t) = \int_0^t K_i(t,\tau) H(\tau) d\tau; \tag{3.21}$$

$$\mathbf{U}_1^i(B_1 H) = \mathbf{U}^i(B_1 H),$$

$$\mathbf{U}_k^i(B_1 H) = \mathbf{U}^i(B_1(\mathbf{U}_{k-1}^i(B_1 H))_{-\omega}), \; k > 1. \tag{3.22}$$

In accordance with (2.19) we can write for $k = n - p + 1$

$$(\mathbf{U}_{\mathbf{k}}^{\mathbf{i}}(B_1H))(t)=\int_0^t\int_0^{t_n-\omega}\ldots\int_0^{t_{p+1}-\omega}[K_i(t,t_n)B_1(t_n)K_i(t_n-\omega,t_{n-1})\ldots$$

$$\ldots B_1(t_{p+1})K_i(t_{p+1}-\omega,t_p)B_1(t_p)H(t_p)]\,dt_pdt_{p+1}\ldots dt_n. \quad (3.23)$$

Let us introduce the constants g, f_q, b_q defined by

$$g=\max\{|G(t)|,\ -\omega\le t\le 0\},$$

$$f_q=\max\{|F(t)|,\ 0\le t\le (q+1)\omega\},$$

$$b_q=\max\{|B_1(t)|,\ 0\le t\le (q+1)\omega\}. \quad (3.24)$$

We shall use the following equality

$$\int_0^{t-\omega}\int_0^{t_n-\omega}\ldots\int_0^{t_{p+1}-\omega}dt_pdt_{p+1}\ldots dt_n=\frac{(t-\omega)[t-(n-p+2)\omega]^{n-p}}{(n-p+1)!} \quad (3.25)$$

In particular, for $t=(n+2)\omega$ we have

$$\int_0^{(n+1)\omega}\int_0^{t_n-\omega}\ldots\int_0^{t_{p+1}-\omega}dt_pdt_{p+1}\ldots dt_n=(n+1)p\omega^{n-p+1}/(n-p+1)!=I_{np}. \quad (3.26)$$

From (2.19) and (3.23) if $k=n-p+1$, we can write

$$(\mathbf{U}_{\mathbf{k}}(B_1H))(t)-(\mathbf{U}_{\mathbf{k}}^{\mathbf{i}}(B_1T))(t)=$$

$$=\int_0^t\int_0^{t_n-\omega}\ldots\int_0^{t_{p+1}-\omega}\{[K(t,t_n)-K_i(t,t_n)]B_1(t_n)K_i(t_n-\omega,t_{n-1})\ldots$$

$$\ldots B_1(t_{p+1})K_i(t_{p+1}-\omega,\ t_p)B_1(t_p)T(t_p)+$$

$$+\sum_{j=p}^{n}K(t,t_n)B_1(t_n)K(t_n-\omega,t_{n-1})\ldots$$

$$\ldots B_1(t_j)[K(t_j-\omega,t_{j-1})-K_i(t_j-\omega,t_{j-1})]B_1(t_{j-1})\ldots$$

$$\ldots B_1(t_{p+1})K_i(t_{p+1}-\omega,t_p)B_1(t_p)T(t_p)+$$

$$+K(t,t_n)B_1(t_n)K(t_n-\omega,t_{n-1})\ldots$$

$$\ldots B_1(t_{p+1})K(t_{p+1}-\omega,t_p)B_1(t_p)[H(t_p)-T(t_p)]\}\,dt_pdt_{p+1}\ldots dt_n. \quad (3.27)$$

If we denote

$$\alpha_1=\max\{|T(t)|,\ p\omega\le t\le (p+1)\omega\},$$

$$\alpha_2=\max\{|H(t)-T(t)|,\ p\omega\le t\le (p+1)\omega\},$$

then from (3.26), (3.27) for $t\in[n\omega,\ (n+1)\omega]$ it follows that

$$\left\| \mathbf{U}_{\mathbf{k}}(B_1H) - \mathbf{U}^i_{\mathbf{k}}(B_1T) \right\| \le 2^k I_{np} \prod_{h=p}^{n} b_h S_h^2 R_h \left\{ \alpha_1 \left[(1+E_i)^{3k} - 1 \right] + \alpha_2 \right\}. \tag{3.28}$$

If $p = 0$ and $H = T = G_{-\omega}$, the expression (3.28) takes on the form

$$\left\| \mathbf{U}_{n+1}(B_1 G_{-\omega}) - \mathbf{U}^i_{n+1}(B_1 G_{-\omega}) \right\| \le \rho_{n,i}, \tag{3.29}$$

where

$$\rho_{n,i} = 2^{n+1} I_{no} g \prod_{h=0}^{n} b_h S_h^2 R_h \left[(1+E_i)^{3(n+1)} - 1 \right] \to 0 \text{ as } i \to \infty. \tag{3.30}$$

If we consider $T = \mathcal{X}_{-\omega}$, $H = \mathcal{X}^i_{-\omega}$, from (2.14), (3.3) and (3.15) for $t_p \in [p\omega, (p+1)\omega]$ it follows that

$$\left| \mathcal{X}(t_p - \omega) - \mathcal{X}^i(t_p - \omega) \right| \le \left(|G(0)| + |G'(0)| \right) S_{p-1} E_i, \tag{3.31}$$

and from (3.28) we can write

$$\left\| \mathbf{U}_{\mathbf{k}}(B_1 \mathcal{X}_{-\omega}) - \mathbf{U}^i_{\mathbf{k}}(B_1 \mathcal{X}^i_{-\omega}) \right\| \le Y_{k,i}, \tag{3.32}$$

where

$$Y_{k,i} = 2^k I_{np} \left(|G(0)| + |G'(0)| \right) \prod_{h=p}^{n} b_h S_h^2 R_h . S_{p-1} \times \left\{ (1+E_i)^{3k} - 1 + E_i \right\} \to 0 \text{ as } i \to \infty. \tag{3.33}$$

Taking $H = (\mathbf{U}^i F)_{-\omega}$, $T = (\mathbf{U}F)_{-\omega}$ and using that $t_p \in [p\omega, (p+1)\omega]$, from (2.9), (3.21) and (3.19) it follows that

$$\left| (\mathbf{U}F)(t_p - \omega) - (\mathbf{U}^i F)(t_p - \omega) \right| \le 2p\omega S_{p-1}^2 R_{p-1} f_{p-1} \gamma_i, \tag{3.34}$$

and from (2.9) and (3.9),

$$\left| \mathbf{U}F(t_p - \omega) \right| \le 2p\omega S_{p-1}^2 R_{p-1} f_{p-1}. \tag{3.35}$$

From (3.28) and (3.34), (3.35) we can write

$$\left| (\mathbf{U}_k(B_1(\mathbf{U}F)_{-\omega}))(t_k) - (\mathbf{U}^i_k(B_1(\mathbf{U}^1 F)_{-\omega}))(t_k) \right| \le U_{k,i}, \tag{3.36}$$

where

$$U_{k,i} = 2^{k+1} I_{np} p\omega f_{p-1} S_{p-1}^2 R_{p-1} \prod_{h=p}^{n} b_h S_h^2 R_h \times \left\{ (1+E_i)^{3k} + (1+E_i)^3 - 2 \right\} \to 0 \text{ as } i \to \infty. \tag{3.37}$$

If we denote by $X^i(t)$ the approximate solution of (1.1) defined in $[n\omega, (n+1)\omega]$ by

$$X^i = \mathcal{X}^i + \mathbf{U}^1 F + \sum_{k=1}^{n} (-1)^k \mathbf{U}^i_{\mathbf{k}}(B_1(\mathbf{U}^1 F)_{-\omega}) + \sum_{k=1}^{n} (-1)^k \mathbf{U}^i_{\mathbf{k}}(B_1 \mathcal{X}^i_{-\omega}) + (-1)^{n+1} \mathbf{U}^i_{n+1}(B_1 G_{-\omega}), \tag{3.38}$$

from (2.18), (3.38) and the above estimates it follows that the error $X(t)-X^i(t)$ for $i\in\mathbf{N}$ and $t\in[n\omega,\ (n+1)\omega]\subset[0,a)$ satisfies

$$\left|X(t)-X^i(t)\right|\le\left(\left|G(0)\right|+\left|G'(0)\right|\right)S_nE_i+2(n+1)\omega S_n^2R_nf_n\gamma_i+\rho_{n,i}+\sum_{k=1}^{n}(U_{k,i}+Y_{k,i}). \quad (3.39)$$

Thus, for a fixed interval $[n\omega,\ (n+1)\omega]$ and an admissible error ε, to construct a finite analytic approximate solution whose error be smaller than ε in $[n\omega,\ (n+1)\omega]$ it suffices to take $i\in\mathbf{N}$ so large that

$$\left(\left|G(0)\right|+\left|G'(0)\right|\right)S_nE_i+2(n+1)\omega S_n^2R_nf_n\gamma_i+\rho_{n,i}+\sum_{k=1}^{n}(U_{k,i}+Y_{k,i})<\varepsilon. \quad (3.40)$$

This is possible by virtue of (3.13), (3.20), (3.30), (3.33) and (3.37).
Hence the following result has been proved.

Theorem 3. *Consider problem* (1.1) *under the hypotheses of Theorem 2 using the previous notation. If* $X^i(t)$ *is the function defined by (3.38) , then* $X^i(t)$ *converges to the exact solution* $X(t)$ *of* (1.1) *as* $i\to\infty$ *for any* $t\in[0,a)$. *If* n *is a positive integer such that* $(n+1)\omega<a$, *then the error of the approximate solution* $X^i(t)$ *with respect to the exact solution* $X(t)$ *satisfies (3.39) for* $t\in[n\omega,\ (n+1)\omega]$. *Given an admissible error* $\varepsilon>0$, *taking* $i\in\mathbf{N}$ *satisfying* (3.40), *one gets an approximation whose error is bounded above by* ε *for* $t\in[n\omega,\ (n+1)\omega]$.

ACKNOWLEDGEMENTS

This work has been supported by a Generalitat Valenciana grant and by the Bulgarian Ministry of Education and Science under Grant MM-8. The third author acknowledges support by the Commission of the European Communities for his stay in Valencia, project No. ERB3510PL90295.

References

1. R. Bellman and K. L. Cooke, *Differential-Difference Equations,* Academic Press, New York, 1962.
2. R. D. Driver, *Ordinary and Delay Differential Equations,* Applied Math. Sciences 20, Springer, Berlin, 1977.
3. G. I. Marchuk, *Mathematical Models in Immunology, Translation Series,* Optimization Software, Springer, New York, 1983.
4. M. Okamoto and K. Hayashi, *Frequency conversion mechanism in enzymatic feedback systems,* J. Theor. Biol., 108(1984), 529-537.
5. A. D. Myshkis, *Linear Differential Equations with a Retarded Argument,* Ed. Gostekhizdat, Moscow-Leningrad, 1951 (in Russian; 2nd Edit. 1972).

6. E. Pinney, *Ordinary Difference-Differential Equations,* Univ. of California Press, Berkeley, Los Angeles, 1958.
7. V. B. Cherepennikov, *Investigation of the solutions of a class of differential-functional equations,* Sb. Asymptotic Methods in System Dynamics, Ed. USF Siberian Filial Acad. Sci., Irkutsk (1985), 38-44 (in Russian).
8. L. Jodar and J. A. Martin, *Explicit solution of systems of second order differential difference equations,* J. Comput. Appl. Maths., 43(1992), 323-334.
9. L. Jodar and J. A. Martin, *Analytical solution of a class of coupled second order differential-difference equations,* Internat. J. Math. & Math. Sci. Vol. 16, No. 2 (1993) 385-396.
10. V. I. Rodinov, *Analytical solution of a linear functional differential equation with a linear deviation of the argument,* Differential Equations, 25(1989), No.4, 616-626 (in Russian).

4th Int. Coll. on Differential Equations, pp. 147-156
D. Bainov and V. Covachev (Eds)

Localization of Global Existence of Holomorphic Solutions of Differential Equations with Complex Parameters

Joji KAJIWARA
Department of Mathematics, Kyushu University 33,
Fukuoka 812, Japan

Abstract. We establish a principle that the global existence of holomorphic solutions to a differential equation $Tf=g$ with complex parameter s, s running over a Stein manifold S, is local with respect to the parameter manifold S .
Keywords: Localization, Global Holomorphic Solutions, Complex Parameters.

1. Introduction

The famed Chinese late philosopher stated that vivid existences have contradictions. The main purpose of the present paper is to unify the oppositive ideas, local and global, by methods of the theory of functions of several complex variables, started from K. Oka [17].

L. Ehrenpreis [2] discoursed on an application of the sheaf theory to differential equations and gave a criterion for the existence of global solutions of differential equations $Tf = g$ in a domain D when the existence of local solutions are assured. The author [3] applied Ehrenpreis' method to linear ordinary differential equationsand gave a necessary and sufficient condition for the global existenc. The condition is that D is either simply connected or doubly connected without non trivial global single-valued holomorphic homogeneous solutions. A soil mechanist asked the author to give a concrete solution of the system of partial differential equations:

$$\frac{\partial u'}{\partial S_\alpha} = (u' \tan\phi + v' \sec\phi)\frac{\partial \theta}{\partial S_\alpha},$$

$$\frac{\partial v'}{\partial S_\beta} = -(u' \sec\phi + v' \tan\phi)\frac{\partial \theta}{\partial S_\beta} \qquad (1)$$

which are concretely solved by a tranformation of independent variables, where S_α and S_β are independent , and u' and v' are dependent variables. In the theory of kinetics of ideal soils, A. J. Spencer [19] gave the following singular system consisting of partial differential equations:

$$\cos\phi \ \frac{\partial p}{\partial \theta} + \frac{2c\cos\phi}{1-\sin\phi} \exp\{(\frac{\pi}{2}+\phi-2\theta)\tan\phi\}\frac{\partial \psi}{\partial \theta} + 2p\ \sin\phi = 0,$$

$$\frac{\partial p}{\partial r} - \frac{2c}{1-\sin\phi}(\frac{\partial \psi}{\partial r}+\frac{2\psi}{r})\exp\{(\frac{\pi}{2}+\phi-2\theta)\tan\phi\} = 0 \qquad (2)$$

A characteristic of the above systems in the soil mechanics are systems of differential equations with repect to each different one independent variable. These systems are investigated in [4], [5], ... and [9] by the author as concrete case studies.

H. Suzuki [22] stated that the global existence of the differential eqution $\partial u/\partial x_1 = f$ with complex parameter depends on the Steinness of the sets of cuts over the parameter space.

Let S be a Stein manifold, D be a Stein domain in the product maniflod of the Stein manifold S and the complex plane **C**, m be a positive integer and $a(z, s)$ be a $m\times m$ matrix valued holomorphic function on D and T be the linear differential operator

$$T = \frac{\partial}{\partial z} + a(z, s) \qquad (3)$$

The author and Y. Mori [13] connected [3] and [22], investigated the condition for the global existence of holomorphic solutions f of the ordinadinary differential equations $Tf = g$ for any given $m\times m$ matrix valued holomorphic function g on D with complex parameters, analyzed the condition into topological conditions for the cuts and into analytical conditions for the coefficient, and proved that the condition for global existence is local with respect to the parameter manifold S which S is Stein. The author and K.H. Shon [14] polished up and generalized those to the case of Stein parameter spaces.

Concerning partial differential equations, the jointworker K. H. Shon [20] reported promptly the main theorem and showed a sheaf-theoretical route of its proof. The aim of the jointwork [15] and [16] is to give complete analytical foundations to it. In the previous paper [11], we gave a necessary and sufficient condition for the global existence for decomposable partial differential operator in polycylindrical domains which are constant as the parameter varies. In the present paper, we discuss the case that the polycylindrical domains vary as the parameter varies.

2. Sheaves and linear differential equations

Let S be a Stein manifold, D be a Stein domain in S and m be a positive integer. Let $\mathbb{O}$ be the structure sheaf of D, i. e., the sheaf of germs of holomorphic functions on D, $T : \mathbb{O}^m \rightarrow \mathbb{O}^m$ be a sheaf homomorphism, and $\operatorname{Ker} T$ and $\operatorname{Im} T$ ($\subset \mathbb{O}^m$) be, respectively, its Kernal and Image. The short exact sequence

$$0 \rightarrow \operatorname{Ker} T \rightarrow \mathbb{O}^m \rightarrow \operatorname{Im} T \rightarrow 0 \qquad (4)$$

of sheaves induces the long sequence

$$0 \to H^0(D, \mathrm{Ker}\, T) \to H^0(D, \mathbb{O}^m) \to H^0(D, \mathrm{Im}\, T) \to$$
$$\to H^1(D, \mathrm{Ker}\, T) \to H^1(D, \mathbb{O}^m) \to H^1(D, \mathrm{Im}\, T) \to \ldots \quad (5)$$

of cohomologies, where $\mathbb{O}^m \to \mathrm{Im}\, T$ and $H^0(D, \mathbb{O}^m) \to H^0(D, \mathrm{Im}\, T)$ are the homomophisms induced by the sheaf homomorphism T and the others are canonical ones. Since $H^1(D, \mathbb{O}^m) = 0$ by the theorem of Cartan-Oka[17]-Serre, we have

$$H^1(D, \mathrm{Ker}\, T) = H^0(D, \mathrm{Im}\, T) / T H^0(D, \mathbb{O}^m) \quad (6)$$

When the sheaf homomorphism T is induced by a linear partial differential operator and when the existence of local solutions are assured so as there holds $\mathrm{Im}\, T = \mathbb{O}^m$, we have

$$H^1(D, \mathrm{Ker}\, T) = H^0(D, \mathbb{O}^m) / T H^0(D, \mathbb{O}^m) \quad (7)$$

and $H^1(D, \mathrm{Ker}\, T)$ amounts to the gap between the destination that, for any given global m- vector valued holomorphic function g on D, we can find a global m-vector valued holomorphic function f on D which is a solution to the partial differential equation $Tf = g$. So, the statement $H^1(D, \mathrm{Ker}\, T) = 0$ is equivant to the global existence stated above.

3. Vanishing the cohomology $H^1(D, \mathbb{O})$

Let Ω be a topological space, S be an open Riemann surface and φ be a local homeomorphism of Ω in S. Then Ω is locally euclidean but Ω is not a complex manifold. However, we can speak of the sheaf $\mathbb{O}$ of germs of holomorphic functions on Ω as if Ω were a complex manifold. In order to reproduce an atomosphere of the proof by the methods of cohomology chases in the simplest case, we give the folowing proposition and prove it:

Proposition 1. *If* $H^1(\Omega, \mathbb{O}) = 0$, *then* Ω *is a Hausdorff space.*

Proof. If Ω were not a Hausdorff space, there would exist two different points w' and w'' in Ω such that, for the filter $\mathbb{V}(w')$ and $\mathbb{V}(w'')$ of neighborhoods of, respectively of w' and w'', the sum $\mathbb{V}(w') \cup \mathbb{V}(w'')$ forms a filter base on S. We may assume that $x_0 = \varphi(w') = \varphi(w'') \in S$. There exist, respectively, open neighborhoods W' and W'' of w' and w'', and an open neighborhood V of x_0 such that the restrictions $\varphi_{W'}$ and $\varphi_{W''}$ of φ on W' and W'' are homeomorphisms of W' and W'' onto V and that there exists a biholomorphic mapping ξ of $V = \varphi(W') = \varphi(W'')$ onto an open disk Δ in the complex plain $\mathbf{C}$. We consider the open neighborhood $\Omega - \{w'\}$ of w''. Then $\mathbb{U} = \{W', \Omega - \{w'\}\}$ is an open covering of Ω. We define a holomorphic function $h = 1/(\xi o \varphi(w) - \xi(x_0))$ for $w \in W' \cap (\Omega - \{w'\}) = W' - \{w'\}$. Then, the function h forms a cocycle $\in Z^1(\Omega, \mathbb{O}) = B^1(\Omega, \mathbb{O})$. Since the canonical mapping $H^1(\mathbb{U}, \mathbb{O}) \to H^1(\Omega, \mathbb{O}) = 0$ is injective, we have

$Z^1(\Omega, \mathbb{O}) = B^1(\Omega, \mathbb{O})$. There exist, respectively, holomorphic functions h' and h'' on W' and $\Omega - \{w'\}$ such that there holds $h = h' - h''$ on $W' - \{w'\}$. By the assumption of reductio ad absurdum, the intersection $W = W' \cap W''$ of the open neighborhoods W' and W'' of two different points w' and w'' with the same image $x_0 = \varphi(w') = \varphi(w'') \in S$ is a non empty open set in Ω, is a schlicht open set over S, bifucates suddenly into two sheets $W' - W' \cap W''$ and $W'' - W' \cap W''$ and we have

$$\frac{1}{\xi \circ \varphi(w) - \xi(x_0)} = h'(w) - h''(w) \tag{8}$$

on the non empty open subset $W' - \{w'\}$ of $\Omega - \{w'\}$. Since the function h'' is holomorphic in the open neighborhood $\Omega - \{w'\}$ of w'', the composite function $h'' \circ \varphi_{w''}^{-1} \circ \xi^{-1}$ is holomorphic in the open neiborhood D of $x_0 = \xi(\varphi(w''))$ and satisfy

$$\frac{1}{z - \xi(x_0)} = = h' \circ \varphi_{w'}^{-1} \circ \xi^{-1}(z) - h'' \circ \varphi_{w''}^{-1} \circ \xi^{-1}(z) \tag{9}$$

in the non empty open subset W of $\xi \circ \varphi(W' - \{w'\})$. By the theorem of identity, (9) holds in $D - x_0$. Since h' is holomorphic in the open neighborhood W' of w', the composite function $h' \circ \varphi_{w''}^{-1} \circ \xi^{-1}(z)$ is holomorphic in the open neighborhood $D = \xi \circ \varphi(W')$. Then, by (9), the second member $h'' \circ \varphi_{w''}^{-1} \circ \xi^{-1}(z)$ of the right hand side of (9) has a simple pole at the point $\xi(x_0)$. This contradicts that the function h'' is holomorphic in the open neighborhood $\Omega - \{w'\}$ of the point w'' of W'' with the image $x_0 = \varphi(w'')$.

Let Ω be a Hausdorff space, S be a two dimensional Stein manifold and φ be a local homeomorphism of Ω in S. Then Ω is locally euclidean and is a complex manifold. Hence, we can speak of the usual sheaf $\mathbb{O}$ of germs of holomorphic functions on Ω. By the argument of the author and H. Kazama [12], we have the following Proposition, an atomosphere of the proof of which we have reproduced in that of Proposition 1.

Proposition 2 *If* $H^1(\Omega, \mathbb{O}) = 0$, *then* Ω *is a Stein manifold.*

Vanishing of the first cohomology of the topological space Ω with coefficient in the structure sheaf $\mathbb{O}$ gives the separability of the topology of Ω when the complex dimension of Ω is one and gives the Steinness of the complex manifold Ω when the complex dimension of Ω is two.

4. Global existence of equations without parameter

Let m be a positive integer, Ω be a domain in the complex plane **C**, $\mathbb{O}$ be the sheaf of germs of holomorphic functions on Ω , $a(z)$ be a $m\times m$ matrix valued holomorphic function on Ω . Then the linear differential operator

$$T = \frac{\partial}{\partial z} + a(z) \tag{10}$$

induces naturally a sheaf homomomorphism $T: \mathbb{O}^m \to \mathbb{O}^m$ and the arguments in the paragraph 2 are applicable. The author [3] gave the following proposition:

Proposition 3. $\mathrm{H}^1(\Omega, \mathrm{Ker}\, T) = 0$ *if and only if* Ω *is either a simply connected domain or* Ω *is a doubly connected domain with* $\mathrm{H}^0(\Omega, \mathrm{Ker}\, T) = 0$.

Proof. Under the assumption $\mathrm{H}^1(\Omega, \mathrm{Ker}\, T) = 0$, assume that Ω were neither simply connected nor doubly connected. There would exist two closed Jordan curves γ_1 and γ_2 in Ω which belong to a homology basis of Ω, and two simply connected subdomains Ω_1 and Ω_2 of Ω such that the intersection $\Omega_1 \cap \Omega_2$ has at least three connected components Δ_0, Δ_1 and so on. We may assume that the two closed curves γ_1 and γ_2 have the counter clock wise orientation, that Δ_0 and Δ_1 are simply connected, that the starting point z_s of γ_1 belongs to Δ_0, the curve γ_1 has a transit point z_t which is the starting point of the curve γ_2 and belongs to Δ_1. We put $\Delta_2 = \Omega - \Delta_0 \cup \Delta_1$. $\mathbb{U} = \{\Omega_1, \Omega_2\}$ is an open covering of Ω and we have $\Omega_1 \cap \Omega_2 = \Delta_0 \cup \Delta_1 \cup \Delta_2$, Δ_0, Δ_1 and Δ_2 being mutually disjoint. Since the domain Δ_0 is simply connected, there exists a $m\times m$ matrix valued holomorphic function $F = (f_{ij})$ on Δ_0 such that the m rows $\{f_{ij}\ ; i = 1, 2, ..., m\}$ are linearly independent holomorphic solutions on the simply connected domain Δ_0 of the homogeneous equation $Tf = 0$. There exists a constant $m\times m$ matrix $\mathrm{Monod}(\gamma_1)$ such that $F\, \mathrm{Monod}(\gamma_1)$ is the result at the terminal point z_s of the analytic continuation of F along γ_1 with the starting point z_s. Let c be any constant $m\times 1$ matrix. Then Fc is a holomorphic solutions on the domain Δ_0 of the equation $Tf = 0$. We associate Fc to the domain Δ_0 and 0 to the domain Δ_1 and the open set Δ_2. Then the above datum give a cocycle of the covering $\mathbb{U}$ with coefficients in the sheaf Ker T. By the assumption, this is a coboundary, i. e., there exist $h' \in \mathrm{H}^1(\Omega_1, \mathrm{Ker}\, T)$ and $h'' \in \mathrm{H}^1(\Omega_2, \mathrm{Ker}\, T)$ satisfying $h' - h'' = Fc$ on Δ_0 and $h' - h'' = 0$ on $\Delta_1 \cup \Delta_2$. There exists a constant $m\times 1$ matrix b such that $h' = Fb$ on Δ_0. Its analytic continuation along the closed curve γ_1 is $F\mathrm{Monod}(\gamma_1)b$ .By the first equation, concerning the difference of before and after analytic continuation along the closed curve γ_1 at the starting and terminal point z_s, we have

$$(\mathrm{I} - \mathrm{Monod}(\gamma_1))b = c \tag{11}$$

Because, for any c there exists b with (10), we have

$$\det\,(\mathrm{I} - \mathrm{Monod}(\gamma_1)) \neq 0 \tag{12}$$

and this implies any non trivial solution f of the homogeneous equation $Tf = 0$ is multiform along the closed curve γ_1. But, the second equation, $h' = h''$ on $\Delta_1 \cup \Delta_2$, implies that any solution f of the homogeneous equation $Tf = 0$ is uniform along the closed curve γ_2. Since the argument is symmetric with respect to the suffix 1 and 2, this is a contradiction.

5. Localization of global existence

Let S be a Stein manifold, D be a Stein domain in the product maniflod of the complex plane $\mathbf{C}$ and the Stein manifold S, $\mathbb{O}$ be the sheaf of germs of holomorphic functions on D, m be a positive integer and $a(z, s)$ be a $m \times m$ matrix valued holomorphic function on D. Then the linear differential operator

$$T = \frac{\partial}{\partial z} + a(z, s) \tag{13}$$

induces naturally a sheaf homomomorphism $T : \mathbb{O}^m \to \mathbb{O}^m$ and the arguments in the paragraph 2 are applicable.

Let $\pi : D \to S$ be the canonical projection. For any point, (z, s) of D, let $D(z, s)$ be the connected component containg the point (z, s) of $\pi^{-1}(s) \subset \mathbf{C} \times \{s\}$. We call $D(z, s)$ the cut of D at the point (z, s) and regard it as a domain in the complex plane $\mathbf{C}$. Let $\widetilde{D}$ be the set of all cuts $D(z, s)$, (z, s) running over D. We introduce in the set D an equivalence relation $\equiv$ by defining two points (z,s) and (z', s') of D to satisfy $(z, s) \equiv (z', s')$ if and only if $D(z, s) = D(z', s')$. Then the set $\widetilde{D}$ is the factor set of the set $D/\equiv$ by this equvalence relation $\equiv$. Let $\psi : D \to \widetilde{D}$ be the canonical mapping. We equip the set $\widetilde{D}$ with the factor topology $D/\equiv$. The topology is the strongest one between those which let ψ be continous. We also define a mapping $\varphi : \widetilde{D} \to S$ by putting canonically $\varphi(D(z, s)) = s \in S$ for $D(z, s) \in \widetilde{D}$. Then we have $\pi = \varphi \circ \psi$ and the mapping φ is a local homoemorphism. Hence, the pair $(\widetilde{D}, \varphi)$ is a target of the arguments of the paragraph 3.

Let A be an analytic subset of S. Then we denote the restriction of T on $\pi^{-1}(A)$ by the same symbol T and regard $T: \mathbb{O}^m \to \mathbb{O}^m$ as a sheaf homomorphism over $\pi^{-1}(A)$. Since any holomorphic function on the analytic subset $\pi^{-1}(A)$ of D is extended to a holomorphic function on D, by the Cartan-Oka[17]-Serre theorem and since the variable $s \in S$ plays only roles of a parameter concerning the differential operator T, the assertion $H^1(D, \operatorname{Ker} T) = 0$ implies $H^1(\pi^{-1}(A), \operatorname{Ker} T) = 0$. Especially, for any $(z, s) \in D$, we consider the analytic set $\{s\} \subset S$ consisting only a single point $s \in S$ and obtain $H^1(D(z, s), \operatorname{Ker} T) = 0$. By Proposition 3, each $D(z, s)$ is either a simply connected domain in the complex plane $\mathbf{C}$ or a doubly connected domain with $H^0(D(z, s), \operatorname{Ker} T) = 0$. By cohomology chases similar to the proofs of Proposition 1 and 3, the author and Y. Mori [13] proved that either all cuts $D(z, s)$ are simutaneously simply connected or all cuts $D(z, s)$ are simutaneously doubly connected with $H^0(D(z, s), \mathrm{Ker}\, T) = 0$. In the latter case, for any $g \in H^0(D(z, s), \mathcal{O}^m)$, there exists uniquely $f \in H^0(D(z, s), \mathcal{O}^m)$ such that $Tf = g$. Hence there exists the inverse mapping $T^{-1}_{(z,s)}: H^0(D(z, s), \mathcal{O}^m) \to H^0(D(z, s), \mathcal{O}^m)$.

In order to reproduce an atomosphere of the proof in the simplest case, we give the following proposition and and prove it:

Proposition 4. *If* $H^1(D, \mathrm{Ker}\, T) = 0$ *and if all cuts of D are doubly connected, then* $\widetilde{D}$ *is a Hausdorf space.*

Proof. If $\widetilde{D}$ were not a Hausdorf space, there would exist two different points $w' = D(z', s')$ and $w'' = D(z'', s'')$ in $\widetilde{D}$ such that, for the filter $\mathbb{V}(w')$ and $\mathbb{V}(w'')$ of neighborhoods of, respectively of w' and w'', the sum $\mathbb{V}(w') \cup \mathbb{V}(w'')$ forms a filter base on S. Since the filter basis $\mathbb{V}(w') \cup \mathbb{V}(w'')$ converges to w' and w'', its images $\varphi(\mathbb{V}(w') \cup \mathbb{V}(w''))$ by the continuous mapping φ converge to $\varphi(w')$ and $\varphi(w'')$ in S. Since S is a Hausdorff space, we have $\varphi(w') = \varphi(w'')$ which is denoted by $s_0 = \varphi(w') = \varphi(w'') \in S$. There exist, respectively, two Jordan curves γ' and γ'' with w' and w'' as starting points in the complex plane $\mathbf{C}$ such that the relatively compact connected components Δ' of $\mathbf{C} - \gamma'$ and Δ'' of $\mathbf{C} - \gamma''$ in the complex plane $\mathbf{C}$, $\Delta' \times \{s_0\}$ and $\Delta'' \times \{s_0\}$ are not contained in the cuts $D(z', s')$ and $D(z'', s'')$. There exists a connected open neighborhood U of s_0 such that $\gamma' \times U \subset D$ and $\gamma'' \times U \subset D$. Then, the subsets $\widetilde{V}' = \{D(z', s); s \in U\}$ and $\widetilde{V}'' = \{D(z'', s); s \in U\}$ are, respectively, open connected neighborhoods U of the points w' and w'' in the space $\widetilde{D}$. By the assumption of this reductio ad absurdum, the intersection $\widetilde{V}' \cap \widetilde{V}''$ is not empty and contains a point s^1. Since $D(z', s^1) = D(z'', s^1)$ is doubly connected, one of the above two domains Δ' and Δ'', for example Δ' satisfies $\Delta' \times \{s^1\} \subset D$. By the Hartogs' phenomenon, any holomorphic function in the domain $(\gamma' \times U) \cup (\Delta' \times \{s_0\})$ contained in D is holomorphically continued to $\Delta' \times U$. Since D is Stein, $\Delta' \times U \subset D$. But this ia a contradiction.

Conversely, if all cuts $D(z, s)$ are simutaneously doubly connected with $H^0(D(z, s), \mathrm{Ker}\, T) = 0$, then by the existence of the inverse mapping $T^{-1}_{(z,s)}: H^0(D(z, s), \mathcal{O}^m) \to H^0(D(z, s), \mathcal{O}^m)$, we can prove that $TH^0(D, \mathcal{O}^m) = H^0(D, \mathcal{O}^m)$, i. e., that $H^1(D, \mathrm{Ker}\, T) = 0$.

In case that $H^1(D, \mathrm{Ker}\, T) = 0$ and that all cuts of Ω are simply connected, sheaf of germs of holomorphic functions on D. We assume moreover

(a) there exists a domain E in $\mathbf{C} \times S$ containing D such that the $m \times m$ matrix valued function $a(z, s)$ can be holomorphically continued in E and the space $\widetilde{E}$ of all cuts is a Hausdorff space.

Then, as is shown in the author and G. H. Shon [14], $\widetilde{D}$ is at first a Hausdorff space and the domain $(\widetilde{D}, \varphi)$ is a Stein domain over the Stein manifold S. The substances of the proofs are firstly to consider any one dimensinal analytic subset A of the parameter manifold S, to derive $H^1(\pi^{-1}(A), \mathrm{Ker}\, T) = 0$ from $H^1(D, \mathrm{Ker}\, T) = 0$, to derive the separation of the toplology of $\widetilde{D}$ as in the proof of Proposition 1 and secondly to consider any two dimensinal analytic subset A of the parameter manifold S, to derive $H^1(\pi^{-1}(A), \mathrm{Ker}\, T) = 0$ from $H^1(D, \mathrm{Ker}\, T) = 0$, to derive the Steinness of the open set $\varphi^{-1}(A)$ as in the author and H. Kazama [12] and the Steinness of $\widetilde{D}$.

Theorem 1. *If* $H^1(D, \mathrm{Ker}\, T) = 0$, *then either all cuts* $D(z, s)$ *are simultaneously simply connected or all cuts* $D(z, s)$ *are simultaneously doubly connected with* $H^0(D(z, s), \mathrm{Ker}\, T) = 0$. *In the latter case, the space* $\widetilde{D}$ *of all cuts is a Hausdorff space. Conversely, if all cuts* $D(z, s)$ *are simutaneously doubly connected with* $H^0(D(z, s), \mathrm{Ker}\, T) = 0$ *and the space* $\widetilde{D}$ *of all cuts is a Hausdorff space, then* $H^1(D, \mathrm{Ker}\, T) = 0$. *In the former case, under the assumption* (a), *the space* $\widetilde{D}$ *of all cuts is a Hausdorff space and, with respect to the natural complex structure,* $\widetilde{D}$ *is a Stein manifold, and conversely, if the space* $\widetilde{D}$ *of all cuts is a Stein manifold, then* $H^1(D, \mathrm{Ker}\, T) = 0$.

Since the property Steinness is local by F. Docquier und H. Grauert [1], the above criterion for the global existence is local with respect to the parameter space S.

6. Decomposable partial differential operators

Let S be a Stein manifold and n be a positive integer. Let D be a Stein domain in the product manifold of the Stein manifold S and the complex space $\mathbf{C}^n$ of n complex vaiables $z = (z_1, z_2, \ldots, z_n)$. Let $\pi : D \to S$ and $\rho : D \to \mathbf{C}^n$ be the canonical projections. For any $j \in \{1, 2, \ldots, n\}$, let $\rho_j : D \to \mathbf{C}$ be the composition of ρ and the projection $\mathbf{C}^n \to \mathbf{C}$ onto the j-th component space. For any $s \in S$, we assume that the open set $D(s) = \pi^{-1}(s)$ is a cylindrical open set of type $D(s) = \prod_{j=1}^{n} D_j(s)$, where each $D_j(s)$ is an open set in the complex plane $\mathbf{C}$.

For any $j \in \{1,2,\ldots,n\}$, let $a_j(z, s)$ be a $m \times m$ matrix valued holomorphic function on D and T_j be the linear differential operator defined by

$$T_j = \frac{\partial}{\partial z_j} + a_j(z, s) \qquad (14)$$

Concerning the differential operator T_j, only the independent variable z_j plays the role of the variable for differentiation and other varaibles $z_1, z_2, \ldots, z_{j-1}, z_{j+1}, \ldots, z_n$ and s play the roles of parameters. For any $(z,s) \in D$, let $D_j(z,s)$ be the connected component containing the point (z,s) of the open set $\rho_j^{-1}(s)$. Let $\widetilde{D}_j$ be the space of cuts $D_j(z,s)$, (z,s) running over D.

Now, we define a linear differential operator T by putting

$$T = T_n \ldots T_2 T_1 \qquad (15)$$

Making use of the preparation of the preceding paper [11], we can prove the following theorem:

Thoerem 2. *The necessary and sufficient condition for* $\mathrm{H}^1(D, \mathrm{Ker}\, T) = 0$ *is that there holds* $\mathrm{H}^1(D, \mathrm{Ker}\, T_j) = 0$ *for any* $j \in \{1,2,\ldots,n\}$.

Thus we can apply Theorem 1 and assert that the condition for the global existence is local with respect to the parameter space S which is a Stein manifold.

References

[1] F. Docquier und H. Grauert, *Levisches Probelem und Rungescher Satz für Teilgebiete Steinscher Mannigfaltigkeit,* Math. Ann., **140**(1960), 94-123.

[2] L. Ehrenpreis, *Sheaves and differential equations*, Proc. Amer. Math. Soc., **7**(1956), 1131-1139.

[3] J. Kajiwara, *On an application of L. Ehrenpreis' method to ordinary differential equations*, Kodai Math. Sem. Rep. **15**(1963), 94-105.

[4] J. Kajiwara, *Some systems of partial differential equations in the theory of soil mechanics*, Mem. Fac. Sci. Kyushu Univ., **24**(1970), 147-230.

[5] J. Kajiwara, *Some systems of partial differential equations in complex domains*, Mem. Fac. Sci. Kyushu Univ., **25**(1971),76-83.

[6] J. Kajiwara, *Holomorphic solutions of a partial differential eqution of a mixed type*, Mathematica Balkanica , **2**(1972), 21-143.

[7] J. Kajiwara, *Cauchy -Kowalewski theorem for a partial differential equatuion of a mixed type*, Rendiconti di Mat. dei Lincei, **6**(1973), 183-192.

[8] J. Kajiwara, *Unique holomorphic solution of a singular partial differential equation with a double pole*, Periodica Mathematica Hungarica **5**-1 (1974) , 61-71.

[9] J. Kajiwara, *Holomorphic solutions of singular Darboux problems*, Rivista di Mathematica Univ. Parma, **4** (1975) , 17-31.

[10] J. Kajiwara, *Solution holomorphes globales des équations différentielles linéaires a` valeurs dans un espace de Hilbert et à parametre complexe*, Jap. J. Math., **2**(1976),91-107.
[11] J. Kajiwara, *Global existence of holomorphic solutions of differential equations with complex variables*, Proceedings of the Third International Colloquium on Differential Equations, Plovdiv, Bulgaria, 18-22 August 1992,119-132.
[12] J. Kajiwara and H. Kazama, *Two dimensional complex manifold with vanishing cohomology*, Math.Ann., **204**(1973),1-12.
[13] J. Kajiwara and Y. Mori, *On the existence of global holomorphic solutions of differential equations with complex parameters*, Czechoslovak Math. J., **24**(99) (1974), 444-454.
[14] J. Kajiwara and K. H. Shon, *Localization of global existence of solutions of ordinary differential equations with parameter in Stein spaces*, Mem. Fac. Sci. Kyushu Univ., **38**(1984), 91-120.
[15] J. Kajiwara and K. H. Shon, *Global existence of holomorphic solutions of differential equations with complex variables -I*, Mem. Fac. Sci. Kyushu Univ., **47**(1993), 147-182.
[16] J. Kajiwara and K. H. Shon, *Global existence of holomorphic solutions of differential equations with complex variables -II*, Pusan Kyongnam Mathematcal Journal, **9**-1(1993), 75-82.
[17] K. Oka, *Sur les fonctions analytiques de plusieurs variables*, Iwanami Shoten (1961).
[18] R. T. Shield, *Mixed type boundary value problems in soil mechanics*, Q. Appl. Math. **11**(1953), 61-75.
[19] R. T. Shield, *Stress and velocity fields in soil mechanics*, J. Math. Phys. **33**(1954), 144-156.
[20] K. H. Shon, *Global holomorphic solutions of differential equations with a parameter in a Stein manifold*, J. Korean Math. Soc. **27**-1 (1990), 820-822.
[21] A. J. Spencer, A theory of the kinetics of ideal soils under plane strain conditions, J. Mech. Phys. Solid, 12(1964), 337-351.
[22] H. Suzuki, *On the global existence of holomorphic solutions of* $\partial u/\partial x_1 = f$, Sci. Rep. Tokyo Kyoiku Daigaku, **11**(1972), 253-258.
[23] I. Wakabayashi, *Non existence of holomorphic solutions of* $\partial u/\partial z_1 = f$, Proc. Jap. Acad., **44**(1968), 820-822.

Phone: +81-92-641-1101 Ext. 4353

FAX: To Prof. Kajiwara +81-92-632-2737

e-mail: kajiwara@math.sci.kyushu-u.ac.jp

4th Int. Coll. on Differential Equations, pp. 157-161
D. Bainov and V. Covachev (Eds)

CONDITIONS FOR THE ABSENCE OF POSITIVE SOLUTIONS OF FIRST ORDER DIFFERENTIAL INEQUALITIES WITH DEVIATING ARGUMENTS

E. Kozakiewicz
Department of Mathematics, Humboldt-University at Berlin
D-10099 Berlin, Germany

ABSTRACT.
Sufficient conditions for the absence of eventually positive solutions of first order differential inequalities with constantly deviating arguments and a coefficient near the critical point are given. In the special case of equality, the results become oscillation criteria.

Keywords: Differential inequality, delay, neutral type, oscillation.

Introduction

In this paper the author continues his investigations on differential inequalities with deviating argument near the critical point $1/e$ from the last year [1].
A solution of the considered differential inequalities is an absolutely continuous function satisfying the inequality almost everywhere. Throughout this paper the function H has to be locally summable without further mentioning.

Delay inequalities

Let R be the set of all real numbers and $T \in R$. For a function $x : [T, \infty) \to (0, \infty)$ we define $G(t) = \ln \dfrac{x(t-1)}{x(t)}$, $T + 1 \leq t < \infty$.

Lemma 1 *Let x be a positive solution of*

$$x'(t) \leq -M(t)x(t-1) \qquad (1\leq)$$

in the interval $T \leq t < \infty$. Assume $M(t) \geq H(t) \geq 0$ and $\int_t^{t+1} H(\sigma)\mathrm{d}\sigma \geq \frac{1}{e}$, $T \leq t < \infty$. Then holds $G(P) \leq 2(1 + \ln 2)$, $T + 3 \leq P < \infty$.

Proof. From ($1\leq$) it follows that $x(t)$ is nonincreasing in $T+1 \leq t < \infty$. Hence, if $T+3 \leq P < \infty$ and $P-1 < Q < P$, we have $x'(t) \leq -H(t)x(P-1)$, $Q \leq t \leq P$. Using $x(P) \geq 0$ and integrating the latter estimation for $x'(t)$ from Q to P we obtain $-x(Q) \leq x(P) - x(Q) \leq -x(P-1)\int_Q^P H(\sigma)d\sigma$ or $\frac{x(Q)}{x(P-1)} \geq \int_Q^P H(\sigma)d\sigma$. Likewise $\frac{x(P)}{x(Q)} \geq \int_P^{Q+1} H(\sigma)d\sigma$. Multiply these inequalities and choose Q such that $\int_Q^P H(\sigma)d\sigma = \frac{1}{2e}$. Then $\int_P^{Q+1} H(\sigma)d\sigma \geq \frac{1}{2e}$ and we obtain $\frac{x(P)}{x(P-1)} \geq \int_Q^P H(\sigma)d\sigma \int_P^{Q+1} H(\sigma)d\sigma \geq \frac{1}{2e}\frac{1}{2e} = \frac{1}{4e^2}$. It follows that $G(P) \leq \ln(4e^2) = 2(1+\ln 2)$, $T+3 \leq P < \infty$. Lemma 1 is proved.

Remark. The value $2(1+\ln 2)$ in Lemma 1 is not the best one. However, this does not matter in the following.

Define a function $g : [T+2,\infty) \to R$ by $g(t) = \min\{G(s); t-1 \leq s \leq t\}$.

Lemma 2 *Under the assumptions of Lemma 1, g is nondecreasing on $[T+2,\infty)$.*

Proof. From ($1\leq$) we conclude $\frac{x'(t)}{x(t)} \leq -H(t)\frac{x(t-1)}{x(t)} = -H(t)e^{G(t)}$, $T+1 \leq t < \infty$. Integration from $t-1$ to t yields $-G(t) \leq -\int_{t-1}^{t} H(\sigma)e^{G(\sigma)}d\sigma$, $T+2 \leq t < \infty$. Using the definition of g we see

$$G(t) \geq e^{g(t)} \int_{t-1}^{t} H(\sigma)d\sigma, \quad T+2 \leq t < \infty. \tag{2}$$

Assume that there exists a point u with $T+2 \leq t < u < \infty$ and $g(t) > G(u)$. Choose $c \neq 1$, $g(t) > c > G(u)$. G is a continuous function. Put $S = \min\{s, G(s) = c, t \leq s < \infty\}$. We have $t < S$ and $G(S) = g(S) = c$. However, due to (2) we would obtain $c = G(S) \geq e^{g(S)}\frac{1}{e} = e^c\frac{1}{e} > ec\frac{1}{e} = c$. This is impossible.

Consequently $g(t) \leq G(u)$, $T+2 \leq t \leq u < \infty$ and, therefore, $g(t) \leq g(v)$, $T+2 \leq t \leq v < \infty$. Lemma 2 is proved.

Define $F : [T+2,\infty) \to R$ by $F(t) = \min\{s; G(s) = g(t), t-1 \leq s \leq t\}$.

Lemma 3 *Under the assumptions of Lemma 1 it holds that $F(t) < t$, $T+2 < t < \infty$.*

Proof. Assume $F(t) = t$ for a point t with $T+2 < t < \infty$. This implies $G(s) > t$, $t-1 \leq s < t$. Since G is continuous, there exists a $\delta > 0$ such that $t - \delta > T+2$ and $G(s) > t$, $t-1-\delta \leq s < t$. Hence $g(t-\delta) > g(t)$ in contradiction to Lemma 2. Lemma 3 is proved.

Clearly, according to Lemma 2, F is nondecreasing, too. Define $F_1(t) = F(t)$, $T+2 \leq t < \infty$ and in case $F_n(t) \geq T+2$, $F_{n+1}(t) = F(F_n(t))$, $n = 1,2,\ldots$.

Lemma 4 *Under the assumptions of Lemma 1 and $F_{2n}(t) > T+2$ it holds that $F_{2n}(t) < t-n$, $n = 1,2,\ldots$.*

Proof. $n = 1$. $F_2(t) > T+2$ implies $F_1(t) > T+2$. In case of $F_1(t) = t-1$, we derive from Lemma 3 that $F_2(t) < t-1$. In case of $F_1(t) > t-1$ it holds that $G(s) > g(t)$, $t-1 \le s < F_1(t)$. From this, Lemma 3 and Lemma 2 we obtain $F_2(t) < t-1$.
Assume that Lemma 4 is true for $n = k$. Then

$$F_{2(k+1)}(t) = F_2(F_{2k}(t)) < F_{2k}(t) - 1 < t - k - 1 = t - (k+1).$$

Lemma 4 is proved by induction.

Lemma 5 *Under the assumptions of Lemma* 1 *it holds that* $g(t) \ge \frac{1}{e}$, $T+4 \le t < \infty$.

Proof. (2) shows that $G(t) \ge 0$, $T+2 \le t < \infty$. Thus, $g(t) \ge 0$, $T+3 \le t < \infty$. Hence, from (2) we obtain

$$G(t) \ge \int_{t-1}^{t} H(\sigma)\mathrm{d}\sigma \ge \frac{1}{e}, \quad T+3 \le t < \infty.$$

Therefore, $g(t) \ge \frac{1}{e}$, $T+4 \le t < \infty$. Lemma 5 is proved.

Theorem 1 *Let x be a solution of* $(1\le)$ *in the interval* $t_o \le t < \infty$, $M(t) \ge H(t) = \frac{1}{e} + \chi(t)$, $\chi(t) \ge 0$ *in the same interval and* $\int_{t_o}^{\infty} \chi(\sigma)\mathrm{d}\sigma = \infty$. *Then x is not eventually positive.*

Proof. Suppose the contrary. Then there exists a $T \ge t_o$ with $x(t) > 0$ for $t \ge T$. Choose t such that $\int_{T+5}^{t-1} \chi(\sigma)\mathrm{d}\sigma > 2(1+\ln 2)$ and a natural number n such that $T+4 \le F_n(t) \le T+5$. By Lemma 3 and Lemma 4 this is possible. Using Lemma 1, the definition of g, the definition of F, (2), $e^x \ge ex$, the definition of χ, $F(t) - 1 \le F_2(t)$ and $\chi(t) \ge 0$, the iteration of the inequality $g(t) \ge g(F_1(t))(1 + e\int_{F_2(t)}^{F_1(t)} \chi(\sigma)\mathrm{d}\sigma)$, $\prod_{\nu=1}^{n}(1+a_\nu) \ge 1 + \sum_{\nu=1}^{n} a_\nu$ for $a_\nu \ge 0$, $F_{n+1}(t) \le F_n(t) \le T+5$ and $t-1 \le F_1(t)$ and $\chi(t) \ge 0$, Lemma 5, the choice of t, we obtain

$$\begin{aligned}
2(1+\ln 2) &\ge G(t) \ge g(t) = G(F(t)) \ge e^{g(F(t))} \int_{F(t)-1}^{F(t)} H(\sigma)\mathrm{d}\sigma \\
&\ge eg(F(t)) \int_{F(t)-1}^{F(t)} H(\sigma)\mathrm{d}\sigma = g(F(t))(1 + e\int_{F(t)-1}^{F(t)} \chi(\sigma)\mathrm{d}\sigma) \\
&\ge g(F_1(t))(1 + e\int_{F_2(t)}^{F_1(t)} \chi(\sigma)\mathrm{d}\sigma) \ge g(F_n(t)) \prod_{\nu=1}^{n}(1 + e\int_{F_{\nu+1}(t)}^{F_\nu(t)} \chi(\sigma)\mathrm{d}\sigma) \\
&\ge g(F_n(t))(1 + e\int_{F_{n+1}(t)}^{F_1(t)} \chi(\sigma)\mathrm{d}\sigma) \ge g(F_n(t))(1 + e\int_{T+5}^{t-1} \chi(\sigma)\mathrm{d}\sigma) \\
&\ge \frac{1}{e} + \int_{T+5}^{t-1} \chi(\sigma)\mathrm{d}\sigma > 2(1+\ln 2).
\end{aligned}$$

This contradiction completes the proof of Theorem 1.

Theorem 2 *In the interval $t_o \le t < \infty$ let x be a solution of* $(1\le)$, $M(t) \ge H(t) \ge 0$, $H(t) = \frac{1}{e} + \chi(t)$, $\int_t^{t+1} \chi(\sigma)\mathrm{d}\sigma$ *a nonincreasing function and* $\int_{t_o}^{\infty} \chi(\sigma)\mathrm{d}\sigma = \infty$. *Then x is not eventually positive.*

Proof. The last estimation in the preceding proof can be changed in the following manner. Choose n such that $T+4 < t-n \le T+5$. Then we obtain

$$\begin{aligned} 2(1+\ln 2) &\ge G(t) \ge g(t) = G(F(t)) \\ &\ge g(F(t))(1+e\int_{F(t)-1}^{F(t)} \chi(\sigma)\mathrm{d}\sigma) \ge g(t-1)(1+e\int_{t-1}^{t} \chi(\sigma)\mathrm{d}\sigma) \\ &\ge g(t-n)\prod_{\nu=1}^{n}(1+e\int_{t-\nu}^{t-\nu+1} \chi(\sigma)\mathrm{d}\sigma) \ge g(t-n)(1+e\int_{t-n}^{t} \chi(\sigma)\mathrm{d}\sigma) \\ &\ge \frac{1}{e} + \int_{t-n}^{t} \chi(\sigma)\mathrm{d}\sigma > \int_{T+5}^{t} \chi(\sigma)\mathrm{d}\sigma > 2(1+\ln 2). \end{aligned}$$

Theorem 2 is proved.

Theorem 3 *Let x be a solution of*

$$x'(t) \le A(t)x(t) - M(t)x(t-1), \quad t_o \le t < \infty, \tag{3$\le$}$$

$A(t) \le a$, $a \in R$, $M(t) \ge H(t) = e^{a-1} + \chi_a(t)$, $\chi_a(t) \ge 0$ *and* $\int_{t_o}^{\infty} \chi_a(\sigma)\mathrm{d}\sigma = \infty$. *Then x is not eventually positive.*

Proof. Assume $x(t) > 0$, $T \le t < \infty$. Then $x'(t) \le ax(t) - M(t)x(t-1)$, $T \le t < \infty$. With $x(t) = e^{at}u(t)$ we get $u'(t) \le -e^{-a}M(t)u(t-1)$, $T \le t < \infty$ and Theorem 3 follows from Theorem 1.

Theorem 4 *Assume that for $t_o \le t < \infty$* $A(t) \le a$, $M(t) \ge H(t) \ge 0$, $H(t) = e^{a-1} + \chi_a(t)$, $\int_t^{t+1} \chi_a(\sigma)\mathrm{d}\sigma$ *is a nonincreasing function and* $\int_{t_o}^{\infty} \chi_a(\sigma)\mathrm{d}\sigma = \infty$. *Then* $(3\le)$ *has no eventually positive solution.*

Proof. Theorem 4 reduces to Theorem 2 as Theorem 3 to Theorem 1.

Inequalities of neutral type

Theorem 5 *Let x be a solution of*

$$x'(t) - cx'(t-\tau) \le -M(t)x(t-1), \quad t_o \le t < \infty,\ 0 \le c \le 1,\ 0 \le \tau < \infty$$

and let $M(t)$ satisfy the conditions of Theorem 1 or 2. Then x is not eventually positive.

Proof. $c = 0$ is known to be true. Suppose $0 < c \le 1$ and assume $x(t) > 0$, $t_o \le T \le t < \infty$. Set $z(t) = x(t) - cx(t-\tau)$. We obtain $z'(t) \le 0$, $T+1 \le t < \infty$. $M(t) \equiv 0$ for sufficiently large t is impossible, therefore also $z'(t) \equiv 0$ and hence $z(t) \equiv 0$ for large t is impossible. Thus we see, unless $z(t) > 0$ in $T+1 \le t < \infty$, then there exists a $T_o \geq T+1$ with $z(t) \leq k < 0$ for $T_o \leq t < \infty$. This implies $x(t+\tau) \leq k + cx(t) \leq k + x(t), T_o \leq t < \infty$ and by iteration $x(T_o+n\tau) \leq nk + x(T_o) < 0$ for n sufficiently large, in contradiction to $x(t) > 0$, $T \leq t < \infty$. Thus, $z(t) > 0$, $T+1 \leq t < \infty$ and $z(t) = x(t) - cx(t-\tau) \leq x(t)$, $T + \tau \leq t < \infty$. Hence, $-M(t)x(t-1) \leq -M(t)z(t-1)$, $T + \tau + 1 \leq t < \infty$. We conclude that z is an eventually positive solution of ($1\leq$). This is impossible. Theorem 5 is proved.

Theorem 6 *Let x be a solution of*

$$x'(t) - cx'(t-\tau) \leq A(t)x(t) - M(t)x(t-1), \quad t_o \leq t < \infty,$$

$A(t) \leq a \leq 0$, $0 \leq ce^{-a\tau} \leq 1$, and let $M(t)$ satisfy the conditions of Theorem 3 or 4. Then x is not eventually positive.

Proof. Assume $x(t) > 0$, $T \leq t < \infty$. Then we obtain

$$x'(t) - cx'(t-\tau) \leq ax(t) - M(t)x(t-1), \quad T \leq t < \infty.$$

Set $x(t) = e^{at}u(t)$, dividing by e^{at} and omitting the non-negative term $-cae^{-a\tau}u(t-\tau)$ on the left-hand side yields

$$u'(t) - ce^{-a\tau}u'(t-\tau) \leq -M(t)e^{-a}u(t-1), \quad T \leq t < \infty.$$

Apply Theorem 5 to this inequality. The proof is complete.

The transformation $x = -u$ shows that Theorem 1′ is also valid.

Theorem 1′ *Let x be a solution of* ($1\geq$) *in the interval $t_o \leq t < \infty$, $M(t) \geq H(t) = \frac{1}{e} + \chi(t)$, $\chi(t) \geq 0$ in the same interval and $\int_{t_o}^{\infty} \chi(\sigma)d\sigma = \infty$. Then x is not eventually negative.*

This holds analogously for the other theorems.
In case of equality Theorem 1 and Theorem 1′ give the following oscillation theorem.

Theorem 1″ *Let x be a solution of* ($1=$) *in the interval $t_o \leq t < \infty$, $M(t) \geq H(t) = \frac{1}{e} + \chi(t)$, $\chi(t) \geq 0$ in the same interval and $\int_{t_o}^{\infty} \chi(\sigma)d\sigma = \infty$. Then x is oscillatory, that means, x has arbitrarily large zeros.*

This holds analogously for the other theorems.

Supplementary remarks. Theorem 1 has been stimulated by a discussion on the Third International Colloquium on Differential Equations in Plovdiv 1992.
Theorem 2″ is a special case of results due to A. Elbert and I.P. Stavroulakis presented at the First International Workshop on Oscillation Theory in Plovdiv, Bulgaria, August 14 - 17, 1993 under the title "Oscillation and Non-oscillation Criteria for Delay Differential Equations". There are also interesting results for the case of $\int_{t_o}^{\infty} \chi(\sigma)d\sigma$ being converging. The transformation $z(t) = x(t) - cx(t-\tau)$ in the proof of Theorem 5 has been taken from the proof of Theorem 3.1 in [2].

References

[1] E. Kozakiewicz, *ZAMM* 73 (1993), T877-T879.

[2] L.H. Erbe and B.G. Zhang, *Differ. Integral Equ.*, 1 (1988), 305-314.

4th Int. Coll. on Differential Equations, pp. 163-166
D. Bainov and V. Covachev (Eds)

On the solution of boundary value problem. The decomposition method

E.S. LEE[(1)] & G. ADOMIAN[(2)]
[(1)] *Yuan-Ze Institute and Kansas State University;* [(2)] *Athens, Georgia**

Abstract. The method developed by Adomian, the decomposition method, supplies approximate solution to the original nonlinear and/or stochastic ordinary, partial, or integral differential equations. The approach may be used to find analytical approximations of the solution of the original equations. Since no linearization or perturbation is used and also since the solution can be approximated as close as needed, this approximate analytical solution forms a very useful tool for analyzing boundary value problems encountered in practice. It should be pointed out that these boundary value problems are usually unstable and very difficult to solve numerically. Furthermore, even for nonlinear problems without the consideration of the boundary value difficulties, linearization or perturbation approaches frequently alter the characteristics of the original problem completely. In this paper, we wish to show the difficulties in handling nonlinear boundary value problems and how the decomposition method can be used to analyze these type of problems.

Keywords and phrases: boundary value problems, decomposition method

1. INTRODUCTION

In most practical applications, the resulting ordinary, partial, or integral differential equations are boundary value problems. Unfortunately, these type of equations are most difficult to solve. Although various approaches have been proposed, only numerical answers are possible [8–10] and analytical solutions are usually impossible to obtain. Since there is a need to analyze the behavior of the system [8], an analytical, or even an approximate representation of the analytical solution would be very useful. It should pointed out that, due to the boundary conditions, the usual linearization or perturbation approaches cannot be used effectively. The decomposition method due to Adomian [1–5] appears to offer such an analytical approximation approach.

The decomposition method decomposes the original problem into different parts: an invertible part and the remainder. If the equation is nonlinear or stochastic, the approach also decomposes the nonlinear and the stochastic parts so that it can be represented separately. The method yields a correct, verifiable approximate analytical solution. The approach does not require linearization, perturbation, or discretization and thus it approximates the original nonlinear and/or stochastic representations directly. Because of this direct approximation, the accuracy can be improved by adding more terms as needed.

*Please address all correspondences to E. S. Lee, Dept. of Industrial Engineering, Yuan-Ze Institute of Technology, 135 Yuan-Tung Road, Nei-Li, Tao-Yuan Shian, Taiwan, 32026 R.O.C.

Furthermore, experiences have shown that the rate of convergence is very fast and highly accurate solutions are obtained with a small number of terms. Cherruault [7] has proved the convergence of this approach and this convergence proof has been extended to partial differential equations recently.

The decomposition method has been proved to be a very useful tool and has been used to solve various difficult nonlinear problems in ordinary, partial, and integral differential equations encountered in practical applications. Some of the examples are: the Navier–Stokes equation, the Duffing oscillator, equations encountered in biomathematics, astrophysics and other applied dynamic systems. The interested reader can consult the literature for details [1–5].

2. NONLINEARITY

To illustrate the sensitivity of nonlinear problems, let us consider the simple example due to Adomian [6]:

$$\frac{dy}{dx} = (y-1)^2, \qquad y(0) = 1\,.$$

The solution of this problem is obviously $y = 1$. However, if we make a 1% change in the initial condition to $y(0) = 1.01$, we obtain a hyperbola:

$$y = 1 - \frac{1}{(x-100)}$$

with one vertical asymptote. If we make a 1% change of the equation as:

$$\frac{dy}{dx} = (y-1)^2 + 0.01$$

we obtain a periodic solution

$$y = 1 + 0.1 \tan \frac{x}{10}$$

with vertical asymptotes at $(2k+1)5\pi$.

Thus, we see how sensitive nonlinear equation can be and a complete change in characteristics resulted even for a very small change in the approximation.

For boundary value problems such as the nonlinear equations resulting from control or optimization, in addition to the sensitivity of nonlinearity, the solution is also very sensitive to the boundary conditions. No useful general approximate analytical solution has be obtained.

3. THE DECOMPOSITION METHOD

Only the essential characteristics of this approach will be given. The reader can consult the literature for more details [1-5].

Consider the deterministic equation $Fu(t) = g(t)$, Where F represents a general nonlinear ordinary differential operator involving both linear and nonlinear terms. The linear term is decomposed into $L + R$, where L is easily invertible and R is the remainder of the linear operator. L is taking as the highest order derivative without involving difficult integrations. Thus the equation is decomposed into:

$$Lu + Ru + Nu = g$$

where Nu represents the nonlinear terms. Solving for Lu and because L is invertible, we have:

$$L^{-1}Lu = L^{-1}g - L^{-1}Ru - L^{-1}N\,.$$

For initial value problems, the integral operator L^{-1} may be regarded as definite integrations. For boundary value problems, indefinite integrations are used and the integration constants are evaluated from the given conditions. Solving for u

$$u = A + Bt + L^{-1}g - L^{-1}Ru - L^{-1}Nu\,.$$

This equation can be expanded as:

$$\sum_{n=0}^{\infty} u_n = u_0 - L^{-1}R\sum_{n=0}^{\infty} u_n - L^{-1}\sum_{n=0}^{\infty} A_n$$

where u_0 represents the terms $(A + Bt + L^{-1}g)$, the first summation on the right hand represents the decomposition of u and the second the decomposition of the nonlinear term Nu, and A_n are special polynomials defined by Adomian. The above equation can be written in separate form as:

$$\begin{aligned} u_1 &= -L^{-1}Ru_0 - L^{-1}A_0 \\ u_2 &= -L^{-1}Ru_1 - L^{-1}A_1 \\ &\vdots \\ u_{n+1} &= -L^{-1}Ru_n - L^{-1}A_n\,. \end{aligned}$$

The A_n polynomials, known as the Adomian polynomials, are given by:

$$\begin{aligned} A_0 &= f(u_0) \\ A_1 &= u_1(d/du_0)f(u_0) \\ A_2 &= u_2(d/du_0)f(u_0) + (u_1^2/2!)(d/du_0^2)f(u_0) \\ A_3 &= u_3(d/du_0)f(u_0) + u_1u_2(d^2/du_0^2)f(u_0) + (u_1^3/3!)d^3/du_0^3f(u_0) \\ &\vdots \end{aligned}$$

These A polynomials can be generated easily from the given problem.

3. BOUNDARY VALUE PROBLEMS

Consider the optimization problem with the system represented by the set of differential equations:

$$\frac{dx}{dt} = f(x_1, x_2, \ldots, x_n; u_1, u_2, \ldots, u_m; t)$$

with the performance functional $J(x, u, t)$ given by

$$J = \Phi\,(x(t_1), t_1) + \int_{t_0}^{t_1} F(x, u, t)\,dt\,.$$

Let the vector p be a vector of Lagrange multipliers. By forming the Lagrange equation and by using the calculus of variations approach, we obtain the adjoin equations

$$\frac{dp_i}{dt} = -\frac{\partial H}{\partial x_i}\,.$$

We now have a system of $2n$ nonlinear differential equations with $2n$ two point boundary conditions. The solution of the $2n$ differential equations can be represented analytically by the decomposition approach. The boundary conditions can be substituted into this representation and the unknown conditions can be obtained by solving this set of equations. This solution can be carried out easily by the use of modern computers. The big advantage is that we have changed a trial-and-error solution method into a straight forward approach which is much more suited to modern computers.

REFERENCES

[1] R. BELLMAN & G. ADOMIAN. Partial Differential Equations — New Methods for Their Treatment and Applications. Reidel, 1986.

[2] G. ADOMIAN. Stochastic Systems. Academic Press, 1983.

[3] G. ADOMIAN. Nonlinear Stochastic Operator Equations. Academic Press, 1986.

[4] G. ADOMIAN. Nonlinear Stochastic Systems, Theory and Applications to Physics. Kluwer, 1990.

[5] G. ADOMIAN. An analytical solution of the stochastic Navier–Stokes system. *Foundation of Physics*, **21**(1991), 831–843.

[6] G. ADOMIAN. Solution of physical problems by decomposition. Presented at the 5th Bellman continuum workshop, Hawaii, January, 1993.

[7] Y. CHERRUAULT. Convergence of Adomian's method. *Kybernetes*, **18**(1989), No. 2, 31–38.

[8] E. S. LEE. Quasilinearization and Invariant Imbedding. Academic Press, 1968.

[9] E. S. LEE. Iterative computation in dynamic optimization. in Approximation, Optimization and Computing: Theory and Applications, A. G. Law & C. L. Wang, (eds.), Elsevier, 1990.

[10] E. S. LEE & H. SUGIYAMA. Solving deterministic problems by stochastic approximation. *J. Math. Analysis Appls.*, **138**(1989), 569.

4th Int. Coll. on Differential Equations, pp. 167-173
D. Bainov and V. Covachev (Eds)

Some uniform estimates for solutions of difference differential equations of elliptic–parabolic type

Masashi Misawa

Department of Information Engineering, Faculty of Engineering, Shinshu University, 500 Wakasato Nagano 380 Japan

Abstract **We establish some uniform estimates for solutions of difference-differential equations and systems of elliptic -parabolic type. It is essential to consider local estimates which are analogous to, but more complicated than those for elliptic and parabolic equations.**

Keywords: **difference differential equation of elliptic-parabolic type, Campanato estimate, a Harnack inequality**

1 Introduction

In the regularity problem for solutions of nonlinear elliptic and parabolic equations or systems, we need to consider the corresponding linear equations or systems(refer to [1,7,8]). It was shown by Moser that a Harnack inequality holds for solutions of elliptic and parabolic equations with bounded and measurable coefficients [10, 11]. We also know that so-called Campanato estimates hold for solutions to linear elliptic and parabolic systems with constant coefficients. These estimates play a fundamental role in obtaining the regularity of solutions to nonlinear elliptic and parabolic equations or systems.
In this paper we report that the above fundamental inequalities also hold for solutions to linear difference-differential equations and systems of elliptic-parabolic type uniformly with respect to approximations(see [6,9]). Previously, such problems had been studied by Kikuchi[4], who treated Hölder estimates for equations independently of approximations. Depending on the relation between the size of a local cube and a time-discrete mesh, the equations show a "parabolic" or "elliptic" behavior, respectively.
Let Ω be a bounded open set in Euclidean space $R^m, m \geq 2$, u be a function: $\Omega \to R$ and $Du = (D_1 u, D_2 u, \dots, D_m u)$, $D_\alpha u = \partial u / \partial x^\alpha$ $(1 \leq \alpha \leq m)$ be the gradient of u. Let T be a positive number arbitrarily given and set $Q = (0,T) \times \Omega$. We use the usual Lebesgue space $L_p(\Omega)$, Sobolev spaces; $W_p^k(\Omega) = W_p^k(\Omega, R)$, $W_{p,0}^k(\Omega) = W_{p,0}^k(\Omega, R)$, $V_2(Q) = L^\infty((0,T); L^2(\Omega)) \cap L^2((0,T); W_2^1(\Omega))$ and $V_{2,0}(Q) = L^\infty((0,T); L^2(\Omega)) \cap L^2((0,T); W_{2,0}^1(\Omega))$. For a positive integer N, $N \geq 2$, we put $h = T/N$ and $t_n = nh$ $(0 \leq n \leq N)$. Let u_0 be a function belonging to $W_2^1(\Omega)$. We shall be concerned with a family of linear elliptic partial differential equations:

$$\frac{u_n - u_{n-1}}{h} = D_\alpha(a_n^{\alpha\beta}(x) D_\beta u_n) \qquad (1 \leq n \leq N). \tag{1.1}$$

In the summation convention over repeated ones, the Greek indices run from 1 to m. The coefficients $a_n^{\alpha\beta}(\cdot)$ $(1 \leq \alpha, \beta \leq m,\ 1 \leq n \leq N)$ are measurable functions defined in Ω

satisfying the uniform ellipticity and boundedness condition with positive constants λ and μ :

$$\lambda|\xi|^2 \le a_n^{\alpha\beta}(x)\xi^\alpha\xi^\beta \le \mu|\xi|^2 \qquad \text{for } \xi = (\xi^\alpha) \in R^m, 1 \le n \le N \text{ and any } \quad x \in \Omega.$$

Consider a family of weak solutions of (1.1) with initial data u_0, that is, a family $\{u_n\}$ $(1 \le n \le N)$ of functions $u_n \in W_2^1(\Omega)$ which satisfy

$$\int_\Omega \frac{u_n - u_{n-1}}{h}\varphi dx + \int_\Omega a_n^{\alpha\beta} D_\beta u_n D_\alpha \varphi dx = 0 \quad \text{for any } \varphi = (\varphi^i) \in W_{2,0}^1(\Omega).$$

For such a family $\{u_n\}(1 \le n \le N)$, define a function $u_h(t,\cdot)$: $t \in [0,T] \to u_h(t,\cdot) \in W_2^1(\Omega)$ as follows:

$$u_h(0,\cdot) = u_0(\cdot), \quad u_h(t,\cdot) = u_n(\cdot) \qquad \text{for } t_{n-1} < t \le t_n \qquad (1 \le n \le N). \tag{1.2}$$

Also define

$$a^{\alpha\beta}(t,\cdot) = a_n^{\alpha\beta}(\cdot), \quad \text{for} \quad t_{n-1} < t \le t_n \ (1 \le n \le N).$$

(1.2) understood, we use the notations u_h and $\{u_n\}$ interchangeably.
Some standard notations: For a point $z_0 = (t_0, x_0) \in Q$, we put

$$\begin{aligned}
B_r(x_0) &= \{x \in R^m : |x^\alpha - x_0^\alpha| < r \ (1 \le \alpha \le m)\}, \\
Q_{r,\tau}(z_0) &= \{t \in R : |t - t_0| < \tau\} \times B_r(x_0), \\
Q_{r,\tau}^+(z_0) &= \{t \in R : t_0 - \tau < t < t_0\} \times B_r(x_0), \\
Q_{r,\tau}^-(z_0) &= \{t \in R : t_0 < t < t_0 + \tau\} \times B_r(x_0).
\end{aligned}$$

These domains are referred to as cubes. For simplicity, we shall use abbreviations:

$$Q_r(z_0) = Q_{r,r^2}(z_0), \ Q_r^+(z_0) = Q_{r,r^2}^+(z_0), \ Q_r^-(z_0) = Q_{r,r^2}^-(z_0).$$

In the above notations, the centre x_0 and z_0 will be omitted when no confusion may arise. For $z_i = (t_i, x_i)$ $(i = 1, 2)$, we introduce the parabolic metric

$$\delta(z_1, z_2) = \max\{|t_1 - t_2|^{1/2}, |x_1^\alpha - x_2^\alpha| \ (1 \le \alpha \le m)\}.$$

For a postive number l we denote by [l] the greatest non-negative integer not greater than l. We also set

$$\tilde{n}_l = \text{the greatest non-negative integer less than } l^2/h.$$

Now let N_0 be a positive integer satisfying $N_0 > \log(1 + m/2)/\log(1 + 2/m)$ and let h_0 be a sufficiently small positive number. From now on, we take N in (1.1) sufficiently large satisfying $N \ge \max\{N_0, T/h_0\}$. We also define a cube $\widetilde{Q_{h_0}}$ as follows:

$$\widetilde{\Omega}_{h_0} = \{x \in \Omega; \operatorname{dist}(x, \partial\Omega) > \sqrt{(N_0+1)h_0}\}, \quad \widetilde{Q_{h_0}} = ((N_0+1)h_0, \ T) \times \widetilde{\Omega}_{h_0}. \tag{1.3}$$

Our main results are the followings:

Theorem 1.1 (Weak Harnack inequality, parabolic regime) *Let u_h be a weak solution of (1.1). If u_h is nonnegative in a cube $Q_r^+(t_{n_0}, x_0) \subset Q$ with $r^2 > h$, then, for any p, $0 < p < 1 + 2/m$, there exists a positive constant γ depending only on λ, μ, m and p such that*

$$\left(\frac{1}{|Q^-_{\sqrt{\tilde{n}_r h}/2}|}\iint_{Q^-_{\sqrt{\tilde{n}_r h}/2}(t_{n_0-\tilde{n}_r}, x_0)} (u_h)^p\,dx dt\right)^{\frac{1}{p}} \leq \gamma \inf_{Q^+_{\sqrt{\tilde{n}_r h}/2}(t_{n_0}, x_0)} u_h. \tag{1.4}$$

Theorem 1.2 (Weak Harnack inequality, elliptic regime) *Let u_h be a weak solution of* (1.1) *satisfying $\iint_Q (u_h)^2 dx dt \leq \gamma_1$ with a uniform constant γ_1. If $u_n \geq 0 (N_0 + 1 \leq n \leq N)$ in $B_{2r}(x_0) \subset \widetilde{\Omega}_{h_0}$ with $r^2 \leq h$, then, for any p, $0 < p < m/(m-2)$, there exist positive constants γ and $\alpha, 0 < \alpha < 1$, depending only on $\lambda, \mu, m, \gamma_1$, and* $\operatorname{dist}(x_0, \partial\Omega)$ *such that, for the same n, $N_0 + 1 \leq n \leq N$ as above, there holds*

$$\left(\frac{1}{|B_{r/2}|}\int_{B_{r/2}(x_0)} (u_n)^p\,dx\right)^{1/p} \leq \gamma \left[\inf_{B_r(x_0)} u_n + r^\alpha\right]. \tag{1.5}$$

Theorem 1.3 (Local boundedness of solutions) *Let u_h be a weak solution of* (1.1) *satisfying $\iint_Q (u_h)^2 dx dt \leq \gamma_1$ with a uniform positive number γ_1. Then, for all $(\bar{t}, \bar{x}) \in \widetilde{Q_{h_0}}$ with* $d = (1/4)\min\{|\bar{t} - N_0 h_0|^{1/2}, \operatorname{dist}(\bar{x}, \partial\Omega)\}$ *and any $p > 1$, there exist positive constants γ and $\alpha, 0 < \alpha < 1$, depending only on λ, μ, γ_1, p and d such that, setting $u_h^\pm = \max\{\pm u_h, 0\}$,*

$$\sup_{Q^+_{r/2}(t_{n_0}, x_0)} u_h^\pm \leq \gamma\left[\left(\frac{1}{|Q_r^+|}\iint_{Q_r^+(t_{n_0}, x_0)} (u_h^\pm)^p\,dx dt\right)^{1/p} + r^\alpha\right] \tag{1.6}$$

holds for any $(t_{n_0}, x_0) \in Q^+_{d/2}(\bar{t}, \bar{x})$ and all $r, 0 < r < d/2$.

For systems of the equations, we have the following results. Consider a family of linear elliptic partial differential equations: For each $1 \leq i \leq M$

$$\frac{u_n^i - u_{n-1}^i}{h} = D_\alpha(a_{ij}^{\alpha\beta} D_\beta u_n^j) \quad (1 \leq n \leq N). \tag{1.7}$$

Suppose that the coefficients $a_{ij}^{\alpha\beta} (1 \leq \alpha, \beta \leq m, 1 \leq i, j \leq M)$ is constant matrix satisfying so-called Legendre-Hadamard condition with a positive constant λ :

$$a_{ij}^{\alpha\beta}\xi_\alpha\xi_\beta\eta^i\eta^j \geq \lambda|\xi|^2|\eta|^2 \qquad \text{for } \xi = (\xi_\alpha) \in R^m \text{ and } \eta = (\eta^i) \in R^M \tag{1.8}$$

and also take N_0 in (1.3) as $N_0 \geq \max\{8(k+2)/3, 4L/(L-2)\} + 1$ with positive numbers $L > 2$, $k \geq [m/2] + 1$.

Then the following holds:

Theorem 1.4 (Campanato estimates) Let u_h be a weak solution of $\widetilde{(1.7)}$ satisfying $\iint_Q |Du_h|^2 dxdt \le \gamma_1$ with a uniform constant γ_1. Then, for all $(\bar{t}, \bar{x}) \in \widetilde{Q_{h_0}}$ with $d = (1/4)\min\{|\bar{t} - N_0 h_0|^{1/2}, \mathrm{dist}(\bar{x}, \partial\Omega)\}$, there exist positive constants γ and $\alpha, 0 < \alpha < 1$, depending only on λ, m, γ_1 and d, such that the estimate

$$\int_{Q_r^+(t_{n_0}, x_0)} |Du_h|^2 dz \le \gamma \left(\frac{r}{\rho}\right)^{m+2} \int_{Q_\rho^+(t_{n_0}, x_0)} |Du_h|^2 dz + \gamma \rho^{m+2\alpha} \tag{1.9}$$

holds for any $(t_{n_0}, x_0) \in Q^+_{d/2}(\bar{t}, \bar{x})$ and $0 < r < \rho < d/2$.

2 Outline of the proof

Proof of Theorem 1.1 - 1.3. We explain only the proof of theorem 1.1. The theorem is obtained by combining some Lemmata. We now state a so-called Caccioppoli inequality.

Lemma 2.1 *Let u_h be a weak solution of (1.1) and let us take $Q^-_{\rho,\tau}(t_{n_0}, x_0), Q^+_{\rho,\tau}(t_{n_0}, x_0) \subset Q$ arbitrarily. Then there exists a positive constant γ depending only on λ, μ and m such that, if u_h is nonnegative in $Q^+_{\rho,\tau}(t_{n_0}, x_0)$ and $u_{n_0-[\tau/h]-1} \ge 0$ in $B_\rho(x_0)$, then*

$$\begin{aligned} \sup_{t_{n_0}-\tau(1-\sigma_2)\le t\le t_{n_0}} \int_{B_{\rho(1-\sigma_1)}(x_0)\times\{t\}} & (u_h + \varepsilon)^p dx \\ &+ \iint_{Q^+_{\rho(1-\sigma_1),\tau(1-\sigma_2)}(t_{n_0}, x_0)} \left| D(u_h+\varepsilon)^{p/2}\right|^2 dxdt \\ &\le \gamma\left((\sigma_1\rho)^{-2} + (\sigma_2\tau)^{-1}\right) \iint_{Q^+_{\rho,\tau}(t_{n_0}, x_0)} (u_h+\varepsilon)^p dxdt \end{aligned} \tag{2.1}$$

holds for any $p < 0$, all $\sigma_1, \sigma_2 \in (0,1)$ and any $\varepsilon > 0$. If u_h is nonnegative in $Q^-_{\rho,\tau}(t_{n_0}, x_0)$ and $u_{n_0} \ge 0$ in $B_\rho(x_0)$,

$$\begin{aligned} \sup_{t_{n_0}\le t\le t_{n_0}+\tau(1-\sigma_2)} \int_{B_{\rho(1-\sigma_1)}(x_0)\times\{t\}} & (u_h + \varepsilon)^p dx \\ &+ \iint_{Q^-_{\rho(1-\sigma_1),\tau(1-\sigma_2)}(t_{n_0}, x_0)} \left| D(u_h+\varepsilon)^{p/2}\right|^2 dxdt \\ &\le \gamma\left((\sigma_1\rho)^{-2} + (\sigma_2\tau)^{-1}\right) \iint_{Q^-_{\rho,\tau}(t_{n_0}, x_0)} (u_h+\varepsilon)^p dxdt \end{aligned} \tag{2.2}$$

holds for any $0 < p < 1$, all $\sigma_1, \sigma_2 \in (0,1)$ and any $\varepsilon > 0$.

Proof. We obtain the desired estimte by testing the identity with appropriate functions. Now we indicate the test functions. To do so, we need to classify two cases.

Let η be an usual cut-off function in $B_\rho(x_0)$.

Case 1 $\sigma_2\tau > 3h$. Then we take $(u_h)^{p-1}\eta^2\sigma(t)$, where $\sigma(t)$ is a special cut-off function defined in $[t_{n_0}-\tau, t_{n_0}]$ as follows(see [6 ,7]):

$$\sigma(t) = \sigma_n \quad \text{for} \quad t_{n-1} < t \le t_n \ (1 \le n \le N)$$

$$\sigma_n = \begin{cases} 1 \ \text{ for } \ n_0 - [(1-\sigma_2)\tau/h] \le n \le n_0, \\ \frac{n-n_0+[\tau/h]-1}{[\tau/h]-1-[(1-\sigma_2)\tau/h]} \ \text{ for } \ n_0 - [\tau/h]+1 \le n \le n_0 - [(1-\sigma_2)\tau/h], \\ 0 \ \text{ for } \ n \le n_0 - [\tau/h]. \end{cases}$$

*Case*2 : $\sigma_2\tau \le 3h$. Then we put $\sigma(t)$ as $\sigma \equiv 1$ on $[t_{n_0}-\tau, t_{n_0}]$.

By Sobolev inequality with Lemma 2.1 and iterating the resulting inequalities, we have

Lemma 2.2 *Let u_h be a weak solution of* (1.1). *If $u_h \ge 0$ in $Q^+_{\rho_0,\tau_0}(t_{n_0}, x_0) \subset Q$ and $u_{n_0-[\tau_0/h]-1} \ge 0$ in $B_{\rho_0}(x_0)$, then there exists a positive constant γ depending only on λ, μ and m such that*

$$\left(\frac{1}{|Q^+_{\rho_0,\tau_0}|}\iint_{Q^+_{\rho_0,\tau_0}(t_{n_0},x_0)} u_h^p(t,x)dtdx\right)^{1/p} \le \gamma(2+\rho_0^{-2}\tau_0)^{-p^{-1}(2+m/2)} \inf_{(t,x)\in Q^+_{\rho_0/2,\tau_0/2}(t_{n_0},x_0)} u_h(t,x) \tag{2.3}$$

holds for any $p<0$. If $u_h \ge 0$ in $Q^-_{\rho_0,\tau_0}(t_{n_0}, x_0) \subset Q$ and $u_{n_0} \ge 0$ in $B_{\rho_0}(x_0)$, then, for any $0 < p < 1+2/m$, there exists a positive constant γ depending only on λ, μ, m and p such that

$$\left(\frac{1}{|Q^-_{\rho_0/2,\tau_0/2}|}\iint_{Q^-_{\rho_0/2,\tau_0/2}(t_{n_0},x_0)} u_h^p(t,x)dtdx\right)^{1/p} \le \gamma\left(\frac{1}{|Q^-_{\rho_0,\tau_0}|}\iint_{Q^-_{\rho_0,\tau_0}(t_{n_0},x_0)} u_h^q(t,x)dtdx\right)^{1/q} \tag{2.4}$$

holds for any q, $0<q<p$.

The following lemma is a variation of the John-Nirenberg estimate for the step function in a time variable. This is the most important in our proof. Let φ be the function defined by $\varphi(s) = \begin{cases} \sqrt{s}, \ s>0, \\ 0, \ s \le 0. \end{cases}$

Lemma 2.3 *Let $v = v_h$ be a time-step function defined in (1.4) and suppose that v satisfies the following relations in a cube $Q^+_R(\bar t, \bar x) \subset Q$ with uniform positive constants γ and κ :*

$$\frac{1}{|Q^+_r||Q^-_r|}\iint_{(t',x')\in Q^+_r}\iint_{(t,x)\in Q^-_r} \varphi(v(t',x')-v(t,x))dtdxdt'dx' \le \gamma$$

for all pairs Q_r^+ *and* Q_r^- *in* $Q_R^+(\bar t,\bar x)$ *with* $3r^2 > h$ *and*

$$|\{x \in B_r(x_0) : |v_n(x) - \overline{v_n}_{B_r}| > \sigma\}| \le e^{\alpha a} e^{-\alpha\sigma\kappa^{-1}} |B_r|$$

for any $r \le \sqrt{h/3}$, *all* $y \in B_R(\bar x)$ *and any* n, $[(\bar t - R^2)/h] \le n \le [\bar t/h] + 1$. *Then there exists a positive constant* ξ *independent of* h *and* v *such that*

$$\frac{1}{|Q^+_{R,R^2/2}||Q^-_{R,R^2/2}|}\iint_{(t',x')\in Q^+_{R,R^2/2}(\bar t,\bar x)}\iint_{(t,x)\in Q^-_{R,R^2/2}(\bar t-2R^2,\bar x)} \Psi(v(t',x') - v(t,x))dtdxdt'dx' \le 1,$$

where $\Psi(s) = \gamma^{-1}e^{\xi s}$.

Now we give the estimtes for $\log u_h$. For details see [9].

Lemma 2.4 *Let* u_h *be a weak solution of* (1.1) *and us take* $B_{2r}(x_0) \subset \Omega$ *arbitrarily. Then there exist positive constants* a *and* α *independent of* h *and* u_h, *depending only on* m, *such that, if* $u_n, u_{n-1} (2 \le n \le N)$ *is nonnegative in* $B_{2r}(x_0)$ *and* $v_n = -\log u_n$ $(1 \le n \le N)$,

$$|\{x \in B_r(x_0) : |v_n(x) - \overline{v_n}_{B_r}| > \sigma\}| \le e^{\alpha a} e^{-\alpha\sigma\kappa^{-1}} |B_r|$$

hold, where $\kappa = \kappa(r) = \gamma\left(16\mu^2/\lambda^2 + 2r^2/\lambda h\right)^{1/2}$.

Lemma 2.5 *Let* u_h *be a weak solution of* (1.1). *Then there exists a positive constant* γ *independent of* h *and* u_h *such that, if* u_h *is nonnegative in* $Q_R^+(\bar t,\bar x) \subset Q$ *and* $u_{[(\bar t - R^2)/h]} \ge 0$ *in* $B_R(\bar x)$ *then, setting* $v = -\log u_h$,

$$\frac{1}{|Q_r^+||Q_r^-|}\iint_{(t',x')\in Q_r^+}\iint_{(t,x)\in Q_r^-} \varphi(v(t',x') - v(t,x))dtdxdt'dx' \le \gamma$$

holds for all pairs Q_r^+ *and* Q_r^- *in* $Q_R^+(\bar t,\bar x)$ *with* $3r^2 > h$.

Once we have Lemma 2.4 and 2.5, we conclude from adopting Lemma 2.3 for $v = -\log u_h$ in $Q_R^+(\bar t,\bar x)$ the following:

Lemma 2.6 *Suppose that the same assumption as Lemma 2.5 is satisfied. Then there exist positive constants* ξ *and* γ *independent of* h *and* u_h *such that*

$$\frac{1}{|Q^+_{R,R^2/2}|}\iint_{Q^+_{R,R^2/2}(\bar t,\bar x)} u^{-\xi}dtdx \frac{1}{|Q^-_{R,R^2/2}|}\iint_{Q^-_{R,R^2/2}(\bar t-2R^2,\bar x)} u^{\xi}dt'dx' \le \gamma. \qquad (2.5)$$

Combining (2.3) and (2.4) with (2.5), we have the assertion in Theorem 1.1.

Proof of Theorem 1.4. Next, for the proof of Theorem 1.4, we note only the followings: Note that Caccioppoli estimates hold for the derivatives of order up to k of solutions of (1.7) in any $Q_\rho^+ \subset \widetilde{Q_{h_0}}$. Combining Caccioppoli inequalities with a variation of Sobolev inequality for step functions in time variable(see [6]), we have the assertion in Theorem 1.4.

References

[1] Giaquinta,M.: *Multiple integrals in the calculus of variations and nonlinear elliptic systems.* Ann.Math.Stud., Princeton Univ. Press,**105.** Princeton, New Jersey, 1983

[2] Gilbarg,D., Trudinger,N.S.: *Elliptic Partial Diffrential Equations of Second Order.* Berlin Heidelberg New York: Springer 1977

[3] John,F., Nirenberg,L.: *On functions of bounded mean oscillation.* Commun.Pure Appl.Math. **14**, 415-426(1961)

[4] Kikuchi,N.: *A construction of solutions to parabolic systems associated with the variational functional of harmonic type.* (preprint)

[5] Kikuchi,N.: *Hölder estimates of solutions for difference-differential equations of elliptic-parabolic type.* Research Report, Keio University 92/002, 1992

[6] Kikuchi,N., Misawa,M.: *Campanato type estimates for solutions of difference-elliptic partial differential equations with constant coefficients.* Rend.Circ.Mat.Palermo,II. Ser.**XXXIX**, 99-131(1990)

[7] Ladyžhenskaya,O.A., Ural'tzeva,N.N.: *Linear and quasilinear elliptic equations.* Providence RI: Am.Math.Soc.1968

[8] Ladyžhenskaya,O.A., Solonnikov,V.A., Ural'tzeva,N.N.: *Linear and quasilinear equations of parabolic type.* (Transl.Math.Monogr.**23**) Providence RI: Am.Math.Soc.1968

[9] Misawa,M.: *A Harnack inequality for solutions of difference differential equations of elliptic-parabolic type.* Math. Z. **213**, 393-424(1993)

[10] Moser,J.: *On Harnack theorem for elliptic differential equations.* Commun.Pure Appl.Math.**14**, 577-591(1961)

[11] Moser,J.: *A Harnack inequality for parabolic differential equations.* Commun.Pure Appl.Math. **17**, 101-134; Errata, ibid.**20**, 231-236(1967)

4th Int. Coll. on Differential Equations, pp. 175-188
D. Bainov and V. Covachev (Eds)

The Eigenvalue Distribution of Elliptic Operators with Hölder Continuous Coefficients

Yôichi MIYAZAKI

School of Dentistry, Nihon University
1-8-13 Kanda-Surugadai, Chiyoda-ku
Tokyo 101, JAPAN

Keywords: asymptotic distribution, eigenvalues, elliptic operators, Hölder continuous, remainder estimate.
Mathematics Subject Classification: 35P20

1. Introduction

The asymptotic distribution of eigenvalues of elliptic operators has been studied by many mathematicians since Weyl's work (1911) for the Laplacian. When the coefficients and the boundary of the domain are sufficiently smooth, we know the best possible estimate for the remainder term in the asymptotic formula for the eigenvalues. But when they are not so smooth, the remainder estimate seems to be unsatisfactory. In this article we will try to improve the remainder estimate when the coefficients are not sufficiently smooth.

First we define an elliptic operator A of order $2m$ whose principal term has Hölder continuous coefficients of exponent $\tau\,(0 < \tau < \infty)$.

Let Ω be a bounded domain of the n-dimensional Euclidean space. For sake of simplicity of statement we assume that the boundary of Ω is Lipschitz continuous. In some cases we may replace this assumption by some weaker assumptions such as the restricted cone property and so on.

In order to treat irregular coefficients it is convenient to define the operator A from a variational triple $\{B, V, L_2(\Omega)\}$ where B is a symmetric integro-differential sesquilinear form:

$$B[u,v] = \int_\Omega \sum_{|\alpha|,|\beta|\leqq m} a_{\alpha\beta}(x) D^\alpha u(x) \overline{D^\beta v(x)}\, dx,$$

and V is a closed subspace of the Sobolev space $H^m(\Omega)$. We assume the following conditions.

(H1) $H_0^m(\Omega) \subset V \subset H^m(\Omega)$ ($H_0^m(\Omega)$ denotes the closure of $C_0^\infty(\Omega)$ in $H^m(\Omega)$).

(H2) There exist $C_0 \geqq 0$ and $\delta_0 > 0$ such that

$$B[u,u] \geqq \delta_0 \|u\|_m^2 - C_0 \|u\|_0^2 \quad \text{for any } u \in V.$$

(H3) The coefficients $a_{\alpha\beta}(x)$ ($|\alpha|+|\beta| \leqq 2m$) are bounded on Ω, and for some $\tau > 0$ the coefficients of top order satisfy

$$a_{\alpha\beta} \in \mathcal{B}^\tau(\Omega) \quad (|\alpha| = |\beta| = m).$$

Here $\mathcal{B}^\tau(\Omega)$ is defined as follows. Let $\tau = k+\sigma$ where k is an integer and $0 < \sigma \leqq 1$. $\mathcal{B}^\tau(\Omega)$ is the space of functions u in Ω such that $\partial^\alpha u$ are bounded and continuous for $|\alpha| \leqq k$ and $|\partial^\alpha u(x) - \partial^\alpha u(y)|/|x-y|^\sigma$ ($x, y \in \Omega$, $x \neq y$) are bounded for $|\alpha| = k$. We set $\mathcal{B}^\infty(\Omega) = \cap_{\tau>0} \mathcal{B}^\tau(\Omega)$.

Under these assumptions we define the elliptic operator A as follows: $u \in V$ belongs to $D(A)$, the domain of definition of A, if and only if there exists $f \in L_2(\Omega)$ such that $B[u,v] = (f,v)$ for any $v \in V$, and we define $Au = f$. Here $(\ ,\ )$ denotes the inner product in $L_2(\Omega)$. Clearly in the distributional sense we have

$$Au = \sum_{|\alpha|, |\beta| \leqq m} D^\beta (a_{\alpha\beta}(x) D^\alpha u)$$

for any $u \in D(A)$.

We give two examples of the operator A.

Example 1. Let us set $V = H_0^m(\Omega)$ and assume that the boundary of Ω and the coefficients are sufficiently smooth. Then A defined in the above is an operator with the Dirichlet boundary conditions:

$$D(A) = H^{2m}(\Omega) \cap H_0^m(\Omega).$$

Example 2. Let A_1 be a self-adjoint realization of a uniformly and strongly elliptic operator of order $2m$ such that $H_0^{2m}(\Omega) \subset D(A_1) \subset H^{2m}(\Omega)$. Generally there does not necessarily exist a variational triple corresponding to A_1. But A_1^2 is a self-adjoint operator associated with the variational triple $\{B_1, V_1, L_2(\Omega)\}$ where

$$B_1[u,v] = (A_1 u, A_1 v)$$

and $V_1 = D(A_1)$. Therefore the investigation of the eigenvalue distribution of the operator such as A_1 is reduced to that of the operator associated with a variational triple.

Since the resolvent of A is compact, the spectrum of A is discrete, that is, the spectrum of A consists only of real eigenvalues accumulating only at ∞. We count the eigenvalues of A in increasing order according to its multiplicity:

$$\lambda_1 \leqq \lambda_2 \leqq \lambda_3 \leqq \cdots \nearrow \infty.$$

For a real number t we denote by $N(t)$, which is called a counting function, the number of eigenvalues of A which do not exceed t.

Generally the asymptotic formula for $N(t)$, which is often called Weyl's formula, is written

$$\text{(WF1)} \qquad N(t) = \mu_A(\Omega)t^{n/2m} + O(t^{(n-\theta)/2m}) \quad \text{as } t \to \infty$$

with an appropriate number θ, $0 < \theta \leqq 1$ where $\mu_A(\Omega)$ is the constant determined by the principal symbol $a(x,\xi)$ of A and Ω as follows:

$$\mu_A(\Omega) = \int_\Omega \mu_A(x)\,dx, \qquad \mu_A(x) = (2\pi)^{-n}\int_{a(x,\xi)<1} d\xi.$$

Weyl obtained the above formula with the remainder term replaced by $o(t^{n/2m})$ for the Laplacian. Many mathematicians have extended Weyl's result to operators of higher order and improved the remainder term.

As will be stated later, the best possible estimate is $\theta = 1$. Our interest is how large we can take θ for fixed τ and what is the minimum value of τ which gives the best possible estimate $\theta = 1$.

For sake of convenience we prepare another asymptotic formula with a log factor:

$$\text{(WF2)} \qquad N(t) = \mu_A(\Omega)t^{n/2m} + O(t^{(n-1)/2m}\log t) \quad \text{as } t \to \infty.$$

Clearly (WF2) is better than (WF1) with $0 < \theta < 1$, and (WF1) with $\theta = 1$ is better than (WF2).

2. Known results and a conjecture

Let us review known results briefly.

First let us assume that the coefficients are infinitely differentiable, that is, $\tau = \infty$, and that the boundary of Ω is sufficiently smooth. Then (WF2) was obtained independently by Brüning [3], who assumed $D(A^k) \subset H^{km}(\Omega)$ for some k with $km > n$, and by Tsujimoto [23] who assumed some estimate for the resolvent kernel.

For the Laplacian ($m = 1$) the log factor was removed, that is, (WF1) with $\theta = 1$ was proved by Seeley [20], [21] and Pham The Lai [19]. For general operators ($m \geqq 1$) defined by boundary conditions Vasil'ev [25] proved (WF1) with $\theta = 1$ (see also Métivier [13]). It is noteworthy that under some assumption on the Hamiltonian flow Vasil'ev [25] obtained the second term:

$$N(t) = \mu_A(\Omega)t^{n/2m} + b\,t^{(n-1)/2m} + o(t^{(n-1)/2m}) \qquad \text{as } t \to \infty,$$

where b is a constant (see also Ivrii [7] for $m = 1$). From this formula we see that the best possible estimate is $\theta = 1$.

Next we assume that $0 < \tau < \infty$.

This case was treated first by Beals [2] and improved by Maruo and Tanabe [9], Maruo [10] and Tsujimoto [24].

The best known result is that (WF1) holds with $\theta = \tau/(\tau+1)$. This was obtained by Métivier [12] when $0 < \tau \leqq 1$, and by the author [14] under the condition of $2m > n$ when $\tau > 1$.

The estimate $\theta = \tau/(\tau+1)$ seems satisfactory for it becomes 1 when τ becomes infinity. But if we look through the proof of (WF2) by Tsujimoto [23] in detail, we see that (WF2) is valid for some finite number τ, the value of which is not so smart. Hence there is a gap between (WF1) with $\theta = \tau/(\tau+1)$ for any fixed $\tau > 0$ and (WF2) for the finite number τ. Therefore we would like to expect that (WF1) holds with θ larger than $\tau/(\tau+1)$.

Considering the simplest case $n = m = 1$, which means that A is a Sturm-Liouville operator, we have reached the following conjecture.

Conjecture. *When $0 < \tau \leqq 1$, (WF1) holds with $\theta = \tau$. In particular, the best possible estimate $\theta = 1$ is attained even when $\tau = 1$. In other words, the best possible estimate does not require $\tau = \infty$.*

3. Main results

If we assume the following condition (H4) or (H4'), the condition of separate type of variables on the coefficients, then the above conjecture can be solved in most cases.

(H4) The coefficients $a_{\alpha\beta}(x)$ ($|\alpha| = |\beta| = m$) can be written in the form $a_{\alpha\beta}(x) = b_{\alpha\beta}p(x)^{\alpha+\beta}$ with some real constants $b_{\alpha\beta}$ and $p(x) = (p_1(x), \cdots, p_n(x))$ where $p_j(x)$ is a function only of x_j for each $1 \leqq j \leqq n$.

(H4') The condition (H4) is satisfied with $a_{\alpha\beta}(x) \equiv 0$ for α and β such that any element of $\alpha + \beta$ does not coincide $2m$.

Theorem 1. *If we assume one of the following conditions (a)–(d), then (WF1) holds with $\theta = \tau$ when $0 < \tau < 1$. In addition, under the condition (a) the formula (WF1) holds with $\theta = 1$ when $\tau = 1$. Under the condition (b) or (c) the formula (WF2) holds.*

(a) *(H4'); Ω is a rectangular parallelepiped whose sides are parallel to one of the axes.*

(b) *(H4'); $V = H_0^m(\Omega)$.*

(c) *(H4); $m = 1$; $V = H_0^m(\Omega)$.*

(d) *(H4); $2m > n$.*

The result for the case (a) shows that the conjecture has been completely solved when $n = 1$.

It is difficult to solve the conjecture without assuming (H4) or (H4'). But for second-order operators we obtain the following.

Theorem 2. *If we assume the following condition (e), then (WF1) holds with $\theta = 3\tau/(2\tau + 3)$ when $0 < \tau < 3$, and (WF2) holds when $\tau = 3$.*

(e) *$m = 1$; the coefficients of top order of A are real-valued.*

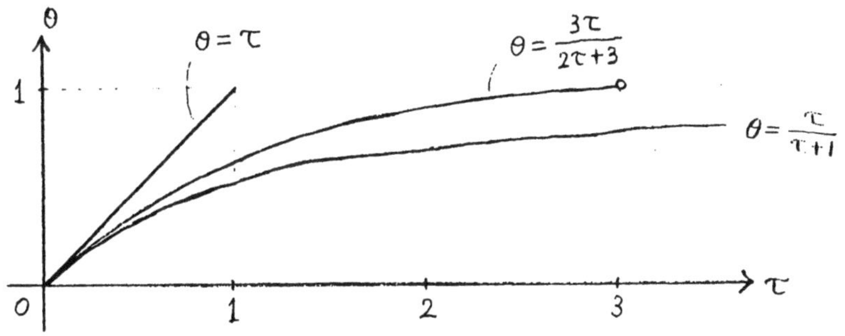

The estimate of Theorem 2 is worse than that of the conjecture, but it is rather better than the known estimate $\theta = \tau/(\tau + 1)$ because $3\tau/(2\tau + 3) = 1$ holds when $\tau = 3$, which is finite.

When the conditions (a)–(e) are not necessarily assumed, we have not improved the estimate $\theta = \tau/(\tau + 1)$ yet. But we try to solve a weak version of conjecture. In place of $N(t)$ we treat the trace of the heat kernel. Let us set

$$U(t) = \operatorname{trace}(e^{-tA}) = \int_{-\infty}^{\infty} e^{-ts} dN(s),$$

and investigate the asymptotic behavior of $U(t)$ as $t \to +0$. Then the following theorem shows that the conjecture is valid in a weak sense.

Theorem 3. *If we assume one of the following conditions (e)–(g), then*

(WF1') $\quad U(t) = c_{n,m}\mu_{\mathcal{A}}(\Omega)t^{-n/2m} + O(t^{(\tau-n)/2m}) \quad$ *as* $t \to +0$

holds when $0 < \tau < 1$, *and*

(WF2') $\quad U(t) = c_{n,m}\mu_{\mathcal{A}}(\Omega)t^{-n/2m} + O(t^{(1-n)/2m}\log t^{-1}) \quad$ *as* $t \to +0$

holds when $\tau = 1$ *where* $c_{n,m} = \Gamma(n/2m + 1)$.

(e) $m = 1$; *the coefficients of top order of* A *are real-valued.*

(f) $2m > n$.

(g) $V = H_0^m(\Omega)$; *the boundary of* Ω *is in* C^{2m}*-class.*

In the following we give the outline of proofs of Theorems 1-3. The details are found in [16], [17].

4. Proof of Theorem 1

At first we give the idea of the proof of Theorem 1 in the case (c) or (d). We may assume $a_{\alpha\beta}(x) \equiv 0 \quad (|\alpha| + |\beta| < 2m)$ without loss of generality.

Step 1 (approximation by operators with smooth coefficients).

We approximate A by a family of operators $\{A_\varepsilon\}_{\varepsilon>0}$ with C^∞ coefficients. Let us take $\varphi \in C_0^\infty(\mathbf{R}^n)$ satisfying

$$\int_{\mathbf{R}^n} \varphi(x)\, dx = 1, \quad \operatorname{supp}\varphi \subset \{x \in \mathbf{R}^n;\ |x| < 1\},$$

and put

$$\varphi_\varepsilon(x) = \varepsilon^{-n}\varphi(x/\varepsilon) \qquad (\varepsilon > 0).$$

We approximate $a_{\alpha\beta}$ by

$$a_{\alpha\beta}^\varepsilon = \varphi_\varepsilon * a_{\alpha\beta} \quad (|\alpha| = |\beta| = m).$$

Then we have

$$(4.1)\qquad |a^{\varepsilon}_{\alpha\beta}(x) - a_{\alpha\beta}(x)| \leqq C\varepsilon^{\tau}, \qquad |D^{\gamma}a^{\varepsilon}_{\alpha\beta}(x)| \leqq C\varepsilon^{\tau-|\gamma|} \quad (|\gamma| \geqq 1).$$

Here and in what follows we denote by C constants independent of x, t and ε. As we define B, A, $N(t)$ for $a_{\alpha\beta}$, we define B_{ε}, A_{ε}, $N_{\varepsilon}(t)$ for $a^{\varepsilon}_{\alpha\beta}$.

Step 2 (difference between $N(t)$ and $N_{\varepsilon}(t)$). Taking (H2) and (4.1) into account and using the min-max principle we get

$$(4.2)\qquad N(t) \leqq N_{\varepsilon}((1 + C\varepsilon^{\tau})t + C\varepsilon^{\tau-2m}).$$

Step 3 (estimate of the spectral function of A_{ε}).

In order to evaluate $N_{\varepsilon}(t)$ we aim at the estimate for the spectral function $e_{\varepsilon}(t, x, y)$ of A_{ε} as follows:

$$(4.3)\qquad |e_{\varepsilon}(t, x, x) - \mu_{A_{\varepsilon}}(x)t^{n/2m}| \leqq C\sum_{j\geqq 1} \varepsilon^{\tau-j}t^{(n-j)/2m}$$

for $t \geqq 1$. Roughly speaking, $\varepsilon^{\tau-j}$ in the right hand side of (4.3) comes from the derivatives $D^{\gamma}a^{\varepsilon}_{\alpha\beta}$ with $|\gamma| \leqq j$. When (H4) is satisfied, we can find the explicit formula for the spectral function for the operator on $L_2(\mathbf{R}^n)$ which has the same principal symbol as A_{ε}. From this explicit formula we get (4.3). Precisely speaking, we must add the term such as $Cd(x)^{-1}t^{(n-1)/2m}$ and so on in the right hand side of (4.3). That is why the log factor remains in the asymptotic formula for $N(t)$ when $\tau = 1$. Here $d(x)$ is the distance from x to the boundary of Ω.

Step 4 (estimate of $N(t)$).

Since it follows that

$$N_{\varepsilon}(t) = \int_{\Omega} e_{\varepsilon}(t, x, x)\, dx,$$

(4.3) yields

$$(4.4)\qquad N_{\varepsilon}(t) \leqq \mu_{A_{\varepsilon}}(\Omega)t^{n/2m} + C\sum_{j\geqq 1} \varepsilon^{\tau-j}t^{(n-j)/2m}.$$

When $\tau = 1$, we must add $Ct^{(n-1)/2m}\log t$ in the right hand side of (4.4). Combining (4.2) and (4.4) we have

$$(4.5)\qquad N(t) \leqq \mu_{A}(\Omega)t^{n/2m} + C\sum_{j\geqq 0} \varepsilon^{\tau-j}t^{(n-j)/2m}.$$

Putting $\varepsilon = t^{-1/2m}$ in (4.5) we get

$$N(t) \leqq \mu_A(\Omega)t^{n/2m} + Ct^{(n-\tau)/2m}.$$

In the same way we get the estimate from below. Thus Theorem 1 has been proved in the case (c) or (d).

The proof in the case (a) or (b) is a little different from the above proof. Without passing through Step 3 we go directly to the estimate (4.4) for $N_\varepsilon(t)$, which is derived from the evaluation of eigenvalues of the Sturm-Liouville operator. Let p be a real-valued function in $C^1(\bar{\Omega})$ satisfying

$$0 < m(p) \leqq |p|_0 < \infty$$

where $|\ \ |_0$ denotes the maximum norm and $m(p)$ denotes the infinimum of $p(x)$ in Ω. Let A be the Sturm-Liouville operator

$$Au = -\frac{d}{dx}\left(p(x)\frac{du}{dx}\right)$$

with $D(A) = \{u \in H_0^1(I);\ Au \in L_2(I)\}$.

Applying a modified Prüfer transform

$$u = \frac{r}{\sqrt{\lambda p}}\sin\theta, \qquad \frac{du}{dx} = \frac{r}{p}\cos\theta$$

to the initial value problem

$$\frac{d}{dx}\left(p(x)\frac{du}{dx}\right) + \lambda u = 0, \quad u(a) = 0, \quad p(a)u'(a) = 1,$$

we get

$$\left|\theta(b,\lambda) - \sqrt{\lambda}\int_a^b \frac{dx}{\sqrt{p(x)}}\right| \leqq \frac{|p'|_0|\Omega|}{4\,m(p)}.$$

Since $\theta(b,\lambda_k) = k\pi$ for the kth eigenvalue λ_k of A, we have

$$|k - \mu_A(\Omega)\sqrt{\lambda_k}| \leqq \frac{|p'|_0|\Omega|}{4\,m(p)},$$

from which it follows that

$$|N(t) - \mu_A(\Omega)t^{1/2}| \leqq \frac{|p'|_0|\Omega|}{4\pi\,m(p)} + \left(\frac{|p|_0}{m(p)}\right)^{1/2} + 1 \tag{4.6}$$

for $t \geqq 0$.

(4.6) is extended to higher-order operators by considering A^m, and to multi-dimensional operators by the method of separate type of variables. Then we get (4.4), from which Theorem 1 follows in the case (a) or (b).

5. Proof of Theorem 2

Theorem 2 is proved in the same way as Theorem 1 in the case (c) or (d). We note that (4.1) remains valid even when $\tau > 1$ if we choose a function φ appropriately. The difficulty lies in obtaining (4.3). In fact we get not (4.3) itself but a worse estimate than (4.3). For example, we must replace $\varepsilon^{\tau-1}t^{(n-1)/2m}$ by $\varepsilon^{\min\{\tau/3-1,0\}}t^{(n-1)/2m}$ in the right hand side of (4.3). If (4.3) itself could be obtained, the conjecture would also hold in the case (e).

When A_ε is a second-order operator, the Euclidean space $\mathbf{R}^n$ is endowed with the Riemannian metric associated with A_ε. Let $r_\varepsilon(x,y)$ denote the geodesic distance between x and y.

In order to evaluate the spectral function we construct a parametrix for the wave equation

$$(5.1) \qquad \begin{cases} \left(\dfrac{\partial^2}{\partial t^2} + A_\varepsilon\right)u(t,x) = 0 \\ u(0,x) = f(x), \quad \partial_t u(0,x) = 0 \end{cases}$$

by the Hadamard-Riesz method. We define an analytic family of distributions $\theta^{(\nu)}(s)$ on $\mathbf{R}$, which is defined by

$$\theta^{(\nu)}(s) = \frac{s_+^{-\nu}}{\Gamma(1-\nu)}, \qquad s_+ = \begin{cases} s & (s \geqq 0) \\ 0 & (s < 0) \end{cases}$$

for $\operatorname{Re}\nu < 1$, and can be analytically continued to the entire plane $\mathbf{C}$ by the first equality of the following:

$$\frac{d}{ds}\theta^{(\nu)}(s) = \theta^{(\nu+1)}(s), \quad s\theta^{(\nu)}(s) = (1-\nu)\theta^{(\nu-1)}(s).$$

Let $\delta_\varepsilon(x)$ be a constant times $\min\{d(x), \varepsilon^{1-\tau/3}\}$. We find the solution of (5.1) in the form

$$(5.2) \qquad u(t,x) = (\cos t\sqrt{A_\varepsilon}\, f)(x) = \sum_{l=0}^{N} I_{\varepsilon,l}(f;t,x) + R_{\varepsilon,N}(f;\,t,x)$$

in the neighborhood of fixed $z \in \Omega$ when $|t| < \delta_\varepsilon(z)$ where

$$I_{\varepsilon,l}(f;t,x) = \int_\Omega a_{\varepsilon,l}(x,y) \cdot \operatorname{sgn} t \cdot \partial_t \theta^{((n-1)/2-l)}(t^2 - r_\varepsilon(x,y)^2) f(y)\, dy$$

and the error term $R_{\varepsilon,N}(f;\, t, x)$ satisfies the Cauchy problem:

$$\begin{cases} \left(\dfrac{\partial^2}{\partial t^2} + A_\varepsilon\right) R_{\varepsilon,N}(f;\, t,x) = Q_{\varepsilon,N}(f;\, t,x) \\ R_{\varepsilon,N}(f;\, 0,x) = \partial_t R_{\varepsilon,N}(f;\, 0,x) = 0 \end{cases}$$

with

$$Q_{\varepsilon,N}(f;\, t,x) = \int_\Omega Q_{\varepsilon,N}(t,x,y) f(y)\, dy,$$
$$Q_{\varepsilon,N}(t,x,y) = -A_{\varepsilon,x} a_{\varepsilon,N}(x,y) \cdot \operatorname{sgn} t \cdot \partial_t \theta^{((n-1)/2-N)}(t^2 - r_\varepsilon(x,y)^2).$$

The amplitudes $a_{\varepsilon,l}$ satisfy the transport equations: $a_{\varepsilon,-1} \equiv 0$ and

$$4r_\varepsilon \frac{da_{\varepsilon,l}}{dr_\varepsilon} - (A_\varepsilon(r_\varepsilon^2) + 2n - 4l) a_{\varepsilon,l} + A_\varepsilon a_{\varepsilon,l-1} = 0 \quad (l \geqq 0).$$

Let us choose $\rho \in \mathcal{S}(\mathbf{R})$ (the set of rapidly decreasing functions) satisfying

$$\widehat{\rho}(0) = 1, \quad \operatorname{supp} \widehat{\rho} \subset (-1,1), \quad \rho(\lambda) \geqq 0,$$
$$\inf_{|\lambda| \leqq 1} \rho(\lambda) = c_0 > 0, \quad \rho : \text{even}$$

where $\widehat{\rho}(t)$ denotes the Fourier transform of $\rho(\lambda)$. For $\delta > 0$ we put $\rho_\delta(\lambda) = \delta\rho(\delta\lambda)$ and have $\widehat{\rho_\delta}(t) = \widehat{\rho}(t/\delta)$. We put

$$E_\varepsilon(\lambda, x, y) = \operatorname{sgn} \lambda \cdot e_\varepsilon(\lambda^2, x, y).$$

For fixed $z \in \Omega$ we set $\delta = \delta_\varepsilon(z)$. Taking the inverse Fourier transform of (5.2) multiplied by $\widehat{\rho_\delta}(t)$ we get

(5.3)
$$\begin{aligned} \frac{1}{2}\rho_\delta * dE_\varepsilon(\lambda, z, z) = &\sum_{0 \leqq l \leqq (n-1)/2} b_l a_{\varepsilon,l}(z,z) \rho_\delta * \left(|\lambda|^{n-2l-1}\right) \\ &+ \sum_{(n-1)/2 < l \leqq N} b_l a_{\varepsilon,l}(z,z) \int_{-\infty}^{\infty} e^{it\lambda} \widehat{\rho_\delta}(t) |t|^{2l-n}\, dt \\ &+ (2\pi)^{-1} \int_{-\infty}^{\infty} e^{it\lambda} \widehat{\rho_\delta}(t) R_{\varepsilon,N}(t,z,z)\, dt \end{aligned}$$

where $R_{\varepsilon,N}(t,x,y)$ is the integral kernel of $R_{\varepsilon,N}(\,\cdot\,;t,x)$ and b_l is a constant depending only on l and n.

Applying elaborated Tauberian argument to (5.3), we get

$$\begin{aligned}\left|E_\varepsilon(\lambda,x,x)-\mu_{A_\varepsilon}(x)\lambda^n\right| \leqq C_N\Big\{ &|a_{\varepsilon,0}|_0\delta_\varepsilon(x)^{-1}\lambda^{n-1}+\sum_{1\leqq l\leqq N}|a_{\varepsilon,l}|_0\lambda^{n-2l}\\ &+R_{\varepsilon,N,N-5n}(x)\delta_\varepsilon(x)^{N+5n-2K+1}\lambda^{-N+5n}\Big\}\end{aligned}$$

with $K=[n/2]+2$ and an appropriate function $R_{\varepsilon,N,N-5n}(x)$ when $\delta_\varepsilon(x)\lambda\geqq 1$.

Since it follows that

$$\begin{aligned}&|a_{\varepsilon,0}|_0\leqq C,\quad |a_{\varepsilon,l}|_0\leqq C\varepsilon^{\min\{\tau-2l,0\}}\quad (l\geqq 1),\\ &0\leqq R_{\varepsilon,N,N-5n}(x)\leqq C\varepsilon^{\tau-(2N+2K+2)},\end{aligned}$$

we obtain the estimate for the spectral function.

6. Proof of Theorem 3

The idea of the proof of Theorem 3 in the case (e) or (g) is also similar to that of Theorem 1 in the case (c) or (d). We replace (4.2) by

$$U(t)\leqq U_\varepsilon((1-C\varepsilon^\tau)t)\exp(C\varepsilon^{-2m+1}t), \tag{6.1}$$

and (4.3) by

$$\begin{aligned}&\left|U_\varepsilon(t,x,x)-c_{n,m}\mu_{A_\varepsilon}(x)t^{-n/2m}\right|\\ &\qquad\leqq Ct^{-n/2m}\exp(C\varepsilon^{-2m}t)(\varepsilon^{\tau-1}t^{1/2m}+\cdots)\end{aligned} \tag{6.2}$$

for $0<t\leqq\min\{\varepsilon^{2m},1/2\}$ where $U_\varepsilon(t,x,y)$ is the heat kernel for A_ε.

In order to derive (6.2) we solve the heat equation

$$\begin{cases}\left(\dfrac{\partial}{\partial t}+A_\varepsilon\right)u(t,x)=0\\ u(0,x)=f(x).\end{cases} \tag{6.3}$$

Taking $\kappa\in C_0^\infty(\mathbf{R}^n)$ such that $\operatorname{supp}\kappa\subset\{x;\,|x|<1\}$ and $\kappa(x)=1$ for $|x|\leqq 1/2$, we set

$$\psi_z(x)=\kappa\Big(\frac{x-z}{d(z)}\Big)$$

for fixed $z \in \Omega$. In the neighborhood of z a parametrix for (6.3) is found in the form

$$(2\pi)^{-n}\psi_z(x)\iint_{\mathbf{R}^{2n}} e^{i(x-y)\xi}e^{-ta(x,\xi)}f(y)\,dy\,d\xi,$$

whose kernel is the principal term of $U_\varepsilon(t,x,y)$.

The remainder term is evaluated by using the inequality

$$\left|\int e^{i(x-y)\xi}\xi^\alpha e^{-ta(x,\xi)}d\xi\right| \leqq Ct^{-(n+|\alpha|)/2m}\exp\left\{-\delta\left(\frac{|x-y|^{2m}}{t}\right)^{1/(2m-1)}\right\}.$$

Then combinig (6.1) and (6.2) and putting $\varepsilon = t^{1/2m}$, we get Theorem 3 in the case (e) or (g).

The proof in the case (f) is a little more complicated.

7. Related topics

It is also interesting to investigate the asymptotic behavior of the spectral function. As far as we know, the best results for the spectral function corresponding to (WF1) and (WF2) are

$$e(t,x,x) = \mu_A(x)t^{n/2m} + O(d(x)^{-\theta}t^{(n-\theta)/2m}) \quad \text{as } t \to \infty$$

with any θ, $0 < \theta < \tau/(\tau+2)$ by Tsujimoto [24] when $0 < \tau < \infty$, and

$$e(t,x,x) = \mu_A(x)t^{n/2m} + O(d(x)^{-1}t^{(n-1)/2m}) \quad \text{as } t \to \infty$$

by Brüning [3] and Tsujimoto [23] when $\tau = \infty$.

We conclude this article with mentioning the case where the boundary of Ω is highly irregular. When the Minkowski dimension of the boundary of Ω is $n-d\,(0 < d < 1)$, Lapidus showed that the remainder term of the asymptotic formula for the operator with constant coefficients is evaluated by $O(t^{(n-d)/2m})$. It is interesting that our conjecture for the case of irregular coefficients is similar to Lapidus' result for the case of irregular boundary.

References

[1] S. Agmon, Lectures on Elliptic Boundary Value Problems, Van Nostrand Mathematical Studies, Princeton, 1965.

[2] R. Beals, *Asymptotic behavior of the Green's function and spectral function of an elliptic operator*, J. Funct. Anal., **5** (1970), 484-503.

[3] J. Brüning, *Zur abschatzung der spektral-function elliptisher operatoren, Math. Z.*, **137** (1974), 75-85.

[4] E. B. Davies: Heat Kernels and Spectral Theory, Cambridge Univ. Press, Cambridge, 1989.

[5] P. Greiner, *An asymptotic expansion for the heat equation*, Arch. Rat. Mech. Anal. **41** (1971), 163-218.

[6] L. Hörmander: *The spectral function of an elliptic operator*, Acta Math. **121** (1968), 193-218.

[7] V. Ivrii, Precise Spectral Asymptotics for Elliptic Operators Acting in Fiberings over Manifolds with Boundary, Lecture Notes in Math., Vol. **1100**, Springer-Verlag, Berlin, 1984.

[8] M. L. Lapidus, *Fractal drum, inverse spectral problems for elliptic operators and a partial resolution of the Weyl-Berry conjecture*, Trans. Amer. Math. Soc. **325** (1991), 465-529.

[9] K. Maruo and H. Tanabe, *On the asymptotic distribution of eigenvalues of operators associated with strongly elliptic sesquilinear forms*, Osaka J. Math., **8** (1971), 323-345.

[10] K. Maruo, *Asymptotic distribution of eigenvalues of non-symmetric operators associated with strongly elliptic sesquilinear forms*, Osaka J. Math., **9** (1972), 547-560.

[11] G. Métivier, *Comportement asymptotique des valeurs propres d'opérateurs elliptiques dégénérés*, Soc. Math. de France, Astérisque, **34-35** (1976), 215-249.

[12] G. Métivier, *Valeurs propres des problèms aux limites elliptiques irréguliers*, Bull. Soc. Math. France Mem., **51-52** (1977), 125-219.

[13] G. Métivier, *Estimation du reste en theorie spectrale*, in Seminar Analysis 1982/1983, Inst. Math., Berlin (1983), 70-96.

[14] Y. Miyazaki, *A sharp asymptotic remainder estimate for the eigenvalues of operators associated with strongly elliptic sesquilinear forms*, Japan. J. Math., **15** (1989), 65-97.

[15] Y. Miyazaki: *On the asymptotic remainder estimate for the eigenvalues of operators associated with strongly elliptic sesquilinear forms*, Proc. Japan Acad. **67** (1991), 299-303.

[16] Y. Miyazaki: *The eigenvalue distribution of elliptic operators with Hölder continuous coefficients*, Osaka J. Math. **28** (1991), 935-973.

[17] Y. Miyazaki: *The eigenvalue distribution of elliptic operators with Hölder con-*

tinuous coefficients II, to appear in Osaka J. Math. **30** (1993).

[18] Y. Miyazaki: *Remarks on irregular open sets and its application to the eigenvalue distribution*, Funkcial. Ekvac. **35** (1992), 571-580.

[19] Pham The Lai, *Meilleures estimations asymptotiques des restes de la fonction spectrale et des valeurs propres relatifs au laplacien*, Math. Scand. **48** (1981), 5-38.

[20] R.T. Seeley, *A sharp asymptotic remainder estimate for the eigenvalues of the Laplacian in a domain of* $\mathbf{R}^3$, Adv. in Math. **29** (1978), 244-269.

[21] R.T. Seeley, *An estimate near the boundary for the spectral function of the Laplace operator*, Amer. J. Math. **102** (1980), 869-902.

[22] N. Shimakura, Ordinary Differential Equations, Shōkabo, Tokyo, in Japanese, 1988.

[23] J. Tsujimoto, *On the asymptotic behavior of spectral functions of elliptic operators*, Japan. J. Math., **8** (1982), 177-210.

[24] J. Tsujimoto, *On the remainder estimates of asymptotic formula for eigenvalues of operators associated with strongly elliptic sesquilinear forms*, J. Math. Soc. Japan, **33** (1981), 557-569.

[25] D. G. Vasil'ev, *Two-term asymptotics of the spectrum of a boundary-value problem under an interior reflection of general form*, Funktsional. Anal. i Prilozhen., **18**, No.4, (1984), 1-13.

4th Int. Coll. on Differential Equations, pp. 189-198
D. Bainov and V. Covachev (Eds)

Applications of an adaptive finite element method

R. MONTENEGRO[1], L. FERRAGUT [2] and A. PLAZA[1]

[1]*Department of Mathematics, University of Las Palmas de Gran Canaria, Campus Universitario de Tafira, 35017-Las Palmas de Gran Canaria, Spain, Fax: +34-28-458811.*

[2]*Department of Applied Mathematics and Informatic Methods, Polytechnic University of Madrid, c/ Rios Rosas 21, 28003-Madrid, Spain, Fax: +34-1-3367051.*

Abstract
An adaptive refinement/derefinement algorithm of nested meshes is described. The main properties of the algorithm are remarked and its efficiency is shown through some numerical examples: a time-dependent convection-diffusion problem and a stationary problem.

Keywords: Numerical analysis for P.D.E., mesh adaptation, finite element method.

1. INTRODUCTION

The traditional approach to the solution of evolution problems using adaptive finite element methods has been revealed very useful to solve this kind of problems [1-3]. However, using local h-refinement by *element subdivision*, if the areas to be refined change with the time, the appearance of a large number of nodes creates a serious difficulty. Thus in the topics of nested meshes it seems necessary to combine a derefinement algorithm able to remove dupe nodes and to get a good approximation of the numerical solution obtained in previous time step. We have used triangular elements with three nodes and the 4-T algorithm of Rivara [4-5] at the refining.

2. THE READAPTIVE PROCESS

A readaptive process is, *in the topics of nested meshes,* the combination of a module of refinement (local) and a derefinement algorithm [6-8]. Figure 1 summarizes such a procedure.

In order to obtain the best piece-wise support to approximate the *initial solution*, some global refinements to achieve the function are performed, after a derefinement application optimizes the mesh with a minimum number of nodes.

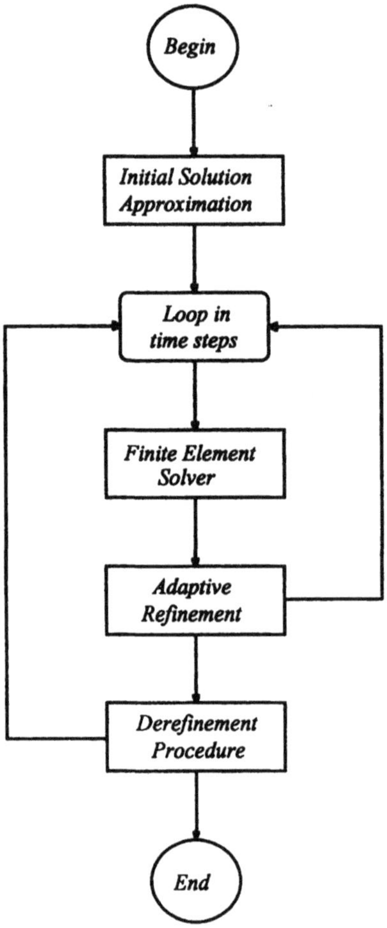

Figure 1. Main modules of a readaptive process in evolutive problems.

The number of *time steps* of each program run is fixed by the user at the beginning. For each time step a *finite element solver* is applied to obtain the numerical solution. In this module we solve the system of equations using a multigrid method [9,10].

Once a numerical solution is obtained, an *adaptive refinement step* is executed using an appropriate error indicator. Several adaptive refinements at the same time step can be thought and after we continue with the *loop in time steps*. This process is executed few times depending on the chosen adaptive strategy.

Afterwards, in order to manage a reasonable number of nodes, a *derefinement procedure* must be executed before the next loop in time steps.

2.1 Definitions and Properties

Let T = { $\tau_1 < \tau_2 < ... < \tau_n$ } be a sequence of nested triangular grids and τ_j any triangulation of T. One node N of τ_j will be called a *proper node of* τ_j if it does not belong to any previous mesh. In other case, N will be called *inherited node in* τ_j. Similarly, edges and elements are named in each level of mesh.

If an edge is divided in two at refining, it is called *father edge* of these, and these are the *son edges* of the former. *Father elements* and *son elements* are defined in the same manner. An edge c is called the *surrounding edge* of the node N if N is in the middle point of c. Figure 2 shows an example of refinement.

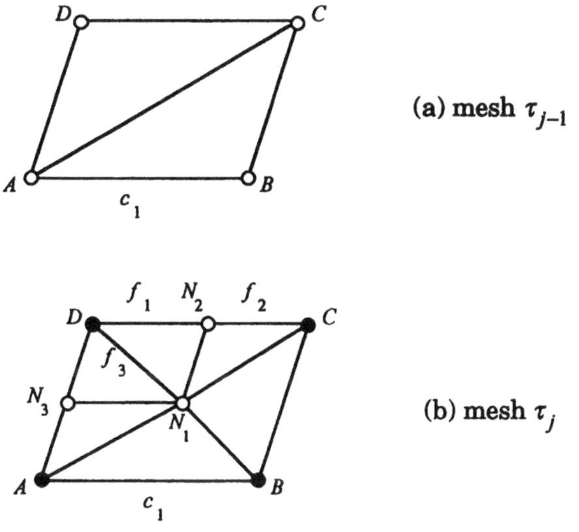

Figure 2. Example of refinement.

Some properties in the sequences of structured triangular grids are:

i) any proper element of some triangulation either is inherited in the following triangulation or has its sons there.

ii) if an element has not sons, it belongs to the finest mesh of the sequence.

At derefining *all levels* of meshes are involved. For this reason, the *family* or *genealogy* of each element must be considered.

The fundamental property of the derefinement algorithm is the following:

iii) *only those elements without successors can be eliminated.*

As derefinement condition the following one has been used: At each node the absolute difference between the numerical solution and the interpolated solution is checked. If this absolute difference is less than a reference value ε, the node can be canceled. However, the relative difference could also be used.

2.2 The derefinement procedure

Let $T = \{ \tau_1 < \tau_2 < ... < \tau_n \}$ be a sequence of nested triangular grids, where τ_1 represents the initial mesh and τ_n the finest mesh. The derefinement algorithm can be shortly described in this form:

INPUT: Sequence $T = \{ \tau_1 < \tau_2 < ... < \tau_n \}$

Loop in levels of T; For j = n to 2, do:

1. For each proper node of τ_j the derefinement condition is evaluated and the nodes and edges able to be eliminated are pointing out.

2. Conformity of the arising new level j is assured.

3.a. If some proper node of τ_j stays then, new nodal connections are defined for the new level j. Genealogy vectors are modified.

3.b. In other case, the current level j is deleted in the data structure. Genealogy vectors are modified.

4. The changes in the mesh are inherited to the following meshes.

OUTPUT: Sequence $T' = \{ \tau_1 < \tau_2' < ... < \tau_m' \}$

Some comments about the derefinement algorithm can be remarked:

We must take into account that only proper nodes are eligible in each mesh-level, and out of these, only those suitable to be canceled out are taken for evaluation. That is because if N is a particular node that cannot be canceled and c is its surrounding edge, then any node of the elements in which c is an edge cannot be canceled either.

The conformity of the arising new level is assured maintaining some nodes that, concerning the derefinement condition, could have been canceled. The conformity procedure can be founded in [8,11].

Once T' is obtained, the number of equation associated to each degree of freedom/node must be redefined, maintaining the global number of the node and the previous numerical solution. At the same time the new number of equations of each level is calculated.

Finally, we remark that our derefinement procedure has several differences with the one proposed by Rivara in [5], see [11].

3. CONVECTION-DIFFUSION PROBLEM

3.1 Semi-implicit formulation

The principal numerical aspects, presented in this section are studied in [12]. We consider the convection-diffusion problem defined in a two-dimensional domain Ω, of boundary Γ:

$$\frac{\partial u}{\partial t}+\vec{v}\cdot\vec{\nabla}u-\vec{\nabla}\cdot\left(k\vec{\nabla}u\right)=f \qquad (1)$$

where $u=u(\vec{x},t)$ is the solution in a fluid element placed in $\vec{x}=x_1\vec{i}+x_2\vec{j}$ at time t; fluid velocity is $\vec{v}=\vec{v}(\vec{x})$; we study the lineal model, $k=k(\vec{x})$; $f=f(\vec{x},t)$ are the external sources. We suppose existence and uniqueness of solution. Equation (1) can be written as:

$$\frac{du}{dt}-\vec{\nabla}\cdot\left(k\vec{\nabla}u\right)=f \qquad (2)$$

Let be a fluid element at point $\underline{P}$, defined by $\underline{\vec{x}}$, at time t_n. After a time step Δt, this fluid element will be in position P, defined by $\vec{x}$, at time t_{n+1} ; such that $t_{n+1}=t_n+\Delta t$ and $\vec{x}=\underline{\vec{x}}+\Delta\vec{x}$. Using the following approximation

$$\frac{du}{dt}\approx\frac{u(\vec{x},t_{n+1})-u(\underline{\vec{x}},t_n)}{\Delta t}=\frac{u^{n+1}(\vec{x})-u^n(\underline{\vec{x}})}{\Delta t} \qquad (3)$$

and applying an Euler implicit scheme to equation (2), we obtain:

$$u^{n+1}(\vec{x})-\Delta t\,\vec{\nabla}\cdot\left[k(\vec{x})\vec{\nabla}u^{n+1}(\vec{x})\right]=\Delta t\,f^{n+1}(\vec{x})+u^n(\underline{\vec{x}}) \qquad (4)$$

In order to evaluate $u^n(\underline{\vec{x}})$ we try to write equations (4) in function of what happens at the point P at every time. For all i = 1,2:

$$\underline{x}_i=x_i(t_{n+1}-\Delta t)=x_i-v_i(\vec{x})\Delta t+\frac{\Delta t^2}{2}\vec{v}(\vec{x})\cdot\vec{\nabla}v_i(\vec{x})+0(\Delta t^3)$$

From this equation we can obtain $\Delta x_i=x_i-\underline{x}_i$ and write:

$$u^n(\underline{\vec{x}})=u^n(\vec{x}-\Delta\vec{x})=u^n(\vec{x})-\sum_{i=1}^{2}\Delta x_i\frac{\partial u^n(\vec{x})}{\partial x_i}+\frac{1}{2}\left[2\Delta x_1\Delta x_2\frac{\partial^2u^n(\vec{x})}{\partial x_1\partial x_2}+\sum_{i=1}^{2}\Delta x_i^2\frac{\partial^2u^n(\vec{x})}{\partial x_i^2}\right]+0(\|\Delta\vec{x}\|^3)$$

and then,

$$u^n(\underline{\vec{x}})=u^n(\vec{x})-\Delta t\sum_{i=1}^{2}v_i(\vec{x})\frac{\partial u^n(\vec{x})}{\partial x_i}+\frac{\Delta t^2}{2}\sum_{i=1}^{2}\left[\vec{v}(\vec{x})\cdot\vec{\nabla}v_i(\vec{x})\right]+\frac{\Delta t^2}{2}\sum_{i=1}^{2}\sum_{j=1}^{2}v_i(\vec{x})v_j(\vec{x})\frac{\partial^2u^n(\vec{x})}{\partial x_i\partial x_j}+0(\Delta t^3)$$

Finally, if we introduce this last expression in (4) we get the following semi-implicit formulation that approximates the evolutive convection-diffusion process, where all the terms are evaluated at the same point P, defined by $\vec{x}$:

$$u^{n+1}-\Delta t\,\vec{\nabla}\cdot\left[k\vec{\nabla}u^{n+1}\right]=\Delta t\,f^{n+1}+u^n-\Delta t\,\vec{v}\cdot\vec{\nabla}u^n+\frac{\Delta t^2}{2}\sum_i(\vec{v}\cdot\vec{\nabla}v_i)\frac{\partial u^n}{\partial x_i}+\frac{\Delta t^2}{2}\sum_{i,j}v_iv_j\frac{\partial^2u^n}{\partial x_i\partial x_j}$$

Now, considering boundary conditions, it is easy to obtain the variational formulation and then apply the finite element method. We have used a consistent integration in all the terms of the final formulation.

3.2 Stability and consistency

Concerning the former formulation, we resume now the principal stability results obtained in one and two dimensions by the Von Neumann method, see[12]. There, in one dimension and over a regular mesh with diameter h, we obtain the following stability condition:

$$C \leq \left[\frac{1}{P_e^2}+\frac{1}{3}\right]^{1/2}+\frac{1}{P_e} \tag{5}$$

where $C=v\Delta t/h$ is the Courant number and $P_e=vh/k$ the Peclet number.

A general stability analysis in two dimensions would be too much complex. For this reason, we take a particular case with a mesh of rectangular triangles whose hypotenuses are oriented following a $\pi/4$-angle and their longitudes are $h\sqrt{2}$. We suppose a velocity field with a direction defined by the angle α with the x_1 axis. We study only two directions of wave propagation: x_1 axis and $3\pi/4$. In these cases we obtain the following stability conditions, respectively:

$$C \leq \frac{1}{\cos^2\alpha}\left[\left(\frac{1}{P_e^2}+\frac{1}{3}\cos^2\alpha\right)^{1/2}+\frac{1}{P_e}\right] \tag{6}$$

$$C \leq \frac{1}{1-\operatorname{sen}2\alpha}\left[\left(\frac{4}{P_e^2}+\frac{1}{3}(1-\operatorname{sen}2\alpha)\right)^{1/2}+\frac{2}{P_e}\right] \tag{7}$$

In the hypothesis of same direction of velocity field and wave propagation, the conditions given by (6) and (7) can be obtained from the condition (5), only taking as h the maximum distance between two points of the two-dimensional element on this direction.

We analyze the consistency following [12] and [13], comparing the improvement, after one time step, of the numerical and the exact solutions. For the analyzed semi-implicit formulation, particularized in one dimension and with constant coefficients and without sources in the second member, it can be proved that it is globally second-order accurate.

4. NUMERICAL RESULTS

4.1 A Convection-diffusion Problem

Let Ω be an unit square domain centered at the point (0.5,0.5), with null Neumann conditions on its boundary Γ. We suppose $f=0$ and a rotating velocity field, with $v_1=\omega x_1(1-x_1)(x_2-0.5)$ and $v_2=\omega x_2(0.5-x_1)(1-x_2)$ such that $\vec{\nabla}\cdot\vec{v}=0$ and tangent to Γ. The maximum value of the velocity field is $\omega/8$ and it is taken in the middle points of the sides of Ω. In this application we have chosen $\omega=2.5\cdot10^4$ and $k=1$. That is, a significant Peclet number of $3.125\cdot10^3$.

The initial solution is a given function that it is approximated in Figure 3(a) using a readaptive strategy. In this paper we have used the following refinement indicator, η_i, for an element Ω_i: $\eta_i=h_i^2\left|\vec{\nabla}u_h\right|$; being h_i the diameter of Ω_i and u_h the lineal numerical solution in the triangular element; and $\varepsilon=0.001$ at derefining.

It is worth to be noted that the number of nodes and their location in the mesh is auto–controlled according to the numerical solution. To get the stationary solution (constant) a total number of 2892 time steps have been calculated. The readaptive strategy was: three refinements followed by the derefinement procedure with increment of the time after each refinement or derefinement. In each time step one multigrid iteration is enough to solve the system of equations. Figure 3 shows some meshes and solutions.

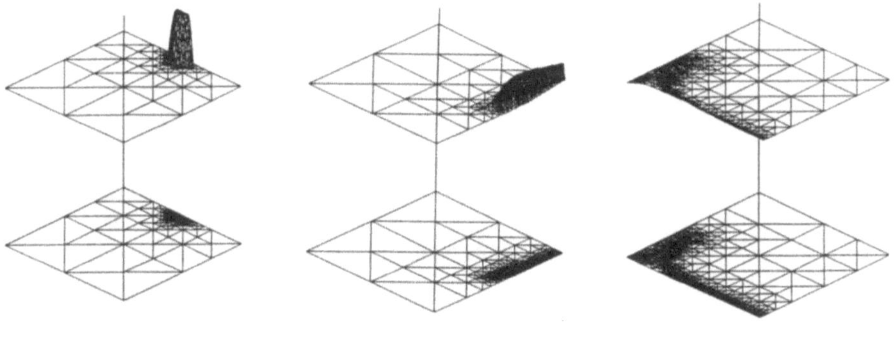

(a) Initial solution, t=0. 234 nodes

(b) t=0.00032, 919 nodes

(c) t=0.00060 863 nodes

Figure 3. Meshes and solutions of an evolutive problem.

4.2 A Stationary Problem

Although the most important applications of the readaptive processes are the evolution problems, the derefinement procedure can be used for local refining. Derefining after global refinement is thereby equivalent to a local refinement procedure, not requiring the use of error indicators.

Other interesting application in stationary problems can be remarked. That is, to stop the execution of the program attending only to the value of the parameter ε in our derefinement condition. That is, when the difference between the number of nodes of two consecutive refined (or derefined) meshes are irrelevant, the execution of the program can be stopped; clearly this is true, if we work with a refinement strategy which introduces a significant number of nodes in each refinement step. This last aspect could be also got with local refinement according to the distribution of the error indicators.

As an example, we present here an elasticity problem studied in [14]. A rectangular plate of 500x600 cm^2 with a fracture of 100 cm in the middle of one of its longest sides is considered. Let us suppose a traction force of 1 Kg/m uniformly distributed in the shorter sides. We are considering Young module $E = 2 \cdot 10^4 Kg/m^2$ and Poisson coefficient $\nu = 0.3$. We do not consider volume forces.

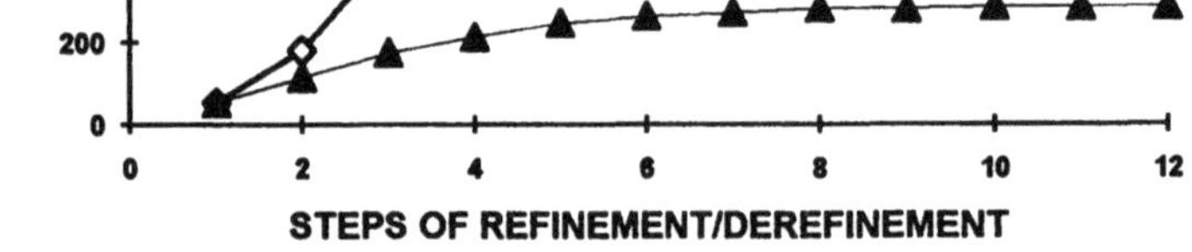

Figure 4. Evolution of number of nodes in a stationary problem.

The readaptive strategy was: after each global refinement, a derefinement with $\varepsilon = 1.5 \cdot 10^{-6}$ and absolute difference as derefinement condition is applied. The initial mesh is built with five elements and seven nodes. The evolution of the number of nodes of the refined and derefined meshes can be observed in Figure 4. It is worth to be noted that the number of nodes in the refined and derefined meshes stabilizes, it means that the solution has been just obtained according to the value of ε. Two refined meshes and their corresponding derefined ones are shown in Figure 5. The multigrid method with conjugate gradient as smoothing gets the solution with a stable number of 12 multigrid iterations.

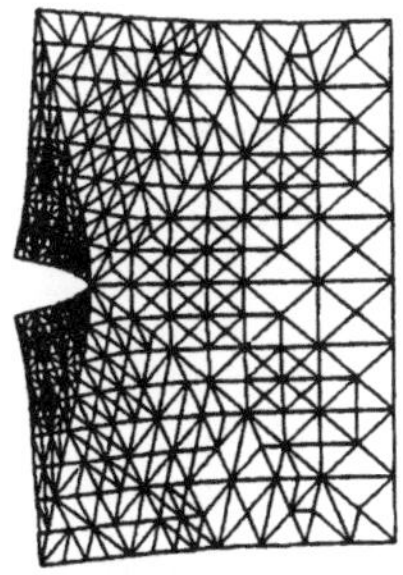

(a) Step 3, refined mesh, 421 nodes

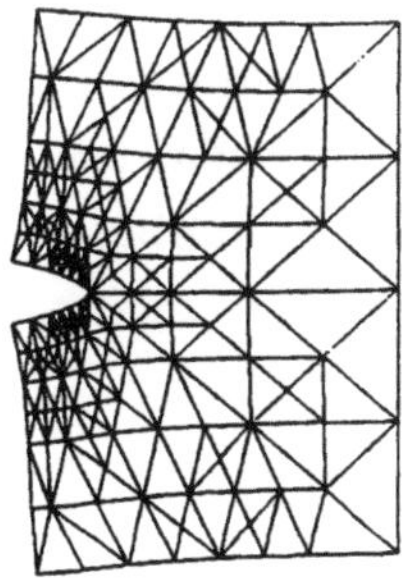

(b) derefined mesh, 175 nodes

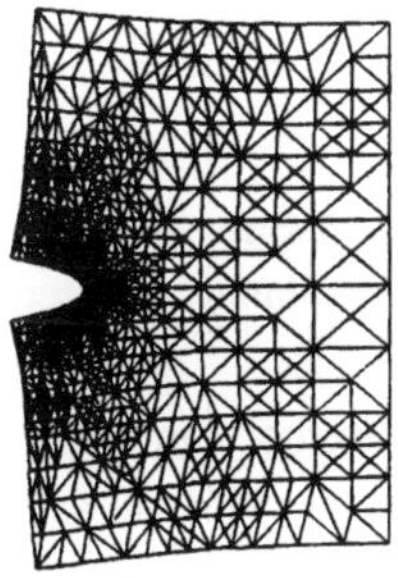

(c) Step 5, refined mesh, 797 nodes

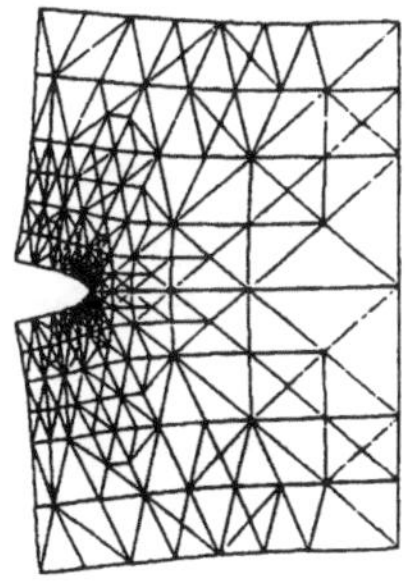

(d) derefined mesh, 246 nodes

Figure 5. Refined and derefined meshes.

5. CONCLUSIONS

The refinement/derefinement combination is very useful to solve time-dependent problems in which moving refinement areas are required. The derefinement algorithm enables us to use the multigrid method in order to solve the system of equations associated to the finite-element method. The additional computation time required by this algorithm amounts less than 1% of the total execution time.

The semi-implicit formulation proposed to solve the convection-diffusion problem shows good results, if it is combined with a readaptive procedure. Without this readaptive procedure, the cost in CPU time to solve evolution problems with Peclet numbers such elevated as the ones managed in this paper should be too high.

As another application, the derefinement algorithm can be used as a local refinement procedure. Derefining after global refinement is thereby equivalent to a local refinement procedure, not requiring the use of error indicators. This implies clear advantages in the class of problems for which error indicators cannot be found or their computation is too costly.

REFERENCES

1. D.W. Kelly, J.P. de S.R. Gago, O.C. Zienkiewicz and I. Babuska, Int. J. Num. Meth. Eng. **19**, 1593-1619 (1983).

2. J.P. de S.R. Gago, D.W. Kelly, O.C. Zienkiewicz and I. Babuska, Int. J. Num. Meth. Eng. **19**, 1621- 1656 (1983).

3. J.Z. Zhu and O.C. Zienkiewicz, J. Comp. Appl. Num. Meth. **4**, 197-204 (1988).

4. M.C. Rivara, Int. J. Num. Meth. Eng. **24**, 1343-1354 (1987).

5. M.C. Rivara, Int. J. Num. Meth. Eng. **28**, 2889-2906 (1989).

6. A. Plaza, L. Ferragut and R. Montenegro, in Algorithms, Software, Architecture, J. van Leeuwen (ed.) IFIP, Elsevier Science Publishers B.V. (North-Holland), Amsterdam (1992) pp. 409-415.

7. A. Plaza, R. Montenegro and L. Ferragut, in Numerical Methods in Engineering, Ch. Hirsch et al. (eds.), Elsevier Science Publishers B.V., Amsterdam (1992) pp. 225-232.

8. A. Plaza, Doctoral Thesis, University of Las Palmas de Gran Canaria (1993).

9. W. Hackbush and V. Trottengurg (eds.), Multigrid Methods, Lectures Notes in Mathematics, Springer-Verlag, Berlin (1982).

10. W. Hackbush and V. Trottengurg (eds.), Multigrid Methods II, Lectures Notes in Mathematics, Springer-Verlag, Berlin, 1986.

11. L. Ferragut, R. Montenegro and A. Plaza, in Com. Num. Meth. Eng., in press.

12. R. Montenegro, G. Montero, L. Ferragut and G. Winter, Rev. Int. Met. Num. Cal. Dis. Ing. **5**, 535-560 (1989).

13. J. Peraire, O.C. Zienkiewicz and K. Morgan, Int. J. Num. Meth. Eng. **22**, 547-574 (1986).

14. G. Montero, R. Montenegro, G. Winter and L. Ferragut, Rev. Int. Met. Num. Cal. Dis. Ing. **6**, 311-332 (1990).

4th Int. Coll. on Differential Equations, pp. 199-203
D. Bainov and V. Covachev (Eds)

Formation of Shocks for Nonlinear Partial Differential Equations of First Order

SHIZUO NAKANE

Tokyo Institute of Polytechnics
1583 Iiyama, Atsugi, Kanagawa 243-02, Japan

Abstract. The Cauchy problems for the equation of conservation law and the Hamilton-Jacobi equation with several space variables are considered. By virtue of the theory of singularities of smooth mappings, local structure of shocks are clarified.

1. INTRODUCTION

In this note, we shall consider the Cauchy problem for single partial differential equations of first order in $\mathbf{R}_t^+ \times \mathbf{R}_x^n$. We treat two kinds of equations, i.e., the equation of conservation law :

$$\text{(CL)} \qquad \frac{\partial u}{\partial t} + \Sigma_{j=1}^n \frac{\partial f_j(u)}{\partial x_j} = 0,$$

and the Hamilton-Jacobi equation :

$$\text{(HJ)} \qquad \frac{\partial u}{\partial t} + f(\frac{\partial u}{\partial x}) = 0,$$

with initial condition $u(0,x) = \varphi(x) \in C_0^\infty(\mathbf{R}^n)$. Here $f_j : \mathbf{R} \to \mathbf{R}$, and $f : \mathbf{R}^n \to \mathbf{R}$ are C^∞-functions.

Because of their nonlinearity, we cannot expect the global smoothness of their solutions even for smooth initial data. That is, singularities appear. We call here this singularity a "shock" since we use later singularity as a "critical point" of smooth mappings. Our purpose is to describe geometrically formation and propagation of their shocks. So we consider them in a weak sense.

Our goal is to construct, under some non-degeneracy hypotheses, "good" weak solutions, namely piecewise C^∞, with smooth shock surfaces, locally near the points of shock formation.

A weak solution for (CL) is a function u satisfying the following

$$\text{(WS)} \qquad \iint_{\mathbf{R}^+ \times \mathbf{R}^n} (u\frac{\partial g}{\partial t} + \Sigma_{j=1}^n f_j(u)\frac{\partial g}{\partial x_j})dtdx + \int_{\mathbf{R}^n} \varphi(x)g(0,x)dx = 0,$$

for any $g \in C_0^\infty(\mathbf{R}^{n+1})$.

We introduce the entropy condition in order to eliminate physically meaningless solutions. A weak solution u is called the entropy solution if it satisfies

$$\text{(E)} \qquad \iint_{\mathbf{R}^+ \times \mathbf{R}^n} sgn(u-k)\{(u-k)\frac{\partial g}{\partial t} + \Sigma_{j=1}^n (f_j(u) - f_j(k))\frac{\partial g}{\partial x_j}\}dtdx \geq 0,$$

for any $g \in C_0^\infty(\mathbf{R}^{n+1})$, $g \geq 0$ and for any $k \in \mathbf{R}$. Existence and uniqueness of entropy solution was established. See Kruzkov [8], for example.

A generalized solution of (HJ) is a Lipschitz continuous function u satisfying (HJ) almost everywhere and having semi-concavity property : there exists $K \geq 0$ such that

$$\text{(SC)} \qquad u(t, x+y) + u(t, x-y) - 2u(t,x) \leq K|y|^2,$$

for any t, x, and y.

Global existence and uniqueness of generalized solutions for (HJ) was established by Douglis [3] and Kruzkov [8], for example.

We are interested in the geometrical desciption of shocks of their solutions. In this context, many authors have studied the global structure of shocks for (CL) in case $n = 1$. See Chen [2], Guckenheimer [4], Jennings [5] and Schaeffer [12]. Recently Tsuji [13], [14] investigated this problem for (HJ) in case $n = 1, 2$. He used the theory of singularities of the plane into the plane. Our work is much inspired by his and aims an extension of his results to the case of general space dimensions. Recently Kossioris [6], [7] has obtained the same result as Theorem 4.

After [14], we shall construct their solutions by the method of characteristics. Generally speaking, they become multi-valued in finite time. By virtue of singularity theory of C^∞-mappings, we shall clarify their structures as multi-valued functions. In fact, we shall show, under some assumptions, their graphs have cusps or swallow's tails as singularities for (CL) or (HJ) respectively.

Such a difference between (CL) and (HJ) is easy to see in case $n = 1$. In fact, in such a case, by differentiating (HJ) by x, it follows that $\partial u/\partial x$ satisfies (CL) for a solution u of (HJ). For $n > 1$, there exists no such a direct relation between them. But our results imply the situation is the same as far as locally near the points of shock formation.

In order to get a single-valued solution, we must give up its smoothness. Shocks appear in this process. As for the shock surfaces, it is more difficult to consider (CL) than (HJ). In fact, we must solve a partial differential equation with non-Lipschitz continuous coefficients in case (CL), which is carried out by the stable manifold theory. On the other hand, in case (HJ), shock surfaces are automatically obtained as "Maxwell's lines" of swallow's tails. For detailed proofs, see Nakane [10], [11].

2. Equation of conservation law

In this section, we consider (CL). Put $a_j(u) = f_j'(u)$. Then, the characteristic line associated with (CL) through $(0, y)$ is the solution of the following :

$$\text{(CE)} \qquad \frac{dx}{dt} = a(u(t,x)), \ x(0) = y.$$

Since $du(t, x(t))/dt \equiv 0$ for smooth solution u of (CL), $u(t, x(t)) \equiv u(0, x(0)) = \varphi(y)$. Hence the solution of (CE) can be written by $x(t) = y + ta(\varphi(y))$.

We define C^∞-mappings Φ_t and Φ by

$$\Phi_t : \mathbf{R}_y^n \to \mathbf{R}_x^n, \quad \Phi_t(y) = y + ta(\varphi(y)),$$
$$\Phi : \mathbf{R}_{(t,y)}^{n+1} \to \mathbf{R}_{(t,x)}^{n+1}, \quad \Phi(t,y) = (t, \Phi_t(y)).$$

Then, the solution u of (CL) is expressed by $u(t,x) = \varphi(\Phi_t^{-1}(x))$ and is of C^∞ at (t, x), which is not the critical value of Φ. Hence our next problem is to study the singularities of mapping Φ.

By a direct calculation, the jacobian $J(\Phi)$ of Φ is written by

$$J(\Phi) = J(\Phi_t) = 1 + t\lambda(y) \equiv 1 + t\Sigma_{j=1}^n a_j'(\varphi(y))\frac{\partial \varphi}{\partial y_j}.$$

If $\lambda(y) \geq 0$, then $J(\Phi) \geq 1$ and there exists a global smooth solution. So we may assume

(A1) $$\min_y \lambda(y) = \lambda(y^0) = -M < 0.$$

Put $t^0 = 1/M$ and $x^0 = \Phi_{t^0}(y^0)$. Then, for $t < t^0$, $J(\Phi) > 0$ and u is of C^∞. From now on, we consider it for $t \geq t^0$ near (t^0, y^0). The following assumption corresponds to the first nondegeneracy condition in [4]:

(A2) $$\lambda'(y) = 0 \text{ implies rank } \lambda''(y^0) = n.$$

From (A1) and (A2), we can easily see that $\lambda''(y)$ is positive definite.

Under (A1) and (A2), we have the following.

Lemma 1. *By an affine transformation of coordinates, we may assume*

(2.1) $$a_j'(\varphi(y^0)) = 0 \quad (1 \leq j \leq n-1),$$

(2.2) $$a_n'(\varphi(y^0)) > 0,$$

(2.3) $$a_j(\varphi(y^0)) = 0 \quad (1 \leq j \leq n-1).$$

Theorem 1. *Under (A1) and (A2), the singularity of Φ must be fold or cusp near (t^0, y^0). Denote Σ^1 (resp. $\Sigma^{1,1}$) the set of fold points (resp. cusp points) of Φ. Then $\Sigma^{1,1}$ (resp. $\Phi(\Sigma^{1,1})$) is a C^∞- submanifold of $\mathbf{R}^{n+1}_{(t,y)}$ (resp. $\mathbf{R}^{n+1}_{(t,x)}$) of codimension two, parametrized by $y' = (y_1, ..., y_{n-1})$ (resp. x').*

Remark. The singularity y^0 of a C^∞-mapping $T : \mathbf{R}^m_y \to \mathbf{R}^m_x$ is said to be fold (resp. cusp) if T is equivalent at y^0 to the following:

$$x_j = y_j \ (1 \leq j \leq m-1), \quad x_m = y_m^2,$$
$$(\text{resp. } x_j = y_j \ (1 \leq j \leq m-1), \quad x_m = y_m^3 - y_1 y_m).$$

Thus, u is triple-valued in the region Ω surrounded by $\Phi(\Sigma^{1,1})$. To get a single-valued solution, it must be discontinuous. Denote the three branches of u in Ω by $u_1 > u_2 > u_3$. We can construct an entropy solution by skipping from u_1 to u_3 on a shock surface $S : \nu(t,x) = 0$. We construct a C^1-shock surface S. From (WS), it satisfies Rankine-Hugoniot condition on S:

(RH) $$n \cdot F(u_1, u_3) = 0,$$

where n is the unit normal of S and $F(u_1, u_3) = ([u], [f_1(u)], ..., [f_n(u)])$, $[u] = u_1 - u_3$, $f_j(u) = f_j(u_1) - f_j(u_3)$. It is equivalent to

(RH') $$\frac{\partial \nu}{\partial t} + \Sigma_{j=1}^n \frac{[f_j(u)]}{[u]}\frac{\partial \nu}{\partial x_j} = 0.$$

The function ν must satisfy also $\nu|_{\Phi(\Sigma^{1,1})} = 0$. The equation (RH') cannot be solved by the classical method of characteristics since its coefficients are not Lipschitz continuous. Nevertheless, we can solve it by the stable manifold theory. Furthermore, we can show it satisfies entropy condition, which is equivalent to

(E') $$n \cdot F(k, u_1) \geq 0 \quad (u_3 \leq k \leq u_1).$$

Thus we obtain

Theorem 2. *Suppose (A1) and (A2). Then, we get the entropy solution of (CL) having C^1-shock surface near (t^0, x^0).*

3. Hamilton-Jacobi equation

In this section, we study (HJ) in the same way as in the previous section. The characteristic equation is as follows :

$$\frac{dx}{dt} = f'(p), \quad x(0) = y,$$
$$\frac{dp}{dt} = 0, \quad p(0) = \varphi'(y),$$
$$\frac{dv}{dt} = -f(p) + p \cdot f'(p), \quad v(0) = \varphi(y),$$

whose solution is written by

$$x = y + tf'(\varphi'(y)) \equiv H_t(y),$$
$$p = \varphi'(y)$$
$$v = \varphi(y) + t\{-f(\varphi(y)) + \varphi'(y) \cdot f'(\varphi'(y))\}.$$

If we define C^∞-mapping H by

$$H : \mathbf{R}^{n+1}_{(t,y)} \to \mathbf{R}^{n+1}_{(t,x)}, \quad H(t,y) = (t, H_t(y)),$$

the solution u of (HJ) is expressed by $u(t,x) = v(H^{-1}(t,x))$.

Example (Arnold [1]). Consider the case $n = 1, f(p) = p^2/2, \varphi(x) = -\frac{3}{4}x^{4/3}$. Then, we have

$$t = t,$$
$$x = y - ty^{1/3},$$
$$v = -\frac{3}{4}y^{4/3} + \frac{1}{2}ty^{2/3}.$$

Instead of v and y, consider $w = v - \varphi'(y)(x - y - tf'(\varphi'(y)))$ and $Y = y^{1/3}$. Then the corresponding mapping $\tilde{H} : \mathbf{R}^3_{(t,x,Y)} \to \mathbf{R}^3_{(t,x,w)}$ is as follows :

$$t = t,$$
$$x = x,$$
$$w = Y^4/4 - tY^2/2 - xY,$$

which is the canonical form of swallow's tail. This example expresses the essential part of the formation of shocks.

Let $\lambda_j(y)$ be the eigenvalues of the matrix $f"(\varphi'(y)) \cdot \varphi"(y)$. Then the jacobian $J(H)$ of H is written by

$$J(H) = \Pi_{j=1}^n (1 + t\lambda_j(y)).$$

We assume

(H0) $\qquad f"(p) \geq cI$ for some $c > 0$,

(H1) $\qquad \lambda_1(y) < \lambda_2(y) \leq \lambda_3(y) \leq \ldots,$

(H2) $\qquad \min_y \lambda_1(y) = \lambda_1(y^0) = -M < 0,$

(H3) $\qquad \lambda'(y) = 0$ implies rank $\lambda_1"(y) = n$.

Put $t^0 = 1/M$. Then, we have

Theorem 3. *Under (H0) - (H3), the singularity of H must be fold or cusp near (t^0, y^0). The set $\Sigma^{1,1}$ (resp. $H(\Sigma^{1,1})$) is a C^∞-submanifold of $\mathbf{R}^{n+1}_{(t,y)}$ (resp. $\mathbf{R}^{n+1}_{(t,x)}$) of codimension two.*

Theorem 4. *Suppose (H0) - (H3). Then the solution u of (HJ) obtained above is triple-valued in the region Ω surrounded by $H(\Sigma^{1,1})$ near $(t^0, x^0) = H(t^0, y^0)$. Its graph has swallow's tails as singularities. From this, we can obtain the single-valued generalized solution of (HJ), whose shocks correspond to "Maxwell's lines."*

Remark. Assumption (H0) assures the semi-concavity of the solution.

REFERENCES

[1] V. I. Arnold, "Mathematical Methods of Classical Mechanics," Springer, 1978.

[2] N. M. Chen, *On types of singularities for solutions of nonlinear hyperbolic systems*, Bull. Inst. Math. Acad. Sinica **10** (1982), 405–416.

[3] A. Douglis, *Solutions in the large for multi-dimensional nonlinear equations of first order*, Ann. Inst. Fourier **15** (1965), 1–35.

[4] J. Guckenheimer, *Solving a single conservation law*, in "Lecture Notes in Math.," Springer, 1975, pp. 108–134.

[5] G. Jennings, *Piecewise smooth solutions of a single conservation law exist*, Adv. in Math. **33** (1979), 192–205.

[6] G. T. Kossioris, *Formation of singularities for viscosity solutions of Hamilton -Jacobi equation in one space variable*, Comm. Partial Differential Equations **18** (1993), 747–770.

[7] ———, *Formation of singularities for viscosity solutions of Hamilton -Jacobi equation in higher dimensions*, Preprint, Carnegie Mellon Univ.

[8] S. N. Kruzkov, *First order quasilinear equations in several independent variables*, Math. USSR Sb. **10** (1970 pages 217–243).

[9] B. Morin, *Formes canoniques de singularites d'une application differentiable*, C. R. Acad. Sci. Paris **260** (1965), 5662–5665.

[10] S. Nakane, *Formation of shocks for a single conservation law*, SIAM j. Math. Anal. **19** (1988), 1391–1408.

[11] ———, *Formation of singularities for Hamilton-Jacobi equation with several space variables*, J. Math. Soc. Japan **43** (1991), 89–100.

[12] D. Schaeffer, *A regularity theorem for conservation laws*, Adv. in Math. **11** (1973), 358–386.

[13] M. Tsuji, *Solution globale et propagation des singularites pour l'equation de Hamilton-Jacobi*, C. R. Acad. Sci. Paris **289** (1979), 397–400.

[14] ———, *Formation of singularities for Hamilton-Jacobi equation II*, J. Math. Kyoto Univ. **26** (1986), 299–308.

Keywords. shocks, singularities of smooth mappings

4th Int. Coll. on Differential Equations, pp. 205-214
D. Bainov and V. Covachev (Eds)

On some degenerate nonlinear parabolic equations

Tokumori NANBU
Department of Mathematics ,Toyama Medical and Phamaceutical University
2630 Sugitani, Toyama, 930-01, Japan

Abstract: We consider the uniqueness, existence and the asymptotic behaviour of solutions of some degenerate nonlinear parabolic equations.
Keywords: degenerate nonlinear parabolic equation; uniqueness; existence; decay estimate; positivity

1. Introduction

We are concerned with the uniqueness, existence and the asymptotic behaviour of solutions of the following initial-boundary value problem $(P)_\psi$:

$$(1) \qquad u_t = \Delta_x(A(u)) + \sum_{k=1}^{N} \frac{dB_k(x,t,u)}{dx_k} + C(x,u) \quad \text{in} \quad Q = \Omega \times R^+,$$

$$(2) \qquad u(x,t) = \psi(x,t) \quad \text{on} \quad \Gamma = \partial\Omega \times R^+,$$

$$(3) \qquad u(x,0) = u_0(x) \quad \text{in} \quad \Omega.$$

Here Ω is a bounded, arcwise connected domain in $R^N (N > 1)$ whose boundary $\partial\Omega$ is of class C^3,$R^+ = (0, +\infty)$, Δ_x denotes the N-dimensional Laplace operator,

$$\frac{dB_k(x,t,u)}{dx_k} = \frac{\partial B_k(x,t,u)}{\partial x_k} + \frac{\partial B_k(x,t,u)}{\partial u}\frac{\partial u}{\partial x_k},$$

$A(u)$ is a function of class $C^1(R)$ satisfying $A(0) = A'(0) = 0$, $A'(u) > 0 (u \neq 0)$, and functions $B_k(x,t,u), C(x,u)$ and $u_0(x)$ satisfy some hypotheses which will be given later.

A lot of researchers have considered the existence of weak solution of the problem $(P)_\Psi$ in more general form ([7]). Also many authors have studied the asymptotic behaviour of solution $u(x,t)$ of the problem $(P)_0$ as $t \to \infty$ in various norms ([2],[5],[7]). For $B_k(x,t,u) = C(x,u) = 0$ and $A(u) = u^m$, when the equation $u_t = \Delta_x(u^m)$ $(m > 1)$

is called the 'porous media equation ', this equation has been investigated by many authors([1],[3],[4]).

Now we mention three examples of the equation (1),where we assume that $m > 1$,$p \geq 1$ and $n_k \geq 1$ are constants.

Example 1.(Porous media equation)

$$(PM) \qquad u_t = \Delta_x(|u|^{m-1}u).$$

Example 2.(Nonlinear diffusion absorption equation)

$$(PMA) \qquad u_t = \Delta_x(|u|^{m-1}u) - \lambda|u|^{p-1}u,$$

where λ is a positive constant.

Example 3.(Nonlinear diffusion absorption convection equation)

$$(PMAC) \qquad u_t = \Delta_x(|u|^{m-1}u) + \sum_{k=1}^{N} b_{0_k}\frac{\partial(|u|^{n_k-1}u)}{\partial x_k} - \lambda|u|^{p-1}u,$$

where $\lambda > 0$ and b_{0_k} are constants.

In [3], Aronson and Peletier proved the following three results on the problem $(P)_0$ of the equation *(PM)* with non-negative initial value $u_0(x)$:

(i) (Positivity of solution)

There exists a $T \in [0,\infty)$ such that

$$\{x \in \Omega \mid u(x,t) > 0\} = \Omega \quad \text{for} \quad \text{any} \quad t \geq T$$

where T depends only on the data,

(ii) (Decay of solution)

There exist positive constants K and $\gamma = 1/(m-1)$ such that

$$0 \leq u(x,t) \leq K(1+t)^{-\gamma} \qquad (t \in [0,\infty))$$

where K depends only on the data,

(iii) (Large time behaviour of solution)

There exist positive constant K and function $f_0(x)$ such that

$$(LB) \qquad \|(1+t)^{\gamma}u(\cdot,t) - f_0\|_{L^\infty(\Omega)} \leq K(1+t)^{-1},$$

where $\gamma = 1/(m-1)$, K is a positive constant depending only on N, m, u_0 and Ω, and $f_0(x)$ is the unique positive solution of the elliptic problem:

$$\Delta(f^m) + \gamma f = 0 \quad \text{in} \quad \Omega,$$

$$f = 0 \quad \text{on} \quad \partial\Omega.$$

In [4], Bertsch, Nanbu and Peletier considered the positivity, the decay and the large time behaviour of solutions for the equation *(PMA)*. In [8] we have done research on those results for the equation *(PMCA)*.

In §2 we give the assumptions and defintions. In §3 we state the uniqueness of solution of the problem $(P)_\psi$ of the equation (1). In §4 and 5 we consider the existence of solution and the decay estimates of solution as $t \to \infty$ of the problem P_0 in the case $B_k(x,t,u) = B_k(x,u)$. In §6 we consider the positivity of solution of the problem P_0 with non-negative initial value for the equation

$$u_t = \Delta_x(u^m) + \sum_{k=1}^{N} \frac{d(b(x)u^{n_k})}{dx_k} - c(x)u^p$$

where $m, n_k, p, b(x)$ and $c(x)$ satisfy certain conditions.

2. Assumption and Definition

We assume the following conditions on $A(u), B_k(x,t,u), C(x,u), u_0(x)$ and $\psi(x,t)$:

(A.1) $A(u) \in C^\infty(R-0) \cap C^1(R), A(0) = A'(0) = 0, a(u) \equiv A'(u) > 0 (u \neq 0).$

(A.2)
$$B_k(x,t,u) \in C^{1+\alpha}(\Omega \times R^+ \times R),$$

$$\frac{dB_k(x,t,u)}{dx_k} = \frac{\partial B_k(x,t,u)}{\partial x_k} + \frac{\partial B_k(x,t,u)}{\partial u}\frac{\partial u}{\partial x_k},$$

we set $b_k(x,t,u) = \frac{\partial B_k(x,t,u)}{\partial u}$, and we assume

$$B_k(x,t,0) = b_k(x,t,0) = 0 \quad \text{for} \quad (x,t) \in \Omega \times R^+.$$

(A.3) For all $M > 0$ there exists a positive constant $\mu_0(M)$ such that

$$b_k^2(x,t,u) \le \mu_0(M)a(u)$$

for $(x,t,u) \in \Omega \times R^+ \times [-M,M] \quad (k = 1,2,..,N)$.

(A.4)
$$C(x,u) \in C^1(\Omega \times R)$$

$$C(x,0) = 0 \quad \text{in} \quad \Omega \quad \text{and} \quad C(x,u) \times sign(u) \le 0 \quad \text{for} \quad (x,u) \in \Omega \times R.$$

(A.5)
$$A(u_0(x)) \in C^1(\bar{\Omega}).$$

(A.6)
$$\psi(x,t) \in L^1_{loc}(\partial\Omega \times [0,\infty)).$$

Definition 1. A *solution* u of the problem (1) - (3) on [0,T] is a function with the following properties :

$$(i) \qquad u \in C([0,T] : L^1(\Omega)) \bigcap L^\infty(Q_T),$$

$$\int_\Omega u(t)\phi(t)dx - \int_0^t \int_\Omega [u\phi_t + A(u)\Delta\phi - \sum_{k=1}^N B_k(x,t,u)\phi_{x_k}]dxd\tau$$

$$(ii) \qquad + \int\int_{\Gamma_t} A(\psi)\frac{\partial\phi}{\partial\nu}dSd\tau = \int_\Omega u_0(x)\phi(x,0)dx + \int\int_{Q_t} C(x,u)\phi dxd\tau$$

for any $t \in (0,T]$ and $\phi \in C^2(\overline{Q_T})$ such that $\phi = 0$ on $\partial\Omega \times [0,T]$.
Here ν denotes the outward unit normal vector on $\partial\Omega$.

Definition 2. A *solution* u of the problem (1)-(3) on $[0,+\infty)$ means a solution on *each* $[0,T]$.

Definition 3. A *subsolution (supersolution)* u of the problem (1) - (3) on $[0,T]$ is a function with the following properties :

$$(i) \qquad u \in C([0,T] : L^1(\Omega)) \bigcap L^\infty(Q_T),$$

$$\int_\Omega u(t)\phi(t)dx - \int_0^t \int_\Omega [u\phi_t + A(u)\Delta\phi - \sum_{k=1}^N B_k(x,t,u)\phi_{x_k}]dxd\tau$$

$$(ii) \qquad + \int\int_{\Gamma_t} A(\psi)\frac{\partial\phi}{\partial\nu}dSd\tau \le (\ge) \int_\Omega u_0(x)\phi(x,0)dx + \int\int_{Q_t} C(x,u)\phi dxd\tau$$

for any $t \in (0,T]$ and $\phi \in C^2(\overline{Q_T})$ such that $\phi \ge 0(\overline{Q_T})$, $\phi = 0$ on $\partial\Omega \times [0,T]$.

3. Uniqueness

Theorem 1 Suppose that all the assumptions (A.1) - (A.6) are satisfied.
(i) (Comparison Principle) Let u be a *subsolution* of the problem (1) - (3) on $[0,T]$ with the data u_0, ψ and let v be a *supersolution* of the problem (1) - (3) on $[0,T]$ with the data v_0, ψ. If $u_0(x) \le v_0(x)$ in Ω, then $u(x,t) \le v(x,t)$ a.e. in Q_T.
(ii) (Uniqueness) The problem (1) - (3) on $[0,+\infty)$ possesses a unique solution.

Proof. Set $M = \max(\|u\|_{L^\infty(Q_T)}, \|v\|_{L^\infty(Q_T)})$.
Let $\mu_0(M)$ be the constant determined by (A.3).
For any $\lambda \ge \mu_0(M)$ and any $t \in [0,T]$, we can show that

$$(3.1) \qquad \int_\Omega (u(x,t) - v(x,t))_+ dx \le e^{-\lambda t}\int_\Omega (u_0(x) - v_0(x))_+ dx +$$

$$\int\int_{Q_t} e^{\lambda(s-t)}[C(x,u) - C(x,v) + \lambda(u-v)]_+ dxds.$$

Here $f_+ = \max\{f, 0\}$. For proof of (3.1), we refer to the proof of Theorem 1 in [8]. From the inequality $|C(x,u) - C(x,v)| \leq K_M|u-v|$, by setting $\lambda = K_0 = \max(K_M, \mu_0(M))$ and $g(t) = e^{K_0 t}\int_\Omega (u(x,t) - v(x,t))_+ dx$, we can obtain

$$g(t) \leq g(0) + 2K_0 \int_0^t g(s)ds \quad (0 \leq t \leq T).$$

Thus we have

$$0 \leq g(t) \leq g(0)e^{2K_0 t} \quad (0 \leq t \leq T).$$

4. Existence

From on now we assume that the functions $B_k(x,t,u)$ are independent of t ,that is , $B_k(x,t,u) = B_k(x,u)$. We consider the following problem $(P)_0$:

(4.1) $$u_t = \Delta_x(A(u)) + \sum_{k=1}^{N} \frac{dB_k(x,u)}{dx_k} + C(x,u) \quad \text{in} \quad Q = \Omega \times R^+,$$

(4.2) $$u(x,t) = 0 \quad \text{on} \quad \Gamma = \partial\Omega \times R^+,$$

(4.3) $$u(x,0) = u_0(x) \quad \text{in} \quad \Omega.$$

To prove the existence and decay of solutions of the problem (4.1) - (4.3), we suppose the additional assumptions (A.7) and (A.8).

(A.7) The function $A(u)$ satisfies the following additional assumption: there exist positive constants $C_0 > 0$ and $m > 1$ such that

$$|A(u)| \leq C_0|u|^m \quad (u \in R),$$

$$(C_0)^{-1}|u|^{m-1} \leq a(u) \leq C_0|u|^{m-1} \quad (u \in R).$$

(A.8) The functions $B_k(x,u) (k = 1, 2, .., N)$ satisfy the following assumptions:

(I) $$\{\sum_{k=1}^{N} \frac{\partial B_k(x,u)}{\partial x_k}\} \times sign(u) \leq 0 \quad (x \in \Omega, u \in R)$$

and

(II) there exist nonnegative constants $C_{k_0}, C_{k_1} \geq 0$ and $n_k > 1$ such that

$$|\frac{\partial B_k(x,u)}{\partial x_k}| \leq C_{k_0}|u|^{n_k} \quad (x \in \Omega, u \in R),$$

$$|\frac{\partial B_k(x,u)}{\partial u}| \leq C_{k_1}|u|^{n_k - 1} \quad (x \in \Omega, u \in R).$$

$$(A.9) \qquad A(u_0(x)) \in C_0^1(\bar{\Omega}).$$

Theorem 2 (Existence) Suppose that (A.1),(A.2),(A.4),(A.8) and (A.9) are satisfied. Then the problem (4.1) - (4.3) has a solution on $[0,+\infty)$ such that

$$(i) \qquad |u(x,t)| \leq M \quad \text{on} \quad \bar{Q} = \bar{\Omega} \times [0,+\infty),$$

$$(ii) \qquad \|\nabla A(u(t))\|_{L^2(\Omega)} \leq c_0(u_0) \quad \text{on} \quad [0,+\infty),$$

$$(iii) \qquad \|(A(u(t))_t\|^2_{L^2(0,T:L^2(\Omega))} \leq (M + \sup_{|s|\leq M} a(s))c_0(u_0).$$

Here $M = \sup_{x\in\Omega} |u_0(x)|$, and $c_0(u_0)$ is a constant depending only on the data.

Proof. For brevity, we assume that $u_0(x)$ is of class $C_0^\infty(\Omega)$.

Let $0 < \epsilon < 1$ and let us consider the following regularized problem:

$$(4.4) \qquad u_t = \Delta_x(\epsilon u + A(u)) + \sum_{k=1}^{N} \frac{dB_k(x,u)}{dx_k} + C(x,u) \quad \text{in} \quad \mathrm{Q} = \Omega \times R^+,$$

$$(4.5) \qquad u(x,t) = 0 \quad \text{on} \quad \Gamma = \partial\Omega \times R^+,$$

$$(4.6) \qquad u(x,0) = u_0(x) \quad \text{in} \quad \Omega.$$

By the standard theory, for each $T \in R^+$ this regularized problem (4.4) - (4.6) has a unique classical solution $u_\epsilon(x,t) \in C^{2,1}(\overline{Q_T})$. We can show the following estimates :

$$(4.7) \qquad |u_\epsilon(x,t)| \leq M \quad \text{on} \quad \bar{Q} = \bar{\Omega} \times [0,+\infty),$$

$$(4.8) \qquad \|\nabla A(u_\epsilon(t))\|_{L^2(\Omega)} \leq c_0(u_0) \quad \text{on} \quad [0,+\infty),$$

$$(4.9) \qquad \|(A(u_\epsilon(t))_t\|^2_{L^2(0,T:L^2(\Omega))} \leq (M + \sup_{|s|\leq M} a(s))c_0(u_0)).$$

We set $v_\epsilon(x,t)$=$A(u_\epsilon(x,t))$. By Ascoli-Arzela's theorem,we can show that there exist a function $v(x,t)$ and a subsequence $\{v_{\epsilon_n}(x,t)\}$ of $\{v_\epsilon(x,t)\}$ such that

$$v_{\epsilon_n}(x,t) \to v(x,t) \quad \text{as} \quad \mathrm{n} \to \infty \quad \text{in} \quad \mathrm{C}([0,T] : \mathrm{L}^2(\Omega)).$$

Put $u(x,t) = A^{-1}(v(x,t))$. Then we can show that $u(x,t)$ is a solution of the problem $(P)_0$.

5. Decay of solution

Theorem 3 (Decay estimates)
Suppose that (A.1),(A.2),(A.4),(A.7),(A.8) and (A.9) are satisfied.
Let u be a solution of the Problem (4.1) - (4.3) on $[0,+\infty)$. Then we have :

$(i)(L^{m+1} - estimate)$ $$\|u(t)\|_{m+1} \le (a_1 t + a_2)^{-\gamma} \qquad (t \in R^+),$$

and

$(ii)(L^{\infty} - estimate)$ $$|u(x,t)| \le C(a_1 t + a_2)^{-\gamma} \quad ((x,t) \in \Omega \times R^+)$$

where $\gamma = 1/(m-1), a_2 = \|u_0\|_{m+1}^{-m+1}$ and $a_1 = (m-1)(C_0)^{-1} S_0^{-2}(mes\Omega)^{(1-m)/(1+m)}$, and C is a constant depending only on the data.

Proof. Let $u_\epsilon(x,t)$ be a solution of the regularized problem (4.4)-(4.6). We denote $u_\epsilon(x,t)$ by $u(x,t)$. Multiply (4.4) by $|u|^{n-1}u \quad (m \le n)$ and integrate in Ω.
By Sobolev's lemma and Hölder's inequality, we have

$$(5.1) \qquad \frac{d}{dt}\{\frac{1}{n+1}\int_\Omega |u|^{n-1}u^2 dx\}$$

$$\le -\frac{4mn}{(n+m)^2}(C_0)^{-1}S_0^{-2}(mes\Omega)^{(1-m)/(1+n)}\{\int_\Omega |u|^{n-1}u^2 dx\}^{(n+m)/(n+1)},$$

where S_0 is a constant of Sobolev inequality. Putting $n = m$ in (5.1), we can show (i).
By the change of variable $\tau = log(a_1 t + a_2)$ and $v(x,\tau) = u(x,t)(a_1 t + a_2)^\gamma$, we can obtain

$$(5.2) \qquad \frac{d}{d\tau}\{\frac{1}{n+1}\int_\Omega |v|^{n-1}v^2 dx\}$$

$$\le -\frac{4mn}{(n+m)^2}(a_1)^{-1}\int_\Omega |\nabla\{|v|^{(n+m-2)/2}v\}|^2 dx + \frac{1}{m-1}\int_\Omega |v|^{n+1} dx$$

for every $n = m + 2^k - 2$ (k=1,2,...).
By Alikakos's argument(Cf.Lemma 3.2 in [1]), we have

$$|v(x,\tau)| \le C \quad \text{in} \quad \Omega \times [0,+\infty).$$

Here the positive constant C depends only on the data.

6. Positivity of solution

We investigate the positivity of solutions of the problem $(P)_0$ on $[0, +\infty)$. At first we consider the following problem with non-negative initial value:

$$u_t = \Delta_x(u^m) + \sum_{k=1}^{N} \frac{dB_k(x,u)}{dx_k} \quad \text{in} \quad Q = \Omega \times R^+, \tag{6.1}$$

$$u(x,t) = 0 \quad \text{on} \quad \Gamma = \partial\Omega \times R^+, \tag{6.2}$$

$$u(x,0) = u_0(x) \geq 0 \quad \text{in} \quad \Omega. \tag{6.3}$$

We assume that

(A.10) $u_0(x)$ is a non-negative continuous function defined on $\bar{\Omega}$ such that the support of $u_0(x)$ is strictly contained in Ω, $u_0(x) \not\equiv 0$ in Ω and $u_0^m \in C^1(\bar{\Omega})$.

Concerning the functions $B_k(x,u)$ of (6.1) ($k = 1, 2, .., N$), we assume that

(A.11) there exist positive constants b_0 and s_0 such that $s_0 < 1$ and for any $(x,s) \in \Omega \times [0, s_0]$

$$|\sum_{k=1}^{N} \frac{\partial B_k(x,s)}{\partial s}| \leq b_0 s^{m-1}$$

and

$$-b_0 s^m \leq \sum_{k=1}^{N} \frac{\partial B_k(x,s)}{\partial x_k} \leq 0$$

Lemma. Suppose that $m > 1$, (A.2), (A.8), (A.10) and (A.11) are satisfied. Let $u(x,t)$ be a solution of the problem (6.1) - (6.3).If $u_0(x_0) > 0$, then $u(x_0,t) > 0$ for any $t > 0$.

Proof. Without generality we can assume x_0=0.

Consider the two-parameter family of functions given by

$$V(x,t;\rho,\tau) = \{\frac{1}{4Nm}(t+\tau)^{-1}(\rho^2 - r^2)_+\}^\gamma$$

where $f_+ = \max\{f, 0\}$, $\gamma = 1/(m-1)$, $r = |x|$, ρ and τ are positive constants. By choosing large enough τ and small enough ρ, we can prove that

$$\Delta_x(V^m) + \sum_{k=1}^{N} \frac{dB_k(x,V)}{dx_k} - V_t \geq 0 \quad \text{in} \quad \{(x,t)| \quad |x| < \rho, 0 \leq t\},$$

$$V(x,0;\rho,\tau) \leq u(x,0) \quad \text{in} \quad \{x| \quad |x| \leq \rho\}$$

Hence we can conclude that $0 < V(0,t;\rho,\tau) \leq u(0,t)$ ($t \in (0, +\infty)$).

Theorem 4. Suppose that $m > 1$, (A.2), (A.8), (A.10) and (A.11) are satisfied. Let $u(x,t)$ be a solution of the problem (6.1) - (6.3).Then we have
(i) there exist positive constants K_1 and K_2 depending only on the data such that

$$0 \leq u(x,t) \leq K_1(t+K_2)^{-1/(m-1)} \quad \text{in} \quad \Omega \times [0,+\infty),$$

and (ii) there exists a time $T > 0$ depending only on the data such that for any $t \geq T$

$$\{x \in \Omega : u(x,t) > 0\} = \Omega \quad \text{holds.}$$

Proof. (i) is obvious by Theorem 3 -(ii).
Consider the two-parameter family of functions given by

$$Z(x,t;\rho,\tau) = \{[\frac{m-1}{m}(t+\tau)^{-1}(\rho^2 - |x|^2 \log^{-1}(t+\tau)]_+\}^{\gamma},$$

where $1 < \tau, 0 < \rho < 1$ and $\gamma = 1/(m-1)$.

Choosing large enough τ and small enough ρ, the function Z satisfies

$$\Delta_x(Z^m) + \sum_{k=1}^{N} \frac{dB_k(x,Z)}{dx_k} - Z_t \geq 0 \quad \text{in} \quad \mathrm{D}_{\rho,\tau},$$

where $\mathrm{D}_{\rho,\tau} = \{(x,t) | \quad |x| < \rho[\log(t+\tau)]^{1/2}\}$.

By the comparison principle of Theorem 1, Lemma and Aronson and Peletier's argument of [3], we can verify (ii).

Finally we discuss the following problem with non-negative initial value:

(6.4) $$u_t = \Delta_x(u^m) + \sum_{k=1}^{N} \frac{d(b(x)u^n)}{dx_k} - c(x)u^p \quad in \quad Q = \Omega \times R^+,$$

(6.5) $$u(x,t) = 0 \qquad on \quad \Gamma = \partial\Omega \times R^+,$$

(6.6) $$u(x,0) = u_0(x) \geq 0 \qquad in \quad \Omega.$$

Here we assume that
(A.12) $b(x) \in C^1(\bar{\Omega}), c(x) \in C(\bar{\Omega}), n, m$ and p in the equation (6.4) satisfy the following conditions :

$$m > 1, \quad m \leq p \qquad \text{and} \quad m \leq n,$$

$$0 \leq c(x) \leq 1 \quad \text{in} \quad \Omega,$$

and there exists a non-negative constant b_0 such that

$$-b_0 \leq \sum_{k=1}^{N} \frac{\partial b(x)}{\partial x_k} \leq 0 \quad \text{for} \quad \text{any} \quad x \in \Omega.$$

Theorem 5. Suppose that the assumptions (A.10) and (A.12) are saisfied.
Let $u(x,t)$ be a solution of the problem (6.4) - (6.6).Then we have
(i) there exist positive constants K_1 and K_2 depending only on the data such that

$$0 \leq u(x,t) \leq K_1(t+K_2)^{-1/(m-1)} \quad \text{in} \quad \Omega \times [0,+\infty),$$

and (ii) there exists a positive time T^* depending only on the data such that if $T^* \leq t$

$$\{x \in \Omega : u(x,t) > 0\} = \Omega.$$

Proof. From Theorem 3 -(ii) ,(i) is obvious.
For proof of (ii) of Theorem 5 ,we refer to the proof of Proposition 3 in [8].
Remark .From Kalashinikov's example([6]) it follows that if m, n and p do not satisfy (A.12), the spread of disturbances may remain localized.

References

1. N.D.Alikakos,L^p bounds of solutions of reaction-diffusion equations, *Comm. Partial Diff.Eq.*, 4,827-868, (1979).

2. D.G,Aronson,M.G.Crandall and L.A.Peletier,Stabilization of solutions of a degenerate nonlinear diffusion problem,*Nonlinear Analysis TMA*, 6,1001-1022,(1982).

3. D.G.Aronson and L.A.Peletier,Large time behaviour of solutions of the porous medium equation in bounded domains,*Jour.Diff.Eq.*, 39,378-412,(1981).

4. M.Bertsch ,T.Nanbu and L.A.Peletier,Decay of solutions of a degenerate nonlinear diffusion equation ,*Nonlinear Analysis TMA*, 6,539-554,(1982).

5. Y.Ebihara and T.Nanbu,Global classical solutions to $u_t - \Delta(u^{2m+1}) + \lambda u = f$, *Jour.Diff.Eq.*, 38,260-277,(1980).

6. A.S.Kalashinikov,On a nonlinear equation appearing in the theory of nonstationary filtration,*Trudy Seminar I.G.Petorvskii*,135-144,(1978).

7. A.S.Kalashinikov,Some problems of the qualitative theory of non-linear degenerate second-order parabolic equations,*Russian Math.Surveys* 42,169-222,(1987).

8. T.Nanbu,Some degenerate nonlinear parabolic equations, *Math.Rep.Coll.Gen.Ed. Kyusyu Univ.* 14,91-110,(1984).

4th Int. Coll. on Differential Equations, pp. 215-228
D. Bainov and V. Covachev (Eds)

EXISTENCE THEORY FOR SINGULAR TWO POINT BOUNDARY VALUE PROBLEMS

Donal O'Regan
Department Of Mathematics, University College Galway
Galway, Ireland.

Abstract. Existence results are presented for $\frac{1}{p}(py')'+\mu qy = f(t,y,py')$ a.e. on $[0,1]$ with y satisfying Sturm Liouville, Neumann, Periodic or mixed boundary data. Here $\mu < \lambda_0$ where λ_0 is the first eigenvalue of $\frac{1}{pq}(pu')' + \lambda u = 0$ a.e. on $[0,1]$ with u satisfying Sturm Liouville, Neumann, Periodic or mixed boundary data.

1. Introduction.

This paper establishes existence results for second order boundary value problems of the form

$$\frac{1}{p(t)}(p(t)y'(t))' + \mu q(t)y(t) = f(t, y(t), p(t)y'(t)) \text{ a.e. on } [0,1] \tag{1.1}$$

with y satisfying either
(i). (Sturm Liouville)

$$(SL) \qquad \begin{cases} -\alpha y(0) + \beta \lim_{t\to 0^+} p(t)y'(t) = 0, \ \alpha > 0, \beta \geq 0 \\ ay(1) + b\lim_{t\to 1^-} p(t)y'(t) = 0, \ a > 0, b \geq 0 \end{cases}$$

(ii). (Neumann)

$$(N) \qquad \lim_{t\to 0^+} p(t)y'(t) = \lim_{t\to 1^-} p(t)y'(t) = 0$$

(iii). (Periodic)

$$(P) \qquad y(0) = y(1), \ \lim_{t\to 0^+} p(t)y'(t) = \lim_{t\to 1^-} p(t)y'(t)$$

or
(iv). (Mixed)

$$(M) \qquad \begin{cases} \lim_{t\to 0^+} p(t)y'(t) = 0 \\ ay(1) + b\lim_{t\to 1^-} p(t)y'(t) = 0, \ a > 0, b \geq 0. \end{cases}$$

Remark. If a function $u \in C[0,1] \cap C^1(0,1)$ with $pu' \in C[0,1]$ satisfies boundary condition (i) we write $u \in (SL)$. A similar remark applies for the other boundary conditions.

Throughout the paper $p \in C[0,1] \cap C^1(0,1)$ together with $p > 0$ on $(0,1)$. Also $pf : [0,1] \times \mathbf{R}^2 \to \mathbf{R}$ is an L^1-Carathéodory function. By this we mean:

Definition 1.1. (i). $t \to p(t)f(t,u,v)$ is measurable for all $(u,v) \in \mathbf{R}^2$
(ii). $(u,v) \to p(t)f(t,u,v)$ is continuous for a.e. $t \in [0,1]$
(iii). for any $r > 0$ there exists $h_r \in L^1[0,1]$ such that $|p(t)f(t,u,v)| \le h_r(t)$ for a.e. $t \in [0,1]$ and for all $|u| \le r$, $|v| \le r$.

For notational purposes let w be a weight function. By $L^1_w[0,1]$ we mean the space of functions u such that $\int_0^1 w(t)|u(t)|\,dt < \infty$. $L^2_w[0,1]$ denotes the space of functions u such that $\int_0^1 w(t)|u(t)|^2\,dt < \infty$; also for $u, v \in L^2_w[0,1]$ define $\langle u,v\rangle = \int_0^1 w(t)u(t)\overline{v(t)}\,dt$. Let $AC[0,1]$ be the space of functions which are absolutely continuous on $[0,1]$.

We next state an existence principle [14], which was established using fixed point methods, for the second order nonresonant boundary value problem

$$(1.2) \qquad \begin{cases} \frac{1}{p}(py')' + \mu q y = f(t,y,py') \text{ a.e. on } [0,1] \\ y \in (SL), (N) \text{ or } (P). \end{cases}$$

By a solution to (1.2) we mean a function $y \in C[0,1] \cap C^1(0,1)$ with $py' \in AC[0,1]$ which satisfies the differential equation in (1.2) a.e. on $[0,1]$ and the stated boundary conditions.

Theorem 1.1. *Let $pf : [0,1] \times \mathbf{R}^2 \to \mathbf{R}$ be an L^1-Carathéodory function with*

$$(1.3) \qquad p \in C[0,1] \cap C^1(0,1) \text{ with } p > 0 \text{ on } (0,1) \text{ and } \int_0^1 \frac{ds}{p(s)} < \infty$$

and

$$(1.4) \qquad q \in L^1_p[0,1] \text{ with } q > 0 \text{ on } (0,1)$$

holding. In addition assume

$$(1.5) \qquad \begin{cases} \frac{1}{p}(py')' + \mu q y = 0 \text{ a.e. on } [0,1] \\ y \in (SL), (N) \text{ or } (P) \end{cases}$$

has only the trivial solution. Also suppose there is a constant M, independent of λ, with

$$\|y\|_* = \max\{\sup_{[0,1]} |y(t)|, \sup_{(0,1)} |p(t)y'(t)|\} \le M$$

for any solution y to

$$(1.6)_\lambda \qquad \begin{cases} \frac{1}{p}(py')' + \mu q y = \lambda f(t,y,py') \text{ a.e. on } [0,1] \\ y \in (SL), (N) \text{ or } (P) \end{cases}$$

for each $\lambda \in (0,1)$. Then (1.2) has at least one solution $y \in C[0,1] \cap C^1(0,1)$ with $py' \in AC[0,1]$.

The following existence principle [5,6,12] for the mixed boundary value problem

$$(1.7) \qquad \begin{cases} \frac{1}{p}(py')' + \mu q y = f(t,y,py') \text{ a.e. on } [0,1] \\ y \in (M) \end{cases}$$

will be used in section 2.

Theorem 1.2. *Let $pf : [0,1] \times \mathbf{R}^2 \to \mathbf{R}$ be an L^1-Carathéodory function with (1.4) and*

$$(1.8) \qquad p \in C[0,1] \cap C^1(0,1) \text{ with } p > 0 \text{ on } (0,1)$$

and

$$(1.9)\quad \begin{cases} \int_0^1 \frac{1}{p(s)} \int_0^s p(x)q(x)\,dx\,ds < \infty \text{ and } \int_0^1 \frac{1}{p(s)} \int_0^s h_r(x)\,dx\,ds < \infty \text{ for any } r > 0; \\ \text{here } h_r \text{ is as described in Definition 1.1} \end{cases}$$

holding. In addition assume there is a constant M, independent of λ, with $\|y\|_\star \le M$ for any solution y to

$$(1.10)_\lambda \quad \begin{cases} \frac{1}{p}(py')' = \lambda[f(t,y,py') - \mu q y] \quad a.e. \text{ on } [0,1] \\ y \in (M) \end{cases}$$

for each $\lambda \in (0,1)$. Then (1.7) *has at least one solution $y \in C[0,1] \cap C^1(0,1)$ with $py' \in AC[0,1]$.*

Next we gather together some results on the Sturm Liouville eigenvalue problem

$$(1.11)\quad \begin{cases} Lu = \lambda u \quad \text{a.e. on } [0,1] \\ u \in (SL), (N) \text{ or } (P) \end{cases}$$

where $Lu = -\frac{1}{pq}(pu')'$. Assume (1.3) and (1.4) hold. Let

$$D(L) = \{w \in C[0,1] : w, pw' \in AC[0,1] \text{ with } w \in (SL), (N) \text{ or } (P)\}.$$

Then L has an countably infinite number [1,9,10,15] of real eigenvalues λ_i with corresponding eigenfunctions $\psi_i \in D(L)$. The eigenfunctions ψ_i may be chosen so that they form a orthonormal set and we may also arrange the eigenvalues so that

$$(1.12)\quad \lambda_0 < \lambda_1 < \lambda_2 < \ldots\ldots .$$

In addition [15] the set of eigenfunctions ψ_i form a basis for $L^2_{pq}[0,1]$ and if $h \in L^2_{pq}[0,1]$ then h has a fourier series representation and h satisfyies Parseval's equality i.e.

$$(1.13)\quad h = \sum_{i=0}^{\infty} \langle h, \psi_i \rangle \psi_i \quad \text{and} \quad \int_0^1 pq|h|^2\,dt = \sum_{i=0}^{\infty} |\langle h, \psi_i \rangle|^2 .$$

In addition we have a Rayleigh-Ritz minimisation theorem:

Theorem 1.3. *Suppose* (1.3) *and* (1.4) *hold. Then*

$$(1.14)\quad \lambda_0 \left(\int_0^1 pqy^2\,dt \right) \le \int_0^1 p(y')^2\,dt + \frac{a}{b}[y(1)]^2 + \frac{\alpha}{\beta}[y(0)]^2$$

for all functions $y \in D^\star(L) = \{w \in C[0,1] : w, pw' \in AC[0,1] \text{ with } w \in (SL)\}$.

Remark. If $b = 0$ then the term $\frac{a}{b}[y(1)]^2$ will not appear in (1.14).

PROOF: First notice for $u \in D(L)$ that

$$\langle Lu, u \rangle = \int_0^1 p(u')^2\,dt + \frac{a}{b}[u(1)]^2 + \frac{\alpha}{\beta}[u(0)]^2.$$

From (1.13) any $u \in D(L)$ has a Fourier series expansion and so

$$\langle Lu, u \rangle = \sum_{i=0}^{\infty} \lambda_i |\langle u, \psi_i \rangle|^2 \ge \lambda_0 \sum_{i=0}^{\infty} |\langle u, \psi_i \rangle|^2 = \lambda_0 \|u\|^2.$$

Consequently $\langle Lu, u \rangle \ge \lambda_0 \|u\|^2$ with equality if $u = \psi_0$. □

Finally consider the singular eigenvalue problem

$$(1.15)\qquad \begin{cases} Lu = \lambda u \text{ a.e. on } [0,1] \\ u \in (M) \end{cases}$$

where $Lu = -\frac{1}{pq}(pu')'$. Assume (1.4), (1.8) and

$$(1.16)\qquad \int_0^1 \frac{1}{p(s)} \left(\int_0^s p(x)q(x)\,dx \right)^{\frac{1}{2}} ds < \infty$$

hold.

Remarks. (i). Notice for example that $\int_{\frac{1}{2}}^1 \frac{ds}{p(s)} < \infty$ since

$$\int_{\frac{1}{2}}^1 \frac{ds}{p(s)} = \int_{\frac{1}{2}}^1 \frac{\int_0^s p(x)q(x)\,dx}{p(s)\int_0^s p(x)q(x)\,dx}\,ds \le \frac{1}{\int_0^{\frac{1}{2}} p(x)q(x)\,dx} \int_{\frac{1}{2}}^1 \frac{1}{p(s)} \int_0^s p(x)q(x)\,dx\,ds.$$

(ii). Notice [12] that (1.16) implies $\int_0^1 p(s)q(s)\left(\int_s^1 \frac{dx}{p(x)}\right)^2 ds < \infty$.

(iii). Now $t = 0$ is a singular point in the limit circle case [15]. To see this we apply Weyl's theorem so all we need show is that all solutions of $Ly = 0$ a.e. on $[0,1]$ lie in $L^2_{pq}[0,1]$. Of course any solution of $Ly = 0$ is of the form

$$y(t) = A + B\int_t^1 \frac{ds}{p(s)}$$

where A and B are constants so (1.16) together with the above remark implies that we have the limit circle case.

Let

$$D_\star(L) = \{w \in C[0,1] : w, pw' \in AC[0,1] \text{ with } \frac{1}{pq}(pw')' \in L^2_{pq}[0,1] \text{ and } w \in (M)\}.$$

Note $L : D_\star(L) \to L^2_{pq}[0,1]$ and for any $h \in L^2_{pq}[0,1]$ the boundary value problem

$$\begin{cases} Ly = h \text{ a.e. on } [0,1] \\ y \in (M) \end{cases}$$

has the unique solution

$$L^{-1}h(t) = y(t) = \frac{b}{a}\int_0^1 p(x)q(x)h(x)\,dx + \int_t^1 \frac{1}{p(s)} \int_0^s p(x)q(x)h(x)\,dx\,ds.$$

Notice $L^{-1} : L^2_{pq}[0,1] \to D_\star(L)$ and it also follows easily [12] that L^{-1} is completely continuous with $\langle L^{-1}u, v\rangle = \langle u, L^{-1}v\rangle$ for $u, v \in L^2_{pq}[0,1]$. Consequently the spectral theorem for compact self adjoint operators [15] implies L has an infinite number of real eigenvalues λ_i (which can be arranged as in (1.12)) with corresponding eigenfunctions $\psi_i \in D_\star(L)$. Now $\{\psi_i\}$ form a basis for $L^2_{pq}[0,1]$ and (1.13) also holds in this case. Essentially the same reasoning as in theorem 1.3 establishes:

Theorem 1.4. *Suppose* (1.4), (1.8) *and* (1.16) *hold. Then*

$$(1.17)\qquad \lambda_0\left(\int_0^1 pqy^2\,dt\right) \le \int_0^1 p(y')^2\,dt + \frac{a}{b}[y(1)]^2$$

for all functions $y \in D_\star(L)$.

2. Existence.

Existence theory is developed for the second order boundary value problem

$$\text{(2.1)} \qquad \begin{cases} \frac{1}{p}(py')' + \mu q y = f(t, y, py') \text{ a.e. on } [0,1] \\ y \in (SL) \end{cases}$$

where $\mu < \lambda_0$; λ_0 (see (1.12)) is the first eigenvalue of

$$\text{(2.2)} \qquad \begin{cases} Lu = \lambda u \text{ a.e. on } [0,1] \\ u \in (SL) \end{cases}$$

where $Lu = -\frac{1}{pq}(pu')'$.

Theorem 2.1. *Let $pf : [0,1] \times \mathbf{R}^2 \to \mathbf{R}$ be an L^1-Carathéodory function with*

$$\text{(2.3)} \qquad q \in L^1_p[0,1] \text{ and } q > 0 \text{ on } (0,1)$$

and

$$\text{(2.4)} \qquad p \in C[0,1] \cap C^1(0,1) \text{ with } p > 0 \text{ on } (0,1) \text{ and } \int_0^1 \frac{ds}{p(s)} < \infty$$

holding. In addition assume f has the decomposition $f(t,u,v) = g(t,u,v) + h(t,u,v)$ with $pg, ph : [0,1] \times \mathbf{R}^2 \to \mathbf{R}$ L^1-Carathéodory and

$$\text{(2.5)} \qquad \begin{cases} |h(t,u,v)| \le \phi_1(t) + \phi_2(t)|u|^\gamma + \phi_3(t)|v|^\theta \text{ a.e. } t \in [0,1] \text{ for} \\ \text{constants } \gamma, \theta \text{ with } 0 \le \gamma, \theta < 1 \end{cases}$$

$$\text{(2.6)} \qquad \begin{cases} ug(t,u,v) \ge c_0 \phi_4(t)|uv| \text{ for a.e. } t \in [0,1] \text{ and } u, v \in \mathbf{R}; \\ \text{here } \phi_4 > 0 \text{ a.e. on } [0,1]. \text{ Also } |g(t,u,v)| \le \phi_5(t) + \phi_6(t)|v|^\omega \\ \text{for } u \text{ in a bounded set; here } \omega \text{ is a constant with } 0 \le \omega < 2 \text{ and} \\ \phi_5, \phi_6 \text{ may depend on the bound of the bounded set} \end{cases}$$

$$\text{(2.7)} \qquad \begin{cases} \phi_1, \phi_2, \phi_5 \in L^1_p[0,1] \text{ with } \int_0^1 (p(t))^{\frac{2+\theta}{2-\theta}} (\phi_3(t))^{\frac{2}{2-\theta}}\, dt < \infty \\ \text{and } \int_0^1 (p(t))^{\frac{2+\omega}{2-\omega}} (\phi_6(t))^{\frac{2}{2-\omega}}\, dt < \infty \end{cases}$$

are satisfied. Then (2.1) has a solution $y \in C[0,1] \cap C^1(0,1)$ with $py' \in AC[0,1]$ in each of the following cases:

(i). $c_0 \ge 0$

(ii). $c_0 < 0$ *and* $\eta_0 + \frac{c_0 Q_0}{\sqrt{\lambda_0}} > 0$ *where in this case we assume* $Q_0 = \sup_{[0,1]} \frac{|\phi_4(t)| p(t)}{\sqrt{q(t)}} < \infty$; *here* $\eta_0 = 1$ *if* $\mu \le 0$ *and* $\eta_0 = 1 - \frac{\mu}{\lambda_0}$ *if* $\mu > 0$.

PROOF: Let y be a solution to

$$\text{(2.8)}_\lambda \qquad \begin{cases} \frac{1}{p}(py')' + \mu q y = \lambda f(t, y, py') \text{ a.e. on } [0,1],\ 0 < \lambda < 1 \\ y \in (SL). \end{cases}$$

Remark. For notational purposes let $\|y'\| = \left(\int_0^1 p|y'|^2\, dt\right)^{\frac{1}{2}}$.

Multiply the differential equation by $-y$ and integrate from 0 to 1 to obtain

$$\text{(2.9)} \qquad \frac{a}{b}[y(1)]^2 + \frac{\alpha}{\beta}[y(0)]^2 + \|y'\|^2 - \mu \int_0^1 pqy^2\, dt = -\lambda \int_0^1 py f(t, y, py')\, dt.$$

Remark. If $\beta = 0$ then the term $\frac{\alpha}{\beta}[y(0)]^2$ does not appear in (2.9).

Apply theorem 1.3 (if $\mu \geq 0$) to obtain

$$(2.10)\quad \eta_0 \left(\frac{a}{b}[y(1)]^2 + \frac{\alpha}{\beta}[y(0)]^2 + \|y'\|^2\right) \leq -\lambda \int_0^1 pyg(t,y,py')\,dt + \int_0^1 p|y||h(t,y,py')|\,dt$$

where

$$\eta_0 = \begin{cases} 1 - \frac{\mu}{\lambda_0} & \text{if } \mu > 0 \\ 1 & \text{if } \mu \leq 0. \end{cases}$$

In addition assumption (2.5) implies

$$(2.11)\quad \int_0^1 p|y||h(t,y,py')|\,dt \leq \int_0^1 p\phi_1|y|\,dt + \int_0^1 p\phi_2|y|^{\gamma+1}\,dt + \int_0^1 p\phi_3|py'|^{\theta}|y|\,dt.$$

For $t \in [0,1]$, Hölder's inequality implies

$$(2.12)\quad |y(t)| \leq |y(0)| + \int_0^t |y'(s)|\,ds \leq |y(0)| + \|y'\| \left(\int_0^1 \frac{ds}{p(s)}\right)^{\frac{1}{2}}.$$

Let $|y|_0 = \sup_{[0,1]} |y(t)|$. Then with assumption (2.7) we have

$$\begin{aligned} \int_0^1 p\phi_1|y|\,dt &\leq |y|_0 \int_0^1 p\phi_1\,dt \leq E_0|y(0)| + E_1\|y'\| \\ \int_0^1 p\phi_2|y|^{\gamma+1}\,dt &\leq |y|_0^{\gamma+1} \int_0^1 p\phi_2\,dt \leq E_2|y(0)|^{\gamma+1} + E_3\|y'\|^{\gamma+1} \\ \int_0^1 p\phi_3|py'|^{\theta}|y|\,dt &\leq |y|_0 \int_0^1 p\phi_3|py'|^{\theta}\,dt \leq E_4|y|_0\|y'\|^{\theta} \\ &\leq E_4|y(0)|\|y'\|^{\theta} + E_5\|y'\|^{\theta+1} \leq (E_4+E_5)\|y'\|^{\theta+1} + E_4|y(0)|^{\theta+1} \end{aligned}$$

for some constants $E_0, \ldots, E_5$. Put these into (2.11) and then (2.10) to obtain

$$(2.13)\quad \begin{aligned} \eta_0 \left(\tfrac{a}{b}[y(1)]^2 + \tfrac{\alpha}{\beta}[y(0)]^2 + \|y'\|^2\right) \leq -\lambda \textstyle\int_0^1 pyg(t,y,py')\,dt + E_0|y(0)| \\ +E_1\|y'\| + E_2|y(0)|^{\gamma+1} + E_3\|y'\|^{\gamma+1} \\ +(E_4+E_5)\|y'\|^{\theta+1} + E_4|y(0)|^{\theta+1}. \end{aligned}$$

There are two cases to consider; $c_0 \geq 0$ and $c_0 < 0$.

Case (i). $c_0 \geq 0$.

Then $yg(t,y,py') \geq 0$ a.e. on $[0,1]$ so (2.13) implies

$$(2.14)\quad \begin{aligned} \eta_0 \left(\tfrac{a}{b}[y(1)]^2 + \tfrac{\alpha}{\beta}[y(0)]^2 + \|y'\|^2\right) \leq E_0|y(0)| + E_1\|y'\| + E_2|y(0)|^{\gamma+1} \\ +E_3\|y'\|^{\gamma+1} + (E_4+E_5)\|y'\|^{\theta+1} + E_4|y(0)|^{\theta+1}. \end{aligned}$$

Now note if $L \geq 0$, $0 \leq k < 2$ are given constants then there exists $N > 0$ such that $Lx^k \leq \frac{1}{2}x^2 + N$ for all $x > 0$. This together with (2.14) implies that there is a constant E_6 with

$$\frac{\eta_0}{2}\left(\frac{a}{b}[y(1)]^2 + \frac{\alpha}{\beta}[y(0)]^2 + \|y'\|^2\right) \leq E_6.$$

Consequently there exists a constant E_7 with

$$(2.15)\quad \|y'\| \leq E_7 \text{ and } |y(0)| \leq E_7$$

for any solution y to $(2.8)_\lambda$.

Remark. If $\beta = 0$ then $y(0) = 0$ so (2.15) is also true in this case.

Case (ii). $c_0 < 0$.

Then with $Q_0 = \sup_{[0,1]} \frac{|\phi_4(t)|p(t)}{\sqrt{q(t)}}$ we have using theorem 1.3

$$\begin{aligned} -\lambda \int_0^1 pyg(t,y,py')\,dt &\le (-c_0)\int_0^1 p\phi_4|ypy'|\,dt \le (-c_0)Q_0\|y'\|\left(\int_0^1 pqy^2\,dt\right)^{\frac{1}{2}} \\ &\le \frac{(-c_0)Q_0}{\sqrt{\lambda_0}}\|y'\|\left(\frac{a}{b}[y(1)]^2 + \frac{\alpha}{\beta}[y(0)]^2 + \|y'\|^2\right)^{\frac{1}{2}} \\ &\le \frac{(-c_0)Q_0}{\sqrt{\lambda_0}}\left(\frac{a}{b}[y(1)]^2 + \frac{\alpha}{\beta}[y(0)]^2 + \|y'\|^2\right). \end{aligned}$$

Put this into (2.13) to obtain

$$\begin{aligned} \left(\eta_0 + \tfrac{c_0Q_0}{\sqrt{\lambda_0}}\right)\left(\tfrac{a[y(1)]^2}{b} + \tfrac{\alpha[y(0)]^2}{\beta} + \|y'\|^2\right) \le{}& E_0|y(0)| + E_1\|y'\| + E_2|y(0)|^{\gamma+1} \\ &+ E_3\|y'\|^{\gamma+1} + (E_4+E_5)\|y'\|^{\theta+1} + E_4|y(0)|^{\theta+1} \end{aligned}$$

so there exists a constant E_6 with

$$\frac{1}{2}\left(\eta_0 + \frac{c_0Q_0}{\sqrt{\lambda_0}}\right)\left(\frac{a}{b}[y(1)]^2 + \frac{\alpha}{\beta}[y(0)]^2 + \|y'\|^2\right) \le E_6$$

and consequently (2.15) is satisfied.

Thus (2.15) is satisfied in both cases. This together with (2.12) implies that there exists a constant E_8 with

$$|y|_0 \le E_8 \tag{2.16}$$

for any solution y to $(2.8)_\lambda$.

This bound together with (2.6) and (2.7) implies that there is a constant E_9 with

$$|(py')'| \le p\phi_1 + E_9p\phi_2 + p\phi_3|py'|^\theta + p\phi_5 + p\phi_6|py'|^\omega.$$

Consequently

$$\begin{aligned} \int_0^1 |(py')'|\,dt &\le \int_0^1 [p\phi_1 + E_9p\phi_2 + p\phi_5]\,dt + \|y'\|^\theta\left(\int_0^1 (p(t))^{\frac{2+\theta}{2-\theta}}(\phi_3(t))^{\frac{2}{2-\theta}}\,dt\right)^{\frac{2-\theta}{2}} \\ &+ \|y'\|^\omega\left(\int_0^1 (p(t))^{\frac{2+\omega}{2-\omega}}(\phi_6(t))^{\frac{2}{2-\omega}}\,dt\right)^{\frac{2-\omega}{2}} \end{aligned}$$

and this together with (2.15) implies that there is a constant E_{10} with $\int_0^1 |(py')'|\,dt \le E_{10}$. Now there exists $t_0 \in (0,1)$ with $y'(t_0) = y(1) - y(0)$ and so (2.16) implies $|y'(t_0)| \le 2E_8$. Thus for $t \in (0,1)$ we have

$$|p(t)y'(t)| \le |p(t_0)y'(t_0)| + |\int_{t_0}^t |(py')'|\,ds| \le 2E_8 \sup_{[0,1]} p(t) + E_{10} \equiv E_{11}$$

so

$$\sup_{(0,1)} |p(t)y'(t)| \le E_{11}. \tag{2.17}$$

The result follows from (2.16), (2.17) and theorem 1.1. □

Remarks. (i). It is possible to let $\omega = 2$ in (2.6) if $p^2(t)\phi_6(t)$ is bounded for $t \in [0,1]$.
(ii). Of course theorem 2.1 is also true if we consider the mixed boundary value problem (1.7).

Next we consider

$$(2.18) \qquad \begin{cases} \frac{1}{p}(py')' + \mu q y = f(t,y,py') \text{ a.e. on } [0,1] \\ y \in (N) \text{ or } (P) \end{cases}$$

with $\mu < \lambda_0 = 0$; $\lambda_0 = 0$ (see (1.12)) is the first eigenvalue of

$$(2.19) \qquad \begin{cases} Lu = \lambda u \text{ a.e. on } [0,1] \\ u \in (N) \text{ or } (P) \end{cases}$$

where $Lu = -\frac{1}{pq}(pu')'$.

Theorem 2.2. *Let $pf : [0,1] \times \mathbf{R}^2 \to \mathbf{R}$ be an L^1-Carathéodory function with* (2.3), (2.4) *and*

$$(2.20) \qquad \int_0^1 \frac{ds}{p(s)q(s)} < \infty$$

holding. In addition assume f has the decomposition $f(t,u,v) = g(t,u,v) + h(t,u,v)$ with $pg, ph : [0,1] \times \mathbf{R}^2 \to \mathbf{R}$ L^1-Carathéodory and suppose (2.5), (2.6) *and* (2.7) *are satisfied. Then* (2.18) *has a solution $y \in C[0,1] \cap C^1(0,1)$ with $py' \in AC[0,1]$ in each of the following cases:*
(i). $c_0 \geq 0$
(ii). $c_0 < 0$ *and there exists an $\epsilon > 0$ with $1 + \frac{c_0 Q_0}{2\epsilon} > 0$ and $\frac{\epsilon c_0 Q_0}{2} - \mu > 0$; in this case we assume $Q_0 = \sup_{[0,1]} \frac{|\phi_4(t)| p(t)}{\sqrt{q(t)}} < \infty$.*

PROOF: Let y be a solution to

$$(2.21)_\lambda \qquad \begin{cases} \frac{1}{p}(py')' + \mu q y = \lambda f(t,y,py') \text{ a.e. on } [0,1],\ 0 < \lambda < 1 \\ y \in (N) \text{ or } (P). \end{cases}$$

Multiply the differential equation by $-y$ and integrate from 0 to 1 to obtain

$$(2.22) \qquad \|y'\|^2 + (-\mu)\int_0^1 pqy^2\,dt \leq -\lambda \int_0^1 pyg(t,y,py')\,dt + \int_0^1 p|y||h(t,y,py')|\,dt.$$

Sobolev's inequality (see remark (i) after proof) implies

$$\begin{aligned}
\int_0^1 p\phi_1|y|\,dt &\leq |y|_0 \int_0^1 p\phi_1\,dt \leq Q_1\left(\|y'\|^2 + \int_0^1 pqy^2\,dt\right)^{\frac{1}{2}} \\
\int_0^1 p\phi_2|y|^{\gamma+1}\,dt &\leq Q_2\left(\|y'\|^2 + \int_0^1 pqy^2\,dt\right)^{\frac{\gamma+1}{2}} \\
\int_0^1 p\phi_3|py'|^\theta|y|\,dt &\leq |y|_0\|y'\|^\theta\left(\int_0^1 (p(t))^{\frac{2+\theta}{2-\theta}}(\phi_3(t))^{\frac{2}{2-\theta}}\,dt\right)^{\frac{2-\theta}{2}} \\
&\leq Q_3\|y'\|^\theta\left(\|y'\|^2 + \int_0^1 pqy^2\,dt\right)^{\frac{1}{2}}
\end{aligned}$$

for some constants Q_1, Q_2 and Q_3. Put these into (2.22) to obtain

$$\begin{aligned}\|y'\|^2 + (-\mu)\textstyle\int_0^1 pqy^2\,dt &\leq -\lambda \textstyle\int_0^1 pyg(t,y,py')\,dt + Q_4\left(\|y'\| + \left(\int_0^1 pqy^2\,dt\right)^{\frac{1}{2}}\right)\\ &+Q_5\left(\|y'\|^{\gamma+1} + \left(\textstyle\int_0^1 pqy^2\,dt\right)^{\frac{\gamma+1}{2}}\right)\\ &+Q_6\left(\|y'\|^{\theta+1} + \left(\textstyle\int_0^1 pqy^2\,dt\right)^{\frac{\theta+1}{2}}\right)\end{aligned}\tag{2.23}$$

for some constants Q_4, Q_5 and Q_6.

Case (i). $c_0 \geq 0$.

Then as in theorem 2.1 there exists a constant Q_7 with

$$\frac{1}{2}\|y'\|^2 + \frac{1}{2}(-\mu)\int_0^1 pqy^2\,dt \leq Q_7.\tag{2.24}$$

Consequently there exists a constant Q_8 with

$$\|y'\| \leq Q_8 \text{ and } \int_0^1 pqy^2\,dt \leq Q_8.\tag{2.25}$$

Case (ii). $c_0 < 0$.

Then with Q_0 as described in the statement of the theorem, we have

$$\begin{aligned}-\lambda\int_0^1 pyg(t,y,py')\,dt &\leq (-c_0)Q_0\|y'\|\left(\int_0^1 pqy^2\,dt\right)^{\frac{1}{2}}\\ &\leq (-c_0)Q_0\left(\frac{\epsilon}{2}\int_0^1 pqy^2\,dt + \frac{1}{2\epsilon}\|y'\|^2\right)\end{aligned}$$

for any $\epsilon > 0$. Put this into (2.23) to obtain

$$\begin{aligned}\left(1+\frac{c_0Q_0}{2\epsilon}\right)\|y'\|^2 + \left(-\mu + \frac{\epsilon c_0 Q_0}{2}\right)\int_0^1 pqy^2\,dt &\leq Q_4\left(\|y'\| + \left(\int_0^1 pqy^2\,dt\right)^{\frac{1}{2}}\right)\\ &+ Q_5\left(\|y'\|^{\gamma+1} + \left(\int_0^1 pqy^2\,dt\right)^{\frac{\gamma+1}{2}}\right)\\ &+ Q_6\left(\|y'\|^{\theta+1} + \left(\int_0^1 pqy^2\,dt\right)^{\frac{\theta+1}{2}}\right)\end{aligned}$$

and (2.25) is again satisfied.

Now (2.25) implies that there exists $t_0 \in [0,1]$ with

$$|y(t_0)| \leq \sqrt{\frac{Q_8}{\int_0^1 p(s)q(s)\,ds}}.$$

Thus for $t \in [0,1]$ we have

$$|y(t)| \leq |y(t_0)| + \int_0^1 |y'(s)|\,ds \leq |y(t_0)| + \|y'\|\left(\int_0^1 \frac{ds}{p(s)}\right)^{\frac{1}{2}}$$

so there exists a constant Q_9 with

$$\sup_{[0,1]}|y(t)| \leq Q_9$$

for any solution y to $(2.21)_\lambda$. The existence of a constant Q_{10} with

$$\sup_{(0,1)} |p(t)y'(t)| \le Q_{10}$$

follows as in theorem 2.1. The result now follows from theorem 1.1. □

Remarks. (i). Let $y \in C[0,1] \cap C^1(0,1)$ with $py' \in AC[0,1]$ and suppose (2.3), (2.4) and (2.20) are satisfied. Then for any $x, r \in [0,1]$ we have that $y(x) = y(r) + \int_r^x y'(t)\,dt$ and this together with Hölder's inequality yields

$$|y(x)| \le |y(r)| + \|y'\| \left(\int_0^1 \frac{ds}{p(s)} \right)^{\frac{1}{2}}.$$

Fix $x \in [0,1]$ and integration with respect to r over $[0,1]$ yields

$$|y(x)| \le \int_0^1 |y(r)|\,dr + \|y'\| \left(\int_0^1 \frac{ds}{p(s)} \right)^{\frac{1}{2}} \le \left(\int_0^1 pqy^2\,dr \right)^{\frac{1}{2}} \left(\int_0^1 \frac{ds}{p(s)q(s)} \right)^{\frac{1}{2}} + \|y'\| \left(\int_0^1 \frac{ds}{p(s)} \right)^{\frac{1}{2}}.$$

Consequently there exists a constant N with

$$|y|_0^2 = \left(\sup_{[0,1]} |y(t)| \right)^2 \le N \left(\|y'\|^2 + \int_0^1 pqy^2\,dt \right)$$

which is the Sobolev inequality referred to in theorem 2.2.
(ii). If we have Periodic (or Dirichlet in theorem 2.1) boundary data and $p \equiv 1$, then a term of the form $k(y)y'$, where $k : \mathbf{R} \to \mathbf{R}$ is continuous say, could be included in the differential equation since $\int_0^1 yk(y)y'\,dt = 0$.

Finally we examine in more detail the mixed boundary value problem

$$(2.26) \qquad \begin{cases} \frac{1}{p}(py')' + \mu qy = f(t, y, py') \text{ a.e. on } [0,1] \\ y \in (M) \end{cases}$$

where $\mu < \lambda_0$; λ_0 is the first eigenvalue (see section 1) of

$$(2.27) \qquad \begin{cases} Lu = \lambda u \text{ a.e. on } [0,1] \\ u \in (M) \end{cases}$$

where $Lu = -\frac{1}{pq}(pu')'$.

Theorem 2.3. *Let $pf : [0,1] \times \mathbf{R}^2 \to \mathbf{R}$ be an L^1-Carathéodory function with* (2.3) *and*

$$(2.28) \qquad \begin{cases} p \in C[0,1] \cap C^1(0,1) \text{ with } p > 0 \text{ on } (0,1) \text{ and} \\ \int_0^1 \frac{1}{p(s)} \left(\int_0^s p(x)q(x)\,dx \right)^{\frac{1}{2}} ds < \infty \end{cases}$$

holding. In addition assume f has the decomposition $f(t,u,v) = g(t,u,v) + h(t,u,v)$ with $pg, ph : [0,1] \times \mathbf{R}^2 \to \mathbf{R}$ L^1-Carathéodory and suppose (2.5) *and*

$$(2.29) \qquad \begin{cases} ug(t,u,v) \ge c_0\phi_4(t)|uv| \text{ for a.e. } t \in [0,1] \text{ and } u, v \in \mathbf{R}. \text{ Here } \phi_4 > 0 \\ \text{a.e. on } [0,1]. \text{ Also } |g(t,u,v)| \le \phi_5(t) + \phi_6(t)|u|^\delta + \phi_7(t)|v|^\omega \\ \text{for a.e. } t \in [0,1] \text{ and } u, v \in \mathbf{R}; \text{ here } \delta \ge 1 \text{ and } 0 \le \omega < 2 \end{cases}$$

$$(2.30) \qquad \begin{cases} \phi_1, \phi_2, \phi_5, \phi_6 \in L^1_p[0,1] \textit{ with } \int_0^1 (p(t))^{\frac{2+\theta}{2-\theta}} (\phi_3(t))^{\frac{2}{2-\theta}}\, dt < \infty \\ \textit{and } \int_0^1 (p(t))^{\frac{2+\omega}{2-\omega}} (\phi_7(t))^{\frac{2}{2-\omega}}\, dt < \infty \end{cases}$$

$$(2.31) \qquad \begin{cases} \int_0^1 p(t)\phi_1(t) \left(\int_t^1 \frac{ds}{p(s)}\right)^{\frac{1}{2}} dt < \infty,\ \int_0^1 p(t)\phi_2(t) \left(\int_t^1 \frac{ds}{p(s)}\right)^{\frac{\gamma+1}{2}} dt < \infty \\ \textit{and } \int_0^1 (p(t))^{\frac{2+\theta}{2-\theta}} (\phi_3(t))^{\frac{2}{2-\theta}} \left(\int_t^1 \frac{ds}{p(s)}\right)^{\frac{1}{2-\theta}} dt < \infty \end{cases}$$

$$(2.32) \qquad \begin{cases} \int_0^1 \frac{1}{p(t)} \int_0^t p(s)\phi_i(s)\, ds\, dt < \infty,\ i = 1,2,3,5,6,7 \textit{ with} \\ \int_0^1 \frac{1}{p(t)} \int_0^t p(s)\phi_2(s) \left(\int_s^1 \frac{dz}{p(z)}\right)^{\frac{\gamma}{2}} ds\, dt < \infty,\ \int_0^1 p(s)\phi_6(s) \left(\int_s^1 \frac{dz}{p(z)}\right)^{\frac{\delta}{2}} ds\, dt < \infty, \\ \int_0^1 \frac{1}{p(t)} \int_0^t p(s)\phi_6(s) \left(\int_s^1 \frac{dz}{p(z)}\right)^{\frac{\delta}{2}} ds\, dt < \infty \textit{ and} \\ \int_0^1 \frac{1}{p(t)} \left(\int_0^t (p(s))^{\frac{2+\theta}{2-\theta}} (\phi_3(s))^{\frac{2}{2-\theta}}\, ds\right)^{\frac{2-\theta}{2}} dt < \infty, \\ \int_0^1 \frac{1}{p(t)} \left(\int_0^t (p(s))^{\frac{2+\omega}{2-\omega}} (\phi_7(s))^{\frac{2}{2-\omega}}\, ds\right)^{\frac{2-\omega}{2}} dt < \infty \end{cases}$$

are satisfied. Then (2.26) *has a solution* $y \in C[0,1] \cap C^1(0,1)$ *with* $py' \in AC[0,1]$ *in each of the following cases:*
(i). $c_0 \geq 0$
(ii). $c_0 < 0$ *and* $\eta_0 + \frac{c_0 Q_0}{\sqrt{\lambda_0}} > 0$ *where in this case we assume* $Q_0 = \sup_{[0,1]} \frac{|\phi_4(t)|p(t)}{\sqrt{q(t)}} < \infty$*; here* $\eta_0 = 1$ *if* $\mu \leq 0$ *and* $\eta_0 = 1 - \frac{\mu}{\lambda_0}$ *if* $\mu > 0$.
PROOF: Let y be a solution to

$$(2.33)_\lambda \qquad \begin{cases} \frac{1}{p}(py')' = \lambda[f(t,y,py') - \mu q y] \ \text{ a.e. on } [0,1],\ 0 < \lambda < 1 \\ y \in (M). \end{cases}$$

Multiply the differential equation by $-y$ and integrate from 0 to 1 to obtain

$$\|y'\|^2 + \frac{a}{b}[y(1)]^2 \leq -\lambda \int_0^1 pyg(t,y,py')\, dt + \int_0^1 p|y||h(t,y,py')|\, dt + \lambda\mu \int_0^1 pqy^2\, dt.$$

Apply theorem 1.4 to obtain

$$(2.34) \qquad \eta_0 \left(\|y'\|^2 + \frac{a}{b}[y(1)]^2\right) \leq -\lambda \int_0^1 pyg(t,y,py')\, dt + \int_0^1 p|y||h(t,y,py')|\, dt$$

where

$$\eta_0 = \begin{cases} 1 - \frac{\mu}{\lambda_0} & \text{if } \mu > 0 \\ 1 & \text{if } \mu \leq 0. \end{cases}$$

Also for $t \in (0,1)$ we have

$$(2.35) \qquad |y(t)| \leq |y(1)| + \|y'\| \left(\int_t^1 \frac{ds}{p(s)}\right)^{\frac{1}{2}}$$

so

$$\int_0^1 p\phi_1 |y|\, dt \leq N_1 |y(1)| + N_2 \|y'\|$$

where $N_1 = \int_0^1 p\phi_1\, dt$ and $N_2 = \int_0^1 p(t)\phi_1(t) \left(\int_t^1 \frac{ds}{p(s)}\right)^{\frac{1}{2}} dt$;

$$\int_0^1 p\phi_2 |y|^{\gamma+1}\, dt \leq N_3 |y(1)|^{\gamma+1} + N_4 \|y'\|^{\gamma+1}$$

where $N_3 = 2^{\gamma+1}\int_0^1 p\phi_2\,dt$ and $N_4 = 2^{\gamma+1}\int_0^1 p(t)\phi_2(t)\left(\int_t^1 \frac{ds}{p(s)}\right)^{\frac{\gamma+1}{2}} dt$;

$$\begin{aligned}\int_0^1 p\phi_3|y||py'|^\theta\,dt &\le \|y'\|^\theta\left(\int_0^1 (p(t))^{\frac{2+\theta}{2-\theta}}(\phi_3(t)|y|)^{\frac{2}{2-\theta}}\,dt\right)^{\frac{2-\theta}{2}}\\ &\le N_5\|y'\|^\theta|y(1)| + N_6\|y'\|^{\theta+1}\end{aligned}$$

where $N_5 = 2\left(\int_0^1 (p(t))^{\frac{2+\theta}{2-\theta}}(\phi_3(t))^{\frac{2}{2-\theta}}\,dt\right)^{\frac{2-\theta}{2}}$ and

$N_6 = 2\left(\int_0^1 (p(t))^{\frac{2+\theta}{2-\theta}}(\phi_3(t))^{\frac{2}{2-\theta}}\left(\int_t^1 \frac{ds}{p(s)}\right)^{\frac{1}{2-\theta}}dt\right)^{\frac{2-\theta}{2}}$. Put this into (2.34) to obtain

$$\begin{aligned}\eta_0\left(\|y'\|^2 + \tfrac{a}{b}[y(1)]^2\right) \le -\lambda\int_0^1 pyg(t,y,py')\,dt + N_7[|y(1)| + \|y'\|\\ +|y(1)|^{\gamma+1} + \|y'\|^{\gamma+1} + |y(1)|^{\theta+1} + \|y'\|^{\theta+1} + 1]\end{aligned} \tag{2.36}$$

for some constant N_7.

Case (i). $c_0 \ge 0$.

Then (2.36) yields

$$\begin{aligned}\eta_0\left(\|y'\|^2 + \tfrac{a}{b}[y(1)]^2\right) \le N_7[|y(1)| + \|y'\| + |y(1)|^{\gamma+1} + \|y'\|^{\gamma+1}\\ +|y(1)|^{\theta+1} + \|y'\|^{\theta+1} + 1]\end{aligned}$$

so there exists a constant M_0 with

$$\|y'\| \le M_0 \text{ and } |y(1)| \le M_0 \tag{2.37}$$

for any solution y to $(2.33)_\lambda$.

Case (ii). $c_0 < 0$.

Then as in theorem 2.1 we have

$$-\lambda\int_0^1 pyg(t,y,py')\,dt \le \frac{(-c_0)Q_0}{\sqrt{\lambda_0}}\left(\frac{a}{b}[y(1)]^2 + \|y'\|^2\right)$$

so (2.34) gives

$$\begin{aligned}\left(\eta_0 + \tfrac{c_0Q_0}{\sqrt{\lambda_0}}\right)\left(\|y'\|^2 + \tfrac{a}{b}[y(1)]^2\right) \le N_7[|y(1)| + \|y'\| + |y(1)|^{\gamma+1} + \|y'\|^{\gamma+1}\\ +|y(1)|^{\theta+1} + \|y'\|^{\theta+1} + 1]\end{aligned}$$

and consequently (2.37) is again satisfied.

Returning to the differential equation we have

$$\int_0^t |(py')'|\,ds \le \int_0^t p|g(s,y,py')|\,ds + \int_0^t p|h(s,y,py')|\,ds + |\mu|\int_0^t pq|y|\,ds. \tag{2.38}$$

Let $k_1(t) = \int_0^t p\phi_1\,ds$, $k_2(t) = 2^\gamma M_0^\gamma[\int_0^t p\phi_2\,ds + \int_0^t p\phi_2\left(\int_s^1 \frac{dz}{p(z)}\right)^{\frac{\gamma}{2}} ds]$ and

$k_3(t) = M_0^\theta\left(\int_0^t (p(s))^{\frac{2+\theta}{2-\theta}}(\phi_3(s))^{\frac{2}{2-\theta}}\,ds\right)^{\frac{2-\theta}{2}}$. Now (2.35) and (2.37) yields

$$\int_0^t p\phi_2|y|^\gamma\,ds \le 2^\gamma\left(|y(1)|^\gamma\int_0^t p\phi_2\,ds + \|y'\|^\gamma\int_0^t p\phi_2\left(\int_s^1 \frac{dz}{p(z)}\right)^{\frac{\gamma}{2}} ds\right) \le k_2(t)$$

and

$$\int_0^t p\phi_3|y||py'|^\theta\,ds \le \|y'\|^\theta\left(\int_0^t (p(s))^{\frac{2+\theta}{2-\theta}}(\phi_3(s))^{\frac{2}{2-\theta}}\,ds\right)^{\frac{2-\theta}{2}} \le k_3(t)$$

so

$$\int_0^t p|h(s,y,py')|\,ds \le k_1(t)+k_2(t)+k_3(t). \tag{2.39}$$

Similarly with $k_4(t)=\int_0^t p\phi_5\,ds$, $k_5(t)=2^\delta M_0^\delta[\int_0^t p\phi_6\,ds+\int_0^t p\phi_6\left(\int_s^1\frac{dz}{p(z)}\right)^{\frac{\delta}{2}}ds]$ and $k_6(t)=M_0^\omega\left(\int_0^t (p(s))^{\frac{2+\omega}{2-\omega}}(\phi_7(s))^{\frac{2}{2-\omega}}\,ds\right)^{\frac{2-\omega}{2}}$ we have

$$\int_0^t p|g(s,y,py')|\,ds \le k_4(t)+k_5(t)+k_6(t). \tag{2.40}$$

Also we have

$$\int_0^t pq|y|\,ds \le M_0\int_0^t pq\,ds + M_0\int_0^t pq\left(\int_s^1\frac{dz}{p(z)}\right)^{\frac{1}{2}}ds \equiv k_7(t). \tag{2.41}$$

Put (2.39), (2.40) and (2.41) into (2.38) to obtain

$$\int_0^t |(py')'|\,ds \le k_1(t)+k_2(t)+k_3(t)+k_4(t)+k_5(t)+k_6(t)+|\mu|k_7(t). \tag{2.42}$$

In addition since

$$y(t)=y(1)-\int_t^1\frac{1}{p(u)}\int_0^u (p(s)y'(s))'\,ds\,du$$

we have for $t\in[0,1]$ that

$$|y(t)|\le M_0+\int_0^1\frac{1}{p(u)}\int_0^u[k_1(x)+k_2+k_3+k_4+k_5+k_6+|\mu|k_7]\,dx\,du\equiv M_1.$$

Thus

$$\sup_{[0,1]}|y(t)|\le M_1 \tag{2.43}$$

for any solution y to $(2.33)_\lambda$. In addition since $|p(t)y'(t)| = |\int_0^t(py')'\,ds|$ we have from (2.42) that

$$\sup_{(0,1)}|p(t)y'(t)|\le k_1(1)+k_2(1)+k_3(1)+k_4(1)+k_5(1)+k_6(1)+|\mu|k_7(1)\equiv M_2. \tag{2.44}$$

The result now follows from theorem 1.2. □

References.

[1]. F.V.Atkinson, Discrete and continuous boundary problems, Academic Press, New York (1964).

[2]. L.E.Bobisud, Existence of solutions for nonlinear singular boundary value problems, Appl. Anal., 35, 46–67 (1990).

[3]. D.R.Dunninger and J.C.Kurtz, A priori bounds and existence of positive solutions for singular nonlinear boundary value problems, SIAM J. Math. Anal., 17, 595–609 (1986).

[4]. A.Fonda and J.Mawhin, Quadratic forms, weighted eigenfunctions and boundary value problems for nonlinear second order ordinary differential equations, Proc. Royal Soc. Edinburgh, 112A, 145–153 (1989).

[5]. M.Frigon and D.O'Regan, Some general existence principles for ordinary differential equations, Topological Methods in Nonlinear Anal, in press.

[6]. M.Frigon and D.O'Regan, Existence results for some initial and boundary value problems without growth restrictions, to appear.

[7]. A.Granas, R.B.Guenther and J.W.Lee, Some general existence principles in the Carathéodory theory of nonlinear differential systems, J. Math. Pures Appl., 70, 153–196 (1991).

[8]. J.Mawhin and W.Omano, Two point boundary value problems for nonlinear perturbations of some singular linear differential equations at resonance, Comm. Math. Univ. Carolinae, 30, 537–550 (1989).

[9]. S.G.Mikhlin, Variational methods in Mathematical Physics, Pergamon Press, Oxford (1964).

[10]. M.A.Naimark, Linear differential operators Part II, Ungar Publishing Co., London (1968).

[11]. D.O'Regan, Boundary value problems for second and higher order differential equations, Proc. Amer. Math. Soc., 113, 761–776 (1991).

[12]. D.O'Regan, Singular Sturm Liouville problems and existence of solutions to singular nonlinear boundary value problems, Nonlinear Anal.,20, 767–779 (1993).

[13]. D.O'Regan, Solvability of some two point boundary value problems of Dirichlet, Neumann or Periodic type, Dynamic Systems and Appl., 2, 163–182 (1993).

[14]. D.O'Regan, Existence principles for second order nonresonant boundary value problems, to appear.

[15]. I.Stakgold, Green's functions and boundary value problems, John Wiley, New York (1979).

4th Int. Coll. on Differential Equations, pp. 229-233
D. Bainov and V. Covachev (Eds)

Microlocal hypoellipticity of some classes of over-determined systems of pseudodifferential operators

P.R. POPIVANOV
Institute of Mathematics of the Bulgarian Academy of Sciences, 1113 — Sofia, BULGARIA

Abstract. The paper deals with the microlocal properties of the solution of some classes of overdetermined pseudodifferential systems having simple and double characteristics.

Keywords and phrases: pseudodifferential operator, hypoellipticity.

1. This paper deals with the microlocal and therefore local hypoellipticity of some classes of pseudodifferential systems with simple and double characteristics. As our systems are consisting of several scalar operators acting on the same complex-valued function u many methods developed in the scalar theory can be used. In fact, we apply some assertions due to Hörmander [1], as well as some ideas coming from [5], [10]. The loss of the smoothness of the corresponding solution is equal to $^1/_2$ or 1 in both classical and microlocal Sobolev spaces $H^s_{\text{loc}}(\Omega)$, H^s_{mcl}. The results here proved enable us to study the C^∞, analytic and G_s (Gevrey) hypoellipticity of complex vector fields system having simple and double characteristics. Some of our results in the simple-characteristics case — local version — reprove in an elementary way theorems due to Baouendi — Trèves [5] and Kwon — Song [8]. Our assertions in the same case are microlocal in contrast with the previous papers and the loss of smoothness here obtained is optimal ($^1/_2$ in the simple-characteristics case, 1 — in the double-characteristics case). The theorems concerning Gevrey hypoellipticity for simple characteristic systems and for double-characteristics seems to be new. Because of the lack of space we omit the standard definitions of classical homogeneous pseudodifferential operators (Ψ.D.O.), C^∞ wave front set (WF(u)), analytic wave front set (WF$_a(u)$), Gevrey wave front set (WF$_s(u), s > 1$). Certainly, the operator P is C^∞-hypoelliptic iff $\operatorname{sing\,supp} Pu = \operatorname{sing\,supp} u$ and P is microlocally hypoelliptic at the point $\rho^0 \in \operatorname{Char} P$ iff $\rho^0 \notin \mathrm{WF}(Pu) \Longrightarrow \rho^0 \notin \mathrm{WF}(u)$. The same is true in Gevrey and analytic categories.

We are going now to formulate the main results of this paper.

2. Consider the overdetermined system of Ψ.D.O.

$$P_\nu u = f_\nu, \quad \nu = 1, \dots, d, \qquad \operatorname{ord} P_\nu = m_\nu, \tag{1}$$

u is a scalar function. In [1] a necessary and sufficient condition for the hypoellipticity of (1) with a loss of the smoothness of the solution u equal to $^1/_2$ was found. To remind it to the reader we introduce the characteristic set of $P = (P_1, \dots, P_d)$ in a standard way:

$$\operatorname{Char} P = \left\{\rho = (x, \xi) \in \Omega \times S^{n-1} : \; p^0_\nu = 0, 1 \le \nu \le d\right\}, \quad \Omega \subset \mathbf{R}^n.$$

Define now at the point $\rho^0 \in \operatorname{Char} P$ the following tensor

$$T(\rho^0) = \sum_{\nu=1}^{n} \frac{\partial p^0(\rho^0)}{\partial \xi_\nu} \otimes_{\mathbf{C}} \frac{\partial p^0(\rho^0)}{\partial x_\nu}$$

where $p^0 = (p_1^0, \dots, p_d^0)$ is the principal symbol of P. There are two possibilities for T:

(i) T is non-symmetric at ρ^0;

(ii) T is symmetric at ρ^0. In this case we introduce the symmetric quadratic form

$$\psi(\nu)(\rho^0) = \operatorname{Re} i \sum_{\nu=1}^{n} \sum_{j,k=1}^{d} \frac{\partial p_j^0(\rho^0)}{\partial \xi_\nu} \overline{\frac{\partial p_k^0(\rho^0)}{\partial x_\nu}} \overline{\nu_j} \nu_k, \quad \nu = (\nu_1, \dots, \nu_d) \in \mathbf{C}^d.$$

It was proved in [1] the next interesting and important result:

Theorem 1. *The operator P is subelliptic with a loss of smoothness $1/2$ in classical Sobolev spaces $H^s_{\mathrm{loc}}(\Omega)$ if and only if at each characteristic point $\rho^0 \in \operatorname{Char} P$ either $T(\rho^0)$ is non-symmetric or $T(\rho^0)$ is symmetric and the quadratic form $\psi(\nu)(\rho^0)$ has at least one positive eigenvalue.*

If T is symmetric at ρ^0 but $\psi(\nu)(\rho^0)$ is negatively definite then P is not subelliptic with a loss of smoothness $\delta \in (0, 1)$.

Remark 1. We can reformulate the condition $\psi(\nu)(\rho^0)$ to have at least one positive eigenvalue in the next way. The symmetric form $\psi(\nu)(\rho^0)$ is not negatively semidefinite.

In order to simplify the proof of our assertions concerning complex vector fields system we shall reformulate Theorem 1 proposing a slight modification with its elementary proof. Thus let $\operatorname{ord} P_j = t$, $p_j^0 = a_j + ib_j$, $a_j = \operatorname{Re} p_j^0$, $b_j = \operatorname{Im} p_j^0$.

Theorem 2. *Consider the system (1) and assume that at the point $\rho^0 \in \operatorname{Char} P$ $\{a_j, b_k\}(\rho^0) \neq \{a_k, b_j\}(\rho^0)$ for some indices $j, k \in [1, d]$. Then (1) is microlocally hypoelliptic at ρ^0 with a loss of smoothness equal to 1, i.e. $Pu \in H^s_{\mathrm{mcl}}(\rho^0) \Longrightarrow u \in H^{s+t-1}_{\mathrm{mcl}}(\rho^0)$.*

Suppose now that $\{a_j, b_k\}(\rho^0) = \{a_k, b_j\}(\rho^0)$ for each $j, k = 1, \dots, d$ and the symmetric matrix $\left[\{a_j, b_k\}\right](\rho^0)$ has at least one positive eigenvalue. Then P is microlocally hypoelliptic at ρ^0 with a loss of smoothness $1/2$, i.e. $Pu \in H^s_{\mathrm{mcl}}(\rho^0) \Longrightarrow u \in H^{s+t-1/2}_{\mathrm{mcl}}(\rho^0)$.

Certainly $\rho^0 \notin \mathrm{WF}(Pu) \Longrightarrow \rho^0 \notin \mathrm{WF}(u)$.

As usual, $\{.,.\}$ stands for the Poisson bracket of two smooth functions and $\{a_k, b_j\} - \{a_j, b_k\}$ is the real part of the principal symbol of the commutator $[L_k, L_j]$.

Remark 2. As a Corollary of Theorem 2 we get a microlocal and local hypoellipticity of (1) in the following cases:

(iii) the matrix $\left[\{a_j, b_k\}\right]_{j,k=1}^{\alpha}(\rho^0)$ is non-symmetric at $\rho^0 \in \operatorname{Char} P$;

(iv) the matrix $\left[\{a_j, b_k\}\right]_{j,k=1}^{\alpha}(\rho^0)$ is symmetric at $\rho^0 \in \operatorname{Char} P$ and not negatively semidefinite.

The validity of either **(iii)** or **(iv)** at each point $\rho^0 \in \operatorname{Char} P$ guarantees the hypoellipticity of P.

Remark 3. Let (1) be a system of differential operators having real-analytic coefficients and $\{a_j, b_k\}(\rho^0) = \{a_k, b_j\}(\rho^0)$, $\rho_0 \in \operatorname{Char} P$ for each j, k and the symmetric matrix $[\{a_j, b_k\}(\rho^0)]$ has at least one positive eigenvalue. Then P is microlocally analytic and microlocally Gevrey hypoelliptic at ρ_0, i.e. $\rho_0 \notin \mathrm{WF}_{\mathrm{G}_s}(Pu) \Longrightarrow \rho_0 \notin \mathrm{WF}_{\mathrm{G}_s}(u)$ for each $s \geq 1$.

Thus the validity of either **(iii)** or **(iv)** at each characteristic point ρ_0 guarantees not only the C^∞ hypoellipticity of (1) but the G_s, $s \geq 1$ hypoellipticity as well.

And now some applications to the (micro)local hypoellipticity of complex vector fields system will be proposed.

Following Baouendi — Trèves we consider a locally integrable m-dimensional Frobenius Lie-algebra L of smooth (real analytic) complex vector fields on an open domain $\Omega \subset \mathbf{R}^{m+1}$, $m+1=n$. That means that

(a) at each point $z_0 = (t_0, x_0) \in \mathrm{R}^m \times \mathrm{R}^1$ the system L is spanned by m linearly independent elements $L_1(z_0), \dots, L_m(z_0)$ of L;

(b) L is closed under the operation $[.,.]$ commutator of two vector fields;

(c) L is locally integrable, i.e. for each $z_0 \in \Omega$ there exists a neighbourhood U ($z_0 \in U$) and a smooth function Z, such that $LZ = 0$, $dZ(z_0) \neq 0$.

As Frobenius theorem is not valid in the complex C^∞ case **(b)** does not imply **(c)**, while **(b)** $\Longrightarrow$ **(c)** in the case of (real) analytic vector fields L.

As it is shown in [5], [6] and without loss of generality in studying hypoellipticity we can assume that we are working in a neighbourhood U of $z_0 = 0$ with local coordinates $z = (t, x)$ for which

$$L_j = \frac{\partial}{\partial t_j} + \lambda_j(t, x)\frac{\partial}{\partial x}, \qquad 1 \leq j \leq m \tag{2}$$

$$Z(t, x) = x + i\phi(t, x), \quad \phi(0, 0) = 0, \quad d\phi(0, 0) = 0,$$

ϕ is a real-valued C^∞ function near the origin. Moreover, then $[L_j, L_k] = 0$ in U. Certainly, then

$$\lambda_j = -i\frac{\frac{\partial\phi}{\partial t_j}}{1 + i\frac{\partial\phi}{\partial x}}.$$

Put

$$\operatorname{Char} L = \left\{(t, x; \tau, \xi) \in U \times S^{n-1} : \tau_j = 0, \frac{\partial\phi}{\partial t_j} = 0, \xi \neq 0, 1 \leq j \leq m\right\},$$

$$C = \operatorname{projection}_z \operatorname{Char} L, \qquad C \neq \{\emptyset\}, U, \quad z = (t, x).$$

The following result is a slight generalization of Theorem 2.1 from [8] but with a better result concerning the loss of smoothness of the solution. The principle of the proof is rather different from this in [8].

Theorem 3. *The system of C^∞ complex-valued vector fields L is microlocally hypoelliptic with a loss of smoothness equal to $1/2$ (i.e. $u \in H^s_{\text{mcl}}(\rho^0)$, $\rho^0 \in \operatorname{Char} L \Longrightarrow u \in H^{s+1/2}_{\text{mcl}}(\rho^0)$) if*

$$\text{(v)} \quad \begin{cases} \rho^0 = (t_0,\, x_0;\, 0,\, -1) \in \operatorname{Char} L \Longrightarrow & H_t\phi(t_0,\, x_0) = \left[\dfrac{\partial^2 \phi}{\partial t_j \partial t_k}(t_0,\, x_0)\right]_{j,k=1}^{m} \\ & \text{has at least one positive eigenvalue} \\ \rho^0 = (t_0,\, x_0;\, 0,\, 1) \in \operatorname{Char} L \Longrightarrow & H_t\phi \text{ has at least one negative eigenvalue.} \end{cases}$$

Let, moreover, L be a Frobenius–Lie algebra of analytic complex vector fields. Then L is microanalytic and microGevrey hypoelliptic for each $s > 1$ at $\rho_0 \in \operatorname{Char} P$ if the condition (**v**) *is valid.*

Corollary. *The system of C^∞ complex-valued vector fields L is C^∞ hypoelliptic with loss of the smoothness $1/2$ if at each point $z_0 = (t_0,\, x_0) \in C$*

$$\text{(vi)} \quad \begin{cases} \text{the Hessian of } \phi \text{ with respect to } t: \ H_t\phi(z_0) \text{ is not} \\ \text{definite, i.e. it is sign} - \text{changeable.} \end{cases}$$

Let, moreover, ϕ be a real-analytic function in U. Then L is analytic and G_s hypoelliptic for all $s > 1$ if at each point $z_0 = (t_0,\, x_0) \in$ **C** *the condition* (**vi**) *is valid.*

At the end of this paper we shall study the C^∞ (micro)hypoellipticity of a second order overdetermined system of complex vector fields, namely

$$(3) \qquad L_j L_k u + \ell_{jk} u = f_{jk}, \qquad 1 \le k \le k \le m,$$

where L_j are the same as in (2) and ℓ_{jk} are arbitrary C^∞ complex vector fields.

Theorem 4. *The system (3) is microlocally hypoelliptic at $\rho^0 \in \operatorname{Char} L$ with a loss of the smoothness equal to 1 if the condition* (**v**) *is fulfilled. Certainly, the system (3) is microelliptic outside* Char L.

Corollary. *The system (3) is C^∞ hypoelliptic in U with loss of the smoothness equal to 1 if the condition* (**vi**) *on ϕ is fulfilled. This hypoellipticity does not depend on the vector fields ℓ_{jk}.*

Moreover, if $\phi(t, x)$ is a real-analytic function and (**vi**) *is valid then the system (3) is analytic and* G_s *hypoelliptic with $1 < s < 2$ for arbitrary analytic vector fields ℓ_{ij}.*

A small modification of the previous result is given for the next system:

$$(4) \qquad L_j^2 u + \ell_j u = f_j, \qquad 1 \le j \le m,$$

where $\{L_j\}$ form a non-degenerate, locally integrable system L of codimension 1 in $\mathbf{R}^{m+1}$ (for the corresponding definitions see [11]) and ℓ_j are arbitrary C^∞ complex-valued vector fields. If the point $z_0 \in C$ then the Levi form L has the signature $\{k,\, m-k\}$, $0 \le k \le m$ which is independent of z_0 provided C is connected.

Proposition. *Let L be a non-degenerate, locally integrable system of codimension 1 in $\mathbf{R}^{m+1}$. Assume that for each $z \in C$ the signature of L is not $\{0,\, n\}$ and for every function Z_1 defined in a neighbourhood U of z and verifying $LZ_1 = 0$ the set $Z_1(U \cap C)$ is a real analytic curve in $\mathbf{C}^1$. Then (4) is C^∞ hypoelliptic with a loss of smoothness equal to 1.*

Certainly in this case L is a Mizohata structure ([11]) in a neighbourhood of any point of C.

The theorems concerning C^∞, analytic and G_s, $s > 1$ hypoellipticities for the system (2) can be enlarged in the C^∞ category by the next result.

Theorem 5. *Consider the C^∞ system (3) and suppose that ϕ depends on t only, i.e. $\phi \equiv \phi(t)$. Let $X(t)$ be a non-degenerate real-valued C^∞ vector field near the origin $t = 0$ and such that $X\phi \geq 0$ near $t = 0$, $x\phi(t) = 0$ iff $t = 0$. Then the system (3) is C^∞ hypoelliptic at $z_0 = 0$ (i.e. $z_0 \notin \operatorname{sing\,supp} Lu \Longrightarrow z_0 \notin \operatorname{sing\,supp} u$).*

3. The proof of Theorem 2 is elementary and Theorem 3 is a simple consequence of Theorem 2. As it concerns Theorem 4 and the Proposition they can be verified by using the theorem on regular microlocal hypoellipticity from [3] ([7]) and thus reducing the problem for the hypoellipticity of (3), (4) to some results of [3], [7]. Theorem 5 is based on an assertion of Hörmander from [9].

REFERENCES

[1] L. Hörmander. Pseudodifferential operators and non-elliptic boundary value problems. *Ann. of Math.*, **83**(1966), No. 1, 129–209.

[2] L. Boutet de Monvel. Hypoelliptic operators with double characteristics and related pseudodifferential operators. *Comm. Pure Appl. Math.*, **27**(1974), 585–637.

[3] P. Popivanov. A new short proof of a theorem on regular hypoellipticity. *Serdica*, **3**(1977), 249–252. (in Russian)

[4] F. Trèves. Analytic hypoellipticity of a class of pseudodifferential operators with double characteristics and application to the $\overline{\partial}$-Neumann problem. *Comm. in Part. Diff. Equations*, **3**(1978), 475–642.

[5] M. S. Baouendi & F. Trèves. A local constancy principle for the solutions of certain overdetermined systems of first order linear partial differential equations. *Adv. in Math. Supplementary studies*, **7A**(1981), 245–262.

[6] M. S. Baouendi & F. Trèves. A property of functions and distributions annihilated by a locally integrable system of complex vector fields. *Ann. of Math.*, **113**(1981), 387–421.

[7] P. Popivanov. Microlocal Analysis of Some Classes of Pseudodifferential Operators. *Doctoral thesis*, Sofia University, 1985.

[8] K. H. Kwon & B. C. Song. Hypoellipticity of a complex vector fields system. *Journal of the Korean Math. Soc.*, **26**(1989), 195–201.

[9] L. Hörmander. The Analysis of Linear Partial Differential Operators. Vol. IV, Springer Verlag, 1985.

[10] L. Rodino. Linear Partial Differential Operators in Gevrey Spaces. World Scientific, 1993.

[11] P. Cordaro & J. Hounie. Local solvability in C_0^∞ of over-determined systems of vector fields. *Journal of Funct. Analysis*, **87**(1989), 231–249.

4th Int. Coll. on Differential Equations, pp. 235-243
D. Bainov and V. Covachev (Eds)

A Two Point Connection Problem For A Certain Third Order Differential Equation

T.K. Puttaswamy

Department of Mathematical Sciences
Ball State University
Muncie, Indiana - 47306, U.S.A.

T.V. Sastry

Department of Mathematics
Bradley University
Peoria, Illinois - 61625, U.S.A.

Abstract

This paper is addressed to examine the two point connection problem for the differential equation

$$z\,\frac{d^3y}{dz^3} + a\,\frac{d^2y}{dz^2} + bz^m\,\frac{dy}{dz} + cz^{m-1}\,y = 0$$

Here m is an arbitrary positive integer, the variable z and the constants a, b, c are complex with $b \neq 0$. We shall also assume that neither $a-1$ nor $a-2$ is congruent to zero modulo $m+1$.

Key Words : Singular Point, Indicial Equation.

1 Introduction

In order to introduce the investigation of this paper, let us consider the linear homogeneous differential equation of third order

$$z\,\frac{d^3y}{dz^3} + a\,\frac{d^2y}{dz^2} + bz^m\,\frac{dy}{dz} + cz^{m-1}\,y = 0 \tag{1.1}$$

Here, m is an arbitrary positive integer, the variable z and the constants a, b, c are complex with $b \neq 0$. Then, in the language of Fuch's theory, the differential equation (1.1) will have a regular singular point $z = 0$ and an irregular singular point at $z = \infty$. We shall assume that neither $a - 1$ nor $a - 2$ is congruent to zero modulo $m + 1$.

If v is the differential operator $z \frac{d}{dz}$, we know that
$z^n \frac{d^n}{dz^n} = v(v-1)(v-2)....(v-n+1)$, $n \geq 1$.

Now, by multiplying (1.1) by z^2, (1.1) can be put in the form

$$f(v)y + z^t g(v) y = 0 \tag{1.2}$$

where

$$f(v) = v(v-1)(v+a-2) \tag{1.3}$$

$$g(v) = bv + c \tag{1.4}$$

$$t = m + 1 \geq 2$$

The indicial equation about the regular singular point $z = 0$ is found to be

$$f(h) = h(h-1)(h+a-2) = 0 \tag{1.5}$$

Hence, the indicial exponents about the regular singular point $z = 0$ are

$$h_j = j - 1, \; j = 1, 2 \tag{1.6}$$

$$and \quad h_3 = 2 - a$$

Since $a - 1$ and $a - 2$ are incongruent to zero modulo $m + 1$, the existing theory of linear differential equations assures us that (1.1) will have three linearly independent solutions $y_i(z)$, $i = 1, 2, 3$ of the form

$$y_i(z) = z^{h_i} \sum_{n=0}^{\infty} g(n) z^{nt} \tag{1.7}$$

which will converge in the whole open plane of the extended complex z plane.

2 Recurrence Relation Satisfied By Each $g_i(n)$, $i = 1, 2, 3$

We substitute $y(z) = z^h \sum_{n=0}^{\infty} g(n) z^{nt}$ into (1.1). Using the power shift $f(v)\, z^a = f(a)\, z^a$ and equating the coefficients of $z^{(n+1)t+h}$, we get

$$p_1(n)\, g(n+1) \;+\; p_0(n)\, g(n) \;=\; 0 \tag{2.1}$$

where

$$p_1(n) \;=\; [h+t(n+1)]\,[h+t(n+1)-1]\,[h+t(n+1)+a-2] \tag{2.2}$$

$$and \;\; p_0(n) \;=\; b[h+tn] \;+\; c \tag{2.3}$$

We arbitrarily take $g(0) = 1$. Since we wish to regard n as complex variable, to emphasize this, we replace n by s.

Our first objective, then, is to find the analytic function $u(s)$ of the complex variable s which is a solution of (2.1) and which takes on the initial value $u(0) = 1$.

3 Solutions $y_i(z), i = 1,2,3$ About The Regular Singular Point $z = 0$

Let α be the root of

$$p_0(s) \;=\; b[h+ts] \;+\; c \;=\; 0 \tag{3.1}$$

then

$$\alpha \;=\; -\,\frac{bh+c}{bt} \tag{3.2}$$

We shall assume for the present that α is not an integer.

Now, from (2.1), we have

$$\begin{aligned}\frac{u(s+1)}{u(s)} &= -\,\frac{p_0(s)}{p_1(s)} \\ &= -\,\frac{b(s-\alpha)}{t^2(s-\beta_1)(s-\beta_2)(s-\beta_3)}\end{aligned} \tag{3.3}$$

where $\beta_j (j = 1,2,3)$ are the roots of $p_1(s) = 0$.

Therefore, the particular solution $u(s)$ which takes on the initial value $u(0) = 1$, is readily found by elementary methods to be

$$u(s) \;=\; (-\frac{b}{t^2})^s\; G(s)$$

where

$$G(s) = \frac{\Gamma(-\beta_1)\Gamma(-\beta_2)\Gamma(-\beta_3)\Gamma(s-\alpha)}{\Gamma(-\alpha)\Gamma(s-\beta_1)\Gamma(s-\beta_2)\Gamma(s-\beta_3)} \tag{3.4}$$

We first observe that in view of the indicial equation (1.5) about the regular singular point $z = 0$, -1 is a root of $p_1(s) = 0$. Let this root be β_3. So, if β_1 and β_2 are the other two roots of

$$p_1(s) = [h + t(s+1)][h + t(s+1) - 1][h + t(s+1) + a - 2] = 0 \tag{3.5}$$

we observe that neither β_1 nor β_2 can be zero or a positive integer, for, if $\beta_i = N - 1 (N = 1, 2, ...)$ for $i = 1, 2$, then

$$p_1(\beta_i) = p_1(N-1) = [h + tN][h + tN - 1][h + tN + a - 2] = 0$$

which would imply that the difference of two roots of the indicial equation about the regular singular point $z = 0$ is congruent to zero modulo $t = m + 1$, which contradicts our hypothesis.

So, (3.4) becomes

$$G(s) = \frac{\Gamma(-\beta_1)\Gamma(-\beta_2)\Gamma(s-\alpha)}{\Gamma(-\alpha)\Gamma(s+1)\Gamma(s-\beta_2)\Gamma(s-\beta_1)} \tag{3.6}$$

Since neither β_1 nor β_2 can be zero or a positive integer, the numerator in (3.6) is necessarily finite.

It follows that each of the solutions $y_i(z)$, $i = 1, 2, 3$ about the regular singular point $z = 0$ is of the form

$$y(z) = z^h \sum_{n=0}^{\infty} G(n) \left(-\frac{b}{t^2}\right)^n . z^{nt}$$

where $G(s)$ is given by (3.6). By rendering precise meaning through certain conventions, the function $G(s)$ as given by (3.6) will be analytic and single valued throughout the finite s-plane except for the poles of first order at the point,

$$s = \alpha - n, \; n = 0, 1, 2, ... \tag{3.8}$$

4 Formal Solution $\overline{y_j}(z)$, $j = 1, 2, 3$ About The Irregular Singular Point $z = \infty$

In (1.1), transform the independent variable z to w by placing $z = w^l$, where $l = \frac{2}{t} = \frac{2}{m+1}$. Then, the differential (1.1) will become

$$\frac{d^3y}{dw^3}+\frac{1}{w}[3-3l+la]\frac{d^2y}{dw^2}+\frac{1}{w^2}[(1-l)(1-2l)+l(1-l)a+bl^2w^2]\frac{dy}{dw}+\frac{cl^3}{w}y=0 \quad (4.1)$$

We now apply the theorem of G.D. Birkoff[1]. For (4.1), the characteristic equation is found to be

$$\rho^3 + \rho l^2 b = 0 \quad (4.2)$$

So, (4.2) will have three distinct roots $\rho_1 = 0$, $\rho_2 = il\sqrt{b}$ and $p_3 = -il\sqrt{b}$. Hence, the differential equation (4.1) will have three linearly independent solutions $\widetilde{y_j}(w)$, $j = 1,2,3$. for which when $|w|$ is large, $\widetilde{y_j}(w) \sim s_j(w)$, where $s_j(w)$ are the divergent series having the forms

$$s_1(w) \sim w^{k_1}[1 + \frac{()}{w} + \frac{()}{w^2} + ...] \quad (4.3)$$

$$s_2(w) \sim e^{\frac{2i\sqrt{b}}{t}w}\, w^{k_2}[1 + \frac{()}{w} + \frac{()}{w^2} + ...] \quad (4.4)$$

and

$$s_3(w) \sim e^{\frac{-2i\sqrt{b}}{t}w}\, w^{k_3}[1 + \frac{()}{w} + \frac{()}{w^2} + ...] \quad (4.5)$$

and k_1, k_2 and k_3 are constants whose values may be determined by formal substitution of these forms into (4.1). The values of these constants are found to be

$$k_1 = \frac{-2c}{bt}$$

$$and \quad k_2 = k_3 = -\frac{3}{2} + \frac{c}{bt} + \frac{3-a}{t} \quad (4.6)$$

It follows, upon making the reverse transformation $w = z^{\frac{1}{l}} = z^{\frac{t}{2}}$, that the original differential equation (1.1) will have three linearly independent solutions $\widetilde{y_j}(z)$, $j = 1,2,3$ such that when $|z|$ is large, we shall have $\widetilde{y_j}(z) \sim s_j(z)$, where $s_j(z)$, $j = 1,2,3$ are given from (4.5) as follows:

$$s_1(z) = z^{-\frac{c}{b}}[1 + \frac{()}{z^{\frac{t}{2}}} + \frac{()}{z^t} + ...] \quad (4.7)$$

$$s_2(z) = e^{\frac{2ib}{t}z^{\frac{t}{2}}}\, z^{\frac{tk_2}{2}}[1 + \frac{()}{z^{\frac{t}{2}}} + \frac{()}{z^t} + ...] \quad (4.8)$$

$$s_3(z) = e^{-\frac{2ib}{t}z^{\frac{t}{2}}}\, z^{\frac{tk_2}{2}}[1 + \frac{()}{z^{\frac{t}{2}}} + \frac{()}{z^t} + ...] \quad (4.9)$$

Here $\frac{tk_2}{2} = \frac{tk_3}{2} = \frac{1}{2}[-\frac{3}{2}t + \frac{c}{b} + 3 - a]$

5 Asymptotic Behavior Of The Solutions $y_i(z)$, $i = 1,2,3$ Corresponding To The Regular Singular Point $z = 0$

If we write $G(s)$ as given by (3.4) in the form

$$G(s) = \frac{\Gamma(-\beta_1)\Gamma(-\beta_2)}{\Gamma(-\alpha)\Gamma(s+1)} \frac{J(s)}{\Gamma(s+\alpha-\beta_1-\beta_2)} \tag{5.1}$$

where

$$J(s) = \frac{\Gamma(s-\alpha)\Gamma(s+\alpha-\beta_1-\beta_2)}{\Gamma(s-\beta_1)\Gamma(s-\beta_2)} \tag{5.2}$$

then, from the well known asymptotic properties of gamma functions, $J(s)$ thus defined, when considered for values of s of large modulus lying in some right half plane is asymptotic to 1. So, for such s, we may write

$$G(s) \sim \frac{\Gamma(-\beta_1)\Gamma(-\beta_2)}{\Gamma(-\alpha)\Gamma(s+1)} \frac{g^*(s)}{\Gamma(s+\alpha-\beta_1-\beta_2)} \tag{5.3}$$

where

$$g^*(s) \sim 1 + \frac{()}{(s+\alpha-\beta_1-\beta_2)} + \frac{()}{(s+\alpha-\beta_1-\beta_2)(s+\alpha-\beta_1-\beta_2+1)} + ... \tag{5.4}$$

We then apply W.B. Ford's VII General Theorem[2] to find the asymptotic development of the function defined in the series (3.7). We obtain

$$y(z) \sim c_3(h)\, s_3(z) - z^h\, R(z) \quad in \quad \frac{\pi}{2} < arg\ (\frac{i\sqrt{b}z^{\frac{t}{2}}}{t}) < \frac{3\pi}{2} \tag{5.5}$$

and

$$y(z) \sim c_2(h)\, s_2(z) - z^h\, R(z) \quad in \quad -\frac{\pi}{2} \leq arg\ (\frac{i\sqrt{b}z^{\frac{t}{2}}}{t}) \leq \frac{\pi}{2} \tag{5.6}$$

where

$$c_3(h) = \frac{\Gamma(-\beta_1)\Gamma(-\beta_2)}{\Gamma(-\alpha)} \frac{1}{2\sqrt{\pi}} (-\frac{ib}{t})^{\beta_1+\beta_2-\alpha+\frac{1}{2}} \tag{5.7}$$

$$c_2(h) = \frac{\Gamma(-\beta_1)\Gamma(-\beta_2)}{\Gamma(-\alpha)} \frac{1}{2\sqrt{\pi}} (\frac{ib}{t})^{\beta_1+\beta_2-\alpha+\frac{1}{2}} \tag{5.8}$$

and $R(z)$ represents the sum of the residues of the function

$$\frac{\pi\, G(s)\, (-\frac{b}{t^2} z^t)^s}{\sin \pi s} \tag{5.9}$$

at the poles (3.8).

Our next problem is to determine $R(z)$. If $R_n(z)$ is the residue of the function (5.9) at the pole $s = \alpha - n$, then

$$R(z) = \sum_{n=0}^{\infty} R_n(z) \tag{5.10}$$

where

$$R_n(z) = \lim_{s \to \alpha - n} \frac{(s - \alpha + n)\pi G(s)(-\frac{b}{t^2} z^t)^s}{\sin \pi s} \tag{5.11}$$

The residues $R_n(z)$, $n = 0, 1, 2, \ldots$ may be obtained, but for our purpose it suffices to note these will be of the forms $c_n z^{t(\alpha - n)}$, c_n being contants. In as much as we are interested only in determining the values of the constants which join the solutions $y_i(z)$, $i = 1, 2, 3$ linearly with $\widetilde{y_j}(z)$, $j = 1, 2, 3$, it will suffice for our purpose to determine $R_0(z)$ at the point $s = \alpha$.

The residue of $\frac{\pi G(s)}{\sin \pi s} (-\frac{b}{t^2} z^t)^s$ at the pole $s = \alpha$ is given by

$$R_0(z) = \lim_{s \to \alpha} \frac{(s - \alpha)\pi G(s)}{\sin \pi s} (-\frac{b}{t^2} z^t)^s \tag{5.12}$$

Since

$$\Gamma(s - \alpha) = \frac{\Gamma(s - \alpha + 1)}{s - \alpha}$$

and

$$\Gamma(-s)\Gamma(s + 1) = \frac{\pi}{\sin \pi s}$$

we obtain that

$$R_0(z) = \frac{\Gamma(-\beta_1)\Gamma(-\beta_2)}{\Gamma(\alpha - \beta_1)\Gamma(\alpha - \beta_2)} (-\frac{b}{t^2} z^t)^\alpha \tag{5.13}$$

Hence

$$\begin{aligned} -z^h R(z) &= -z^h \sum_{n=0}^{\infty} R_n(z) \\ &= c_1(h)\, s_1(z) \end{aligned} \tag{5.14}$$

where

$$c_1(h) = \frac{-\Gamma(-\beta_1)\Gamma(-\beta_2)}{\Gamma(\alpha-\beta_1)\Gamma(\alpha-\beta_2)}\left(-\frac{b}{t^2}\right)^{\alpha} \tag{5.15}$$

We observe that the expression for $c_i(h)$, $i = 1, 2, 3$ preserve meaning whether α is an integer or not. Thus we have arrived at the following theorem:

Theorem :

If in the differential equation $b \neq 0$ and if neither $a - 1$ nor $a - 2$ is congruent to zero modulo $m + 1$, then the solutions $y_i(z)$, $i = 1, 2, 3$ about the regular singular point $z = 0$, when considered for values of z of large modulus may be developed asymptotically in the form

$$y_i(z) \sim c_1(h_i)s_1(z) + c_3(h_i)s_3(z)$$
$$in \quad \frac{\pi}{2} < arg\ \left(\frac{i\sqrt{b}}{t}z^{\frac{t}{2}}\right) < \frac{3\pi}{2}$$

and

$$y_i(z) \sim c_1(h_i)s_1(z) + c_2(h_i)s_2(z)$$
$$in \quad -\frac{\pi}{2} \leq arg\ \left(\frac{i\sqrt{b}}{t}z^{\frac{t}{2}}\right) \leq \frac{\pi}{2}$$

where $c_1(h)$, $c_2(h)$, $c_3(h)$ are respectively given by (5.15), (5.8), (5.7) and $s_i(z)$, $i = 1, 2, 3$ are given respectively by (4.7), (4.8), and (4.9).

References

[1] G.D. Birkoff, 'Collected Mathematical Papers', Dover (1968).

[2] W.B. Ford, 'Studies in Divergent Series and Summability and the Asymptotic Developments of Functions Defined by Maclaurin Series', Chelsa Publishing Company, New york (1960).

[3] T.K. Puttaswamy, 'Solution in The Large of a Certain n^{th} Order Differential Equations', Proceedings of Ramanuja Birth Centinary Year International Symposium on Analysis, Macmillan India Limited (1989),451- 465.

[4] T.K. Puttaswamy, 'A Connection Problem for a nth Order Homogeneous Differential Equation'. Proceedings of the Internation al Conference on Theory and Applications of Differential Equations, Ohio University Press (1989), 317-321.

[5] T.K. Puttaswamy, 'A Generalization of a Theorem of W.B. Ford', Proceedings of the International Conference on New Trends in Geometry Function Theory and Applications, World Scientific (1991), 90-95.

[6] T.K. Puttaswamy, 'Stokes Multipliers for a Certain Third Order Differential Equation of W.B. Ford's type', Jour. of Ramanujan Math. Soc. 8. No's 1 and 2 (1993), 139-166.

[7] T.K. Puttaswamy, 'Stokes Multipliers for a Certain n^{th} Order Differential Equation', Proceedings of the Third International Colloquium on Differential Equations, International Science Publishers (1993), 157-168.

[8] T.K. Puttaswamy, 'Connection Coefficients for a Certain n^{th} Order Differential Equation', to appear in the Proceedings of the International Symposium on Ordinary Differential Equations held at Western Michigan University during May 20-22, 1993,World Scientific (1993).

[9] T.K. Puttaswamy and T.V. Sastry, 'Solution in the Large of a Certain Third Order Differential Equation of Langer', to appear in the Proceedings of the International Symposium on Ordinary Differential Equations, held at Western Michigan University during May 20-22, 1993,World Scientific (1993).

[10] H.L. Turrittin, 'Convergent Solution of Ordinary Linear Homogeneous Differential Equations in the Neighborhood of an Irregular Singular Point', Acta. Math. 93 (1955), 27-66.

[11] H.L. Turrittin, 'Stokes Multiplier for Asymptotic Solutions of a Certain Differential Equation', Trans. Amer. Math. Soc., 68, 1950,304-329.

4th Int. Coll. on Differential Equations, pp. 245-251
D. Bainov and V. Covachev (Eds)

Strong-electric-field eigenvalue asymptotics for the perturbed magnetic Schrödinger operator

G.D.RAIKOV

Institute of Mathematics, Bulgarian Academy of Sciences,

P.O.B. 373, 1090 Sofia, Bulgaria

Abstract.–We consider the Schrödinger operator $H_g^{\pm} := (i\nabla + A)^2 + F \mp gV$, $V \geq 0$. We study the asymptotic behaviour for large g of the eigenvalues situated in the gaps of the essential spectrum of the unperturbed operator H_0.

INTRODUCTION

For $u \in C_0^{\infty}(\mathbf{R}^m)$, $m \geq 2$, introduce the quadratic form

$$h_0[u] := \int_{\mathbf{R}^m} \left\{ |i\nabla u + Au|^2 + F|u|^2 \right\} dx.$$

Here $A \in L^2_{\rm loc}(\mathbf{R}^m; \mathbf{R}^m)$ is the magnetic potential and $F = \bar{F}$ is a scalar function. It is supposed that the multiplier by $|F|$ is $-\Delta$-form-bound with zero form bound, i.e. we have

$$\int_{\mathbf{R}^m} |F||u|^2 \, dx \leq \varepsilon \int_{\mathbf{R}^m} |\nabla u|^2 \, dx + c(\varepsilon) \int_{\mathbf{R}^m} |u|^2 \, dx, \quad \forall \varepsilon > 0, \quad \forall u \in C_0^{\infty}(\mathbf{R}^m). \tag{1}$$

Under these assumptions the quadratic form h_0 is lower-bounded and closable in $L^2(\mathbf{R}^m)$ (see [Av.Her.Sim, Theorem 2.5]). Denote by H_0 the selfadjoint operator generated in $L^2(\mathbf{R}^m)$ by the closed quadratic form h_0.

Further, assume that the multiplier by the function $V \geq 0$ is $-\Delta$-form-compact. Set

$$H_g^{\pm} := H_0 \mp gV$$

where $g > 0$ is the coupling constant of the perturbation $\mp V$. Since the multiplier by V is -Δ-form-compact (and, consequently, H_0-form-compact), we have

$$\sigma_{ess}(H_g^{\pm}) = \sigma_{ess}(H_0), \forall g > 0. \tag{2}$$

Fix a real number $\lambda \in \varrho(H_0)$ where $\varrho(H_0)$ denotes the resolvent set of H_0. By (2), λ is not in $\sigma_{ess}(H_g^{\pm})$ for all $g > 0$. Denote by $\mathcal{N}_g^{\pm}(\lambda)$ the number of the eigenvalues of the operator $H_t^{\pm}$ crossing λ as the parameter t grows from 0 to the value $g > 0$. The paper is devoted to the study of the asymptotic behaviour of $\mathcal{N}_g^{\pm}(\lambda)$ as $g \to \infty$, the value $\lambda = \bar{\lambda} \in \varrho(H_0)$ being fixed. It summarizes the results of the author published in [Bir.Rai] and [Rai 1-2].

Before we state our results we introduce some necessary notations and formulate some identities concerning $\mathcal{N}_g^{\pm}(\lambda)$.

Let T be a selfadjoint operator, and $P_I(T)$ be the spectral projection of T corresponding to the interval $I \subset \mathbf{R}$. Set

$$N(\lambda; T) := \text{rank } P_{(-\infty,\lambda)}(T), \quad \lambda \in \mathrm{R},$$

$$n_{\pm}(\mu; T) := \text{rank } P_{(\mu,\infty)}(\pm T), \quad \mu > 0.$$

If T is a linear compact operator, we put

$$\nu(\mu; T) := n_{\pm}(\mu^2; T^*T), \quad \mu > 0.$$

Then the identity

$$\mathcal{N}_g^{\pm}(\lambda) = n_{\pm}(g^{-1}; W(H_0 - \lambda)^{-1}W), \; \forall g > 0, \tag{3}$$

holds for $W := V^{1/2}$ and $\lambda = \bar{\lambda} \in \varrho(H_0)$. If, in addition, $\lambda < \inf \sigma_{ess}(H_0)$, we have

$$\mathcal{N}_g^{+}(\lambda) = N(\lambda; H_g^{+}), \; \forall g > 0. \tag{4}$$

1. THE CASE OF ARBITRARY A AND F, AND RAPIDLY DECAYING V

The main result of the present section is contained in the following theorem.

Theorem 1 ([Bir.Rai, Theorem 1.1]) *Let $m \geq 3$. Assume $A \in L^m_{\text{loc}}(\mathbf{R}^m; \mathbf{R}^m)$. Suppose that (1) holds and $V \in L^{m/2}(\mathbf{R}^m; \mathbf{R}_+)$. Fix $\lambda = \bar{\lambda} \in \varrho(H_0)$. Then we have*

$$\lim_{g\to\infty} g^{-m/2}\mathcal{N}_g^{+}(\lambda) = \frac{(4\pi)^{-m/2}}{\Gamma(1+m/2)} \int_{\mathbf{R}^m} V(X)^{m/2}\, dX, \tag{5}$$

$$\mathcal{N}_g^{-}(\lambda) = o(g^{m/2}), \; g \to \infty. \tag{6}$$

Sketch of the proof. In the case $\lambda < \inf \;\; \sigma_{ess}(H_0)$, and $F \equiv 0$, the asymptotics (5) have been established in [Rai 1, Theorem 1.1]. In this case we make use of the identity (4) and the Lieb estimate

$$N(0; H_1^{+}) \leq c_m \int_{\mathbf{R}^m} V(X)^{m/2}\, dX, \tag{7}$$

valid for $m \geq 3$, $F \equiv 0, A \in L^2_{\text{loc}}(\mathbf{R}^m; \mathbf{R}^m)$ and a constant c_m which depends only on the dimension m (see [Av.Her.Sim, Theorem 2.15]). Now (5) follows from a standard variational argument. In the case of arbitrary $\lambda = \bar{\lambda} \in \varrho(H_0)$, and general F satisfying (1), we write the weighted resolvent identity

$$W(H_0 - \lambda)^{-1}W = W(H_0 - F - \lambda_0)^{-1}W - W(H_0 - \lambda)^{-1}(F + \lambda_0 - \lambda)(H_0 - F - \lambda_0)^{-1}W \quad (8)$$

with $\lambda_0 < \inf \sigma(H_0 - F) \leq \inf \sigma_{ess}(H_0 - F)$, The next step is to establish the asymptotic estimates

$$\nu(s; W(H_0 - F - \lambda_0)^{-1}) = o(s^{-m}), s \downarrow 0, \quad (9)$$

$$\nu(s; W(H_0 - F - \lambda_0)^{-1/2}) = O(s^{-m}), s \downarrow 0, \quad (10)$$

Making use of (3), (7)-(10) and the result for $F \equiv 0$ and $\lambda = \lambda_0$, we easily come to (5) (see [Bir, Theorem 1.2]). The relation (6) is obtained in a similar way; we have just to take into account (3), and the fact that the operator $H_0 - F - \lambda_0$ is nonnegative, so that we have $n_-(s; W(H_0 - F - \lambda_0)^{-1}W) = 0$ for each $s > 0$.

The intrinsic reason for the banal Weylian type of the asymptotics (5) independent of quite general A, F and λ, is the fact that the perturbation V decays rapidly at infinity (i.e. $V \in L^{m/2}(\mathbf{R}^m)$). In the next section we shall see that the picture changes in a drastic way if V decays slowly at infinity.

2. THE CASE OF CONSTANT MAGNETIC FIELD B AND SLOWLY DECAYING V

In the sequel we assume $F \equiv 0$.

Introduce the magnetic-field

$$B \equiv \text{curl } A := \{B_{ij}\}_{ij=1}^{m}, B_{ij} = \partial_{X_i} A_j - \partial_{X_j} A_i, \; i,j = 1, \dots, m.$$

Throughout the section we assume that the entries $B_{i,j}, i,j = 1, \dots, m$, are independent of $X \in \mathbf{R}^m$, and, besides, B does not vanish identically. Let $b_1 \geq ... \geq b_d > 0$ be such numbers that the eigenvalues of the skew-symmetric matrix B are equal together with the multiplicities to the imaginary numbers $-ib_j$ and ib_j, $j = 1, \dots, d$. Thus we have rank $B = 2d$. Set $k := m - 2d = \dim \text{Ker } B$.

Introduce the selfadjoint operator

$$\mathrm{h} = \sum_{j=1}^{d} b_j \left\{ -\frac{\partial^2}{\partial x_j^2} + x_j^2 \right\} \quad (11)$$

defined originally on $C_0^\infty(\mathbf{R}^d)$, and then closed in $L^2(\mathbf{R}^d)$. The spectrum of h is purely discrete. Let $\{\Lambda_q\}$, $q \geq 1$, be the nondecreasing sequence of the eigenvalues of h counted with the multiplicities. The eigenvalues Λ_q have the form

$$\Lambda_q = \sum_{j=1}^{d} b_j(2n_j - 1)$$

where n_j, $j = 1, \ldots, d$, are positive integers.

Under our assumptions the spectrum of H_0 can be described explicitly. Namely, we have

$$\sigma(H_0) = \sigma_{ess}(H_0) = [\Lambda_1, \infty), \quad k \geq 1,$$

$$\sigma(H_0) = \sigma_{ess}(H_0) = \bigcup_{q=1}^{\infty} \Lambda_q, \quad k = 0.$$

Moreover, if $k \geq 1$, the spectrum of H_0 is purely absolutely continuous, while in the case $k = 0$ the spectrum of H_0 is purely point and each Λ_q is an eigenvalue of infinite multiplicity.

Let $t \in \mathbf{R}$. For $k \in \mathbf{Z}$, $k \geq 1$, set $\theta_k(t) := t_+^{k/2}$; respectively, $\theta_0(t)$ is the Heaviside function continuous from the left. For $t \in \mathbf{R}$, and a constant magnetic field B introduce the quantity

$$\Theta(t; B) := b_1 \ldots b_d \sum_{q=1}^{\infty} \theta_k(t - \Lambda_q)/2^{k+d}\pi^{m/2}\Gamma(1 + k/2).$$

Next, we define some classes of real-valued functions over $\mathbf{R}^m$ which decay at infinity. Let $f = \bar{f} \in C^l(\mathbf{R}^m), l \in \{1, 2, \ldots\}$. We shall write $f \in \mathcal{D}_{\alpha,l}, \alpha > 0$, if and only if the estimates

$$|D_\beta f(X)| \leq C_\beta < X >^{-\alpha-|\beta|}, < X >:= (1 + |X|^2)^{1/2}, \forall X \in \mathbf{R}^m,$$

hold for each multiindex β such that $|\beta| \leq l$. Moreover, set $\mathcal{D}_{\alpha,\infty} = \bigcap_{l=1}^{\infty} \mathcal{D}_{\alpha,l}$. Further, we shall write $f \in \mathcal{D}_{\alpha,l}^+$, $\alpha > 0$, $l \in \{1, \ldots, \infty\}$, if and only if $f \in \mathcal{D}_{\alpha,l}$, and the estimate

$$C < X >^{-\alpha} \leq f(X), \ |X| > R,$$

holds for some $C > 0, R > 0$. Finally, we shall write $f \in \mathcal{D}_{\alpha,l}^{++}$, $\alpha > 0$, $l \in \{1, \ldots, \infty\}$, if and only if $f \in \mathcal{D}_{\alpha,l}^+$, and the estimate

$$C < X >^{-\alpha} \leq -X.\nabla f(X), \ |X| > R,$$

holds for some $C > 0$, $R > 0$.

Assume $V \in \mathcal{D}_{\alpha,1}^+, \alpha > 0$, and for $g > 0$, $\lambda < \Lambda_1$, set

$$\nu_g^+(\lambda) := \int_{\mathbf{R}^m} \Theta(gV(X) + \lambda)\, dX.$$

Note that the estimates

$$C_\alpha^{-1} g^{m/\alpha} \leq \nu_g^+(\lambda) \leq C_\alpha g^{m/\alpha}, \ \alpha \in (0, 2),$$

$$C_2^{-1} g^{m/2} \log g \leq \nu_g^+(\lambda) \leq C_2 g^{m/2} \log g, \ \alpha = 2,$$

hold with some constants $C_\alpha \geq 1$, $\alpha \in (0, 2]$, for g large enough. Assume that the potential $V \in \mathcal{D}_{\alpha,1}^+, \alpha > 0$, obeys the asymptotics

$$V(X) = v(\hat{X})|X|^{-\alpha}(1 + o(1)), \hat{X} := X/|X|, \ |X| \to \infty, \tag{12}$$

with $v \in C(\mathbf{S}^{m-1})$, $v = \bar{v} > 0$. Then we have

$$\lim_{g\to\infty} g^{-m/\alpha}\nu_g^+(\lambda) =$$

$$\frac{b_1\dots b_d}{2^{k+d}\pi^{m/2}}\frac{\Gamma(m/\alpha-k/2)}{\alpha\Gamma(1+m/\alpha)}\sum_{q=1}^{\infty}(\Lambda_q-\lambda)^{k/2-m/\alpha}\int_{\mathbf{S}^{m-1}} v(\omega)^{m/\alpha}\,dS(\omega), \alpha \in (0,2), \tag{13}$$

$$\lim_{g\to\infty}(g^{m/2}\log g)^{-1}\nu_g^+(\lambda) = \frac{(4\pi)^{-m/2}}{2\Gamma(1+m/2)}\int_{\mathbf{S}^{m-1}} v(\omega)^{m/2}\,dS(\omega), \alpha = 2. \tag{14}$$

Let $V \in \mathcal{D}^+_{\alpha,1}$, $\alpha > 0$. For $s > 0$ set

$$\Phi_V(s) := \text{vol}\,\{X \in \mathbf{R}^m : V(X) > s\}.$$

We shall say that the potential V satisfies the condition $\mathcal{T}$ if and only if we have

$$\lim_{\delta\downarrow 0}\limsup_{s\downarrow 0}\Phi_V((1-\delta)s)/\Phi_V(s) = 1,$$

A sufficient condition that V satisfies $\mathcal{T}$ is $V \in \mathcal{D}^{++}_{\alpha,1}$, $\alpha > 0$. Another sufficient condition is that V obeys (12) any with $\alpha > 0$.

Theorem 2 ([Rai 1, Theorem 2.1]) *Let* $V \in \mathcal{D}^+_{\alpha,1}$, $\alpha \in (0,2]$. *If* $k \equiv \dim\operatorname{Ker} B = 0$ *and* $\alpha \in (0,2)$, *assume in addition that* V *satisfies the condition* $\mathcal{T}$. *Let* $\lambda < \Lambda_1$. *Then we have*

$$\mathcal{N}_g^+(\lambda) = \nu_g^+(\lambda)(1+o(1)), \quad g \to \infty. \tag{15}$$

Remark. Note that if $V \in \mathcal{D}_{\alpha,1}$, with $\alpha > 2$, then $V \in L^{m/2}(\mathbf{R}^m)$. Hence, in this case Theorem 1 is valid (provided $m \geq 3$).

The proof of Theorem 2 is based on the following important technical result.

Lemma 1 ([CdV, Theorem 3.1]) *Let* $Q_R \subset \mathbf{R}^m$, $m \geq 2$, *be a cube of side-length* R, *and* χ_R *be the selfadjoint operator generated in* $L^2(Q_R)$ *by the closure of the quadratic form* $\int_{Q_R}|i\nabla u + Au|^2\,dx$, $u \in C_0^\infty(Q_R)$, *with constant* $B \equiv \operatorname{curl} A$. *Then we have*

$$N(\lambda;\chi_R) \leq R^m\Theta(\lambda;B),$$

$$N(\lambda;\chi_R) \geq (R-R_0)^m\Theta(\lambda - C/R_0^2;B),$$

for each $\lambda \in \mathbf{R}$, $R > 0$, *any* $R_0 \in (0,R/2)$, *and a constant* C *which depends only on the dimension* m.

Besides, the proof of Theorem 2 employs the minimax principle, an appropriate covering of $\mathbf{R}^m$ by disjoint cubes introduced essentially in [Roz, Lemma 4], and a suitable version of the argument in the proof of [Tam, Theorem 1 (i)].

Here it should be noted that an important moment in the proof of (15) is the application of the relation

$$\lim_{\delta\downarrow 0}\limsup_{g\to\infty}\nu^+_{(1+\delta)g}(\lambda)/\nu^+_g(\lambda)=1,\quad \forall\lambda<\Lambda_1. \tag{16}$$

If $k\geq 1$ and $\alpha\in(0,2]$, or $k=0$ and $\alpha=2$, the relation (16) is implied just by $V\in\mathcal{D}^+_{\alpha,1}$. However, if $k=0$, and $\alpha\in(0,2)$ it is necessary to impose the requirement that V satisfies the condition $\mathcal{T}$ (see [Rai 3, Lemma 2.3]) in order to guarantee the validity of (16). Unfortunately, the assumption that V satisfies $\mathcal{T}$ if $k=0$ and $\alpha\in(0,2)$ has been omitted in the hypotheses of [Rai 1, Theorem 2.1].

If $k\geq 1$, the relation $\lambda=\bar\lambda\in\varrho(H_0)$ entails $\lambda<\Lambda_1$; hence, this case has been exhausted in Theorem 2. If $k=0$, however the point $\lambda=\bar\lambda\in\varrho(H_0)$ can be situated between two subsequent distinct eigenvalues Λ_q and Λ_{q+1}. In the remaining part of the paper we deal with this case.

Note that for $k=0$ we have

$$\nu^+_g(\lambda)=\frac{b_1\dots b_d}{(2\pi)^{m/2}}\sum_{q=1}^{\infty}\Phi_V(g^{-1}(\Lambda_q-\lambda)),\lambda<\Lambda_1.$$

We generalize this formula for arbitrary $\lambda=\bar\lambda\in\varrho(H_0)$ by setting

$$\nu^+_g(\lambda)=\frac{b_1\dots b_d}{(2\pi)^{m/2}}\sum_{q:\Lambda_q>\lambda}\Phi_V(g^{-1}(\Lambda_q-\lambda)),\lambda=\bar\lambda\in\varrho(H_0).$$

Moreover, we put

$$\nu^-_g(\lambda)=\frac{b_1\dots b_d}{(2\pi)^{m/2}}\sum_{q:\,\Lambda_q<\lambda}\Phi_V(g^{-1}(\lambda-\Lambda_q)),\lambda=\bar\lambda\in\varrho(H_0).$$

If $V\in\mathcal{D}^+_{\alpha,1}$, satisfies (12), then, similarly to (13), we have

$$\lim_{g\to\infty}g^{-m/\alpha}\nu^+_g(\lambda)=\frac{b_1\dots b_d}{m(2\pi)^{m/2}}\sum_{q:\,\Lambda_q>\lambda}(\Lambda_q-\lambda)^{k/2-m/\alpha}\int_{\mathbf{S}^{m-1}}v(\omega)^{m/\alpha}\,dS(\omega),\alpha\in(0,2);$$

if $\alpha=2$, then (14) is valid for arbitrary $\lambda=\bar\lambda\in\varrho(H_0)$. Besides, we have

$$\lim_{g\to\infty}g^{-m/\alpha}\nu^-_g(\lambda)=\frac{b_1\dots b_d}{m(2\pi)^{m/2}}\sum_{q:\,\Lambda_q<\lambda}(\lambda-\Lambda_q)^{k/2-m/\alpha}\int_{\mathbf{S}^{m-1}}v(\omega)^{m/\alpha}\,dS(\omega),\forall\alpha>0.$$

Theorem 3 ([Rai 2, Theorems 1.1 - 1.3]) *Let* $V\in\mathcal{D}^{++}_{\alpha,\infty}$, $k\equiv\dim\operatorname{Ker}B=0$, $\lambda=\bar\lambda\in\varrho(H_0)$.

(i) Assume $\alpha\in(0,2]$. *Then we have*

$$\mathcal{N}^+_g(\lambda)=\nu^+_g(\lambda)(1+0(1)),\quad g\to\infty.$$

(ii) Assume $\alpha>0$. *Then we have*

$$\mathcal{N}^-_g(\lambda)=\nu^-_g(\lambda)(1+0(1)),\quad g\to\infty.$$

Recall the operator h acting in $L^2(\mathbf{R}_x^d)$ (see (11)) and introduce the operator

$$\mathcal{H}_0 = \int_{\mathbf{R}_y^d}^{\oplus} \mathrm{h}\, dy$$

acting in $L^2(\mathbf{R}_{x,y}^{2d})$. Let $\mathcal{V}$ be a pseudodifferential operator (ΨDO) with Weyl symbol $V_b(x - \eta, y - \xi)$, $(x, y; \xi, \eta) \in T^*\mathbf{R}^{2d}$, where

$$V_b(x,y) := V(b_1^{-1/2}x_1, \ldots, b_d^{-1/2}x_d, b_1^{-1/2}y_1, \ldots, b_d^{-1/2}y_d).$$

The proof of Theorem 3 is based on the unitary equivalence of the operators $H_g^{\pm}$ and $\mathcal{H}_g^{\pm} := \mathcal{H}_0 \mp g\mathcal{V}$. Moreover, we employ the general theory of ΨDOs with Weyl symbols (see [Shu, Ch. IV]) and, in particular, the results in [Dau.Rob] on spectral asymptotics for ΨDOs of negative order acting on $L^2(\mathbf{R}^m)$.

Acknowledgments The work has been partly supported by the Bulgarian Science Foundation under Grant MM 8/91.

References

[Av.Her.Sim] J.Avron, I.Herbst, B.Simon, *Schrödinger operators with magnetic fields. I. General interactions*, Duke. Math. J. **45** (1978), 847-883.

[Bir] M.Š.Birman, *Discrete spectrum in the gaps of a continuous one for perturbations with large coupling constant.* In: Estimates and Asymptotics for Discrete Spectra of Integral and Differential Equations. Advances in Soviet Mathematics, **7** (1991), 57-74, AMS, Providence.

[Bir.Rai] M.Š.Birman, G.D.Raikov, *Discrete spectrum in the gaps for perturbations of the magnetic Schrödinger operator*, In: Estimates and Asymptotics for Discrete Spectra of Integral and Differential Equations. Advances in Soviet Mathematics, **7** (1991), 75-84, AMS, Providence.

[CdV] Y. Colin de Verdière, *L'asymptotique de Weyl pour les bouteilles magnétiques*, Commun. Math. Phys. **105** (1986), 327-335.

[Dau.Rob] M.Dauge, D.Robert, *Weyl's formula for a class of pseudodifferential operators with negative order on $L^2(\mathbf{R}^n)$*, In: Proceedings of the Conference on Pseudo-Differential Operators, Oberwolfach, Lect.Notes Math. **1256** (1987), 91-122.

[Rai 1] G.D,Raikov, *Strong electric field eigenvalue asymptotics for the Schrödinger operator with electromagnetic potential*, Lett.Math.Phys. **21** (1991), 41-49

[Rai 2] G. D. Raikov, *Strong-electric-field eigenvalue asymptotics for the perturbed magnetic Schrödinger operator*, Commun.Math.Phys. **155** (1993), 415-428.

[Rai 3] G.D.Raikov, *Semiclassical and weak-magnetic-field eigenvlaue asymptotics for the Schrödinger operator with electromagnetic potential*, to appear in Ann. Inst. H. Poincaré, Phys. Théorique.

[Roz] G.V.Rozenbljum, *An asymptotic of the negative discrete spectrum of the Schrödinger operator*, Math.Notes USSR **21** (1977), 222-227.

[Shu] M.A.Shubin, *Pseudodifferential Operators and Spectral Theory*, Berlin, Heidelberg, New York: Springer, 1987

[Tam] H.Tamura *Asymptotic distribution of eigenvalues for Schrödinger operators with homogeneous magnetic fields*, Osaka J. Math. **25** (1988) 633-647

4th Int. Coll. on Differential Equations, pp. 253-264
D. Bainov and V. Covachev (Eds)

Stability Theory for a Class of 2D Linear Systems Described by a Set of Recursive Differential Equations

E Rogers* and D H Owens+

* *Advanced Systems Research Group, Department of Aeronautics & Astronautics, University of Southampton, UK*

+ *Centre for Systems and Control Engineering, School of Engineering, University of Exeter, UK*

Abstract

A rigorous stability theory for the class of 2D linear systems known as repetitive processes is described using an abstract model in a Banach space setting. This is then applied to the sub-class of differential linear processes to yield stability tests which can be implemented using standard software. Some extensions of these basic results are also noted together with some currently open research problems.

Introduction

The essential unique characteristic of a repetitive, or multipass, process is a series of sweeps, or passes, through a set of dynamics defined over a finite fixed duration - the pass length. On each pass an output - the pass profile - is produced which acts as a forcing function on, and hence contributes to, the dynamics of the next pass profile. This is the source of the unique general control problem in that the output sequence can contain oscillations which increase in amplitude from pass to pass.

In the simplest case, therefore, the output at any point on the current pass is a function of the inputs/disturbances at this point and the output at the same point on the previous pass. This is the so-called 'point' assumption and here an abstract model in a Banach space setting is used to present a rigorous stability theory. Application of this theory to the sub-class of differential linear processes then yields conditions which can be tested using standard software. Some extensions of these basic results are also noted together with some currently open research problems.

Preliminaries

A study of known examples [1] indicates that repetitive, or multipass, processes have a number of common features which distinguish them from more familiar dynamic processes. These are detailed below

(i) A number of passes through a set of (assumed) known dynamics, indexed here by $k = 1, 2, \ldots$

(ii) Each pass is characterised by a pass length $\alpha_k \in (0, +\infty)$ and a pass profile $Y_k(t)$ defined on $0 \leq t_k \leq \alpha_k$. Note that $Y_k(t)$ need not be a scalar quantity.

(iii) An initial pass profile $Y_0(t)$, $0 \leq t \leq \alpha_0$, where α_0 is the initial pass length and, in effect, Y_0 is an initial condition for the process.

(iv) Each pass is subject to its own boundary conditions, disturbance and control inputs.

(v) The process has the unit memory property, i.e. the dynamics of pass k evolve from the independent inputs to that pass and the pass profile on pass $k - 1$.

An abstract setting for the development of a rigorous stability theory for repetitive processes can be obtained by regarding the pass profile $Y_k(t)$ on pass k as a point in a suitably chosen function space. In particular, regard the pass profile Y_k as a point in a Banach space E_{α_k}, i.e. $Y_k \in E_{\alpha_k}$, $k \geq 0$. Then a general abstract model of the process dynamics has the recursion relation structure

$$Y_{k+1} = f_{k+1}(Y_k),\ k \geq 0 \tag{1}$$

(where f_{k+1} is an abstract mapping of E_{α_k} into $E_{\alpha_{k+1}}$) together with a method of updating the pass length α_k

$$\alpha_{k+1} = g_{k+1}(\alpha_k, Y_k, Y_{k+1}),\ k \geq 0 \tag{2}$$

Note also that this structure has a natural extension to the case where it is the previous $M > 1$ pass profiles which contribute to the current one. The integer M is termed the memory length, and these so-called non-unit memory processes are not considered further here since, see [1] for a complete treatment, all of the results presented generalise in a natural maner.

Clearly the analysis of the abstract model of (1) would be a formidable, if not impossible, task. Hence attention is restricted to processes with linear dynamics and a constant pass length, i.e. $\alpha_k = \alpha$, $k \geq 0$. The following abstract model includes [1] all such processes as special cases.

Definition 1: A linear repetitive process $S(E_\alpha, W_\alpha, L_\alpha)$ of constant pass length $\alpha > 0$ consists of a Banach space E_α, a linear subspace W_α of E_α, and a bounded linear operator L_α of E_α into itself. The system dynamics are described by linear recursion relations of the form

$$Y_{k+1} = L_\alpha Y_k + b_{k+1}, k \geq 0 \tag{3}$$

where $Y_k \in E_\alpha$ is the pass profile on pass k and $b_{k+1} \in W_\alpha$, $k \geq 0$. The term $L_\alpha Y_k$ represents the contribution of pass k to pass $k+1$ and b_{k+1} represents known initial conditions, disturbances and control input effects.

In this paper only one sub-class of $S(E_\alpha, W_\alpha, L_\alpha)$ is considered. This consists of so-called differential linear repetitive processes which in the simplest case have the state-space model

$$\dot{X}_{k+1}(t) = AX_{k+1}(t) + BU_{k+1}(t) + B_0 Y_k(t)$$

$$Y_{k+1}(t) = CX_{k+1}(t) + D_1 Y_k(t)$$

$$X_{k+1}(0) = d_{k+1}, 0 \leq t \leq \alpha, k \geq 0 \tag{4}$$

Here $Y_k(t)$ is the $m \times 1$ vector pass profile on the kth pass, $X_k(t)$ is the $n \times 1$ state vector describing the dynamics on the kth pass and $U_k(t)$ is an $\ell \times 1$ vector control input to the kth pass. The matrices A, B, B_0, C and D_1 have dimensions $n \times n$, $n \times \ell$, $n \times m$, $m \times n$ and $m \times m$ respectively. Note also again that the pass length α is finite in all cases.

To write (4) in the form of definition 1 first note that

$$Y_{k+1}(t) = C\int_0^t e^{A(t-t^1)}\left\{B_0 Y_k(t^1) + BU_{k+1}(t^1)\right\}dt^1$$

$$+ Ce^{At}d_{k+1} + D_1 Y_k(t),\ 0 \leq t \leq \alpha, k \geq 0 \tag{5}$$

Further, consider the problem in the context of the Banach space $E_\alpha = C_m(0, \alpha)$ of bounded continuous mappings of the interval $0 \leq t \leq \alpha$ into the vector space of complex m-vectors C^m with norm

$$\|Y\| = \sup_{0 \leq t \leq \alpha} \|Y(t)\|_m \tag{6}$$

where $\|\,.\,\|_m$ is any convenient norm in C^m. Then L_α is defined by the relation

$$(L_\alpha Y)(t) = D_1 Y(t) + C\int_0^t e^{A(t-t^1)} B_0 Y(t^1)\,dt^1, 0 \leq t \leq \alpha \tag{7}$$

and the disturbance b_{k+1} by

$$b_{k+1}(t) = Ce^{At} d_{k+1} + C \int_0^t e^{A(t-t^1)} BU_{k+1}(t^1)\, dt^1 \tag{8}$$

Suppose also that the initial conditions d_{k+1}, $k \geq 0$, of interest lie in a subspace W of R^n. Further, consider the case of piecewise continuous control inputs $U_{k+1}(t)$, $k \geq 0$. Then $W_\alpha \subset E_\alpha$ is obtained by evaluating (8) for all such d_{k+1} and U_{k+1}.

The model of (4) is the simplest possible in the sense that at any point on the current pass the only previous pass contribution is from the same point. Examples also exist, however, where this point assumption is no longer valid due to dynamic interaction between passes, termed interpass smoothing, which severely distorts the previous pass profile prior and/or during the production of the current one. For example, in long-wall coal cutting, the interpass smoothing is caused by the machines weight (up to 5 tonnes) as it is 'pushed over' to rest on the newly cut floor profile.

One feasible means of enhancing the 'point' model in such cases is to assume that the output at $t \in [0, \alpha]$ on the current pass is a function of the complete previous pass profile. Edwards and co-workers [2] have carried out extensive modelling studies on this problem of interpass smoothing for the long-wall coal cutting example. This has led to a number of possible models but here attention is restricted to

$$\dot{X}_{k+1}(t) = AX_{k+1}(t) + BU_{k+1}(t) + B_0 \int_0^\alpha K(t,\tau)\, Y_k(\tau)\, d\tau$$

$$Y_{k+1}(t) = CX_{k+1}(t),\; X_{k+1}(0) = d_{k+1},\; 0 \leq t \leq \alpha,\; k \geq 0 \tag{9}$$

In effect, the term $\int_0^\alpha K(t, \tau)\, Y_k(\tau)\, d\tau$ here represents a 'smoothing out' of the previous pass profile in a maner governed by the properties of the kernel $K(t, \tau)$. Further, the choice of $K(t, \tau) = \delta(t - \tau)\, I_m$, where $\delta(.)$ denotes the Dirac delta function, reduces (9) to (4). It is also easy to show that (9) is a linear repetitive process $S(E_\alpha, W_\alpha, L_\alpha)$, $E_\alpha = C_m(0, \alpha)$, with

$$(L_\alpha Y)(t) = C \int_0^t e^{A(t-\tau)} B_0 \int_0^\alpha K(\tau, t^1)\, Y(t^1)\, dt^1\, d\tau \tag{10}$$

Abstract Stability Theory

The basic unique control problem for repetitive processes is that the output sequence $\{Y_k\}_{k \geq 1}$ can contain oscillations which increase in amplitude from pass to pass. Hence the natural definition of asymptotic stability is to demand that, given any initial profile Y_0 and any disturbance sequence $\{b_k\}_{k\geq 1}$ which 'settles down' to a 'steady' disturbance b_∞ as $k \to +\infty$, $\{Y_k\}_{k\geq 1}$ converges to a 'steady' profile Y_∞ as $k \to +\infty$.

In practical applications it is clearly desirable that any definition of stability is retained if the model is perturbed slightly due to modelling errors or simulation approximations. Hence the following definition of asymptotic stability is used since it ensures that the 'set of stable systems' is (in a well defined sense) open in the class of all linear repetitive processes.

Definition 2: $S(E_\alpha, W_\alpha, L_\alpha)$ is said to be asymptotically stable if $\exists$ a real scalar $\delta > 0$ such that, given any Y_0 and any strongly convergent disturbance sequence $\{b_k\}_{k\geq 1} \subset W_\alpha$, $\{Y_k\}_{k\geq 1}$ generated by the perturbed process

$$Y_{k+1} = (L_\alpha + \gamma)\,Y_k + b_{k+1},\ k \geq 0 \tag{11}$$

converges strongly to a 'limit profile' $Y_\infty \in E_\alpha$ whenever $\|\gamma\| \leq \delta$.

Note: Y_∞ does, of course, depend on the model perturbation γ, Y_0, and $\{b_k\}_{k\geq 1}$.

The following result, proved in [1], gives the necessary and sufficient condition for asymptotic stability of $S(E_\alpha, W_\alpha, L_\alpha)$.

Theorem 1: $S(E_\alpha, W_\alpha, L_\alpha)$ is asymptotically stable if, and only if,

$$r(L_\alpha) < 1 \tag{12}$$

where $r(.)$ denotes the spectral radius.

In practical terms, this result provides little information on transient behaviour. To remove this difficulty, the following definition and theorem (again see [1] for complete details) have been introduced.

Definition 3: Suppose that $S(E_\alpha, W_\alpha, L_\alpha)$ is asymptotically stable and $\{b_k\}_{k\geq 1}$ converges strongly to b_∞. Then the strong limit

$$Y_\infty := \operatorname*{limit}_{k \to +\infty} Y_k \tag{13}$$

is termed the limit profile corresponding to $\{b_k\}_{k\geq 1}$.

Theorem 2: Suppose that the conditions of definition 3 hold. Then Y_∞ is the unique solution of the linear equation

$$Y_\infty = L_\alpha Y_\infty + b_\infty \tag{14}$$

Note again here that the pass length α is finite and fixed. Further, what is really required in all practical cases is a 'robust' stability theory which is independent of α. (This requirement will be clearly demonstrated in the next section which applies the results of this section to differential processes). This is achieved by defining it in terms of the so-called extended process, denoted by $S(E_\alpha, W_\alpha, L_\alpha)_{\alpha \geq \alpha_0}$, where α_0 is some nominal value of interest. The formal definition is as follows

Definition 4: $S(E_\alpha, W_\alpha, L_\alpha)_{\alpha \geq \alpha_0}$ is termed stable along the pass if, and only if, $\exists$ real numbers $M_\infty > 0$ and $\lambda_\infty \in (0, 1)$, independent of α, such that

$$\|L_\alpha^k\| \leq M_\infty \lambda_\infty^k, \; k \geq 0 \tag{15}$$

where $\|\,.\,\|$ denotes the norm on E_α

Necessary and sufficient conditions for stability along the pass are given in the following result.

Theorem 3: $S(E_\alpha, W_\alpha, L_\alpha)_{\alpha \geq \alpha_0}$ is stable along the pass if, and only if,

a) $$r_\infty := \sup_{\alpha \geq \alpha_0} r(L_\alpha) < 1 \tag{16}$$

and

b) $$M_0 := \sup_{\alpha \geq \alpha_0} \sup_{|z| \geq \lambda} \|(zI - L_\alpha)^{-1}\| < +\infty \tag{17}$$

for some $\lambda \in (r_\infty, 1)$.

Proof: This is based on that of theorem 3.3.2 in [1]. To prove necessity, first (using the basic definition) note that

$$r(L_\alpha) = \lim_{k \to +\infty} \|L_\alpha^k\|^{1/k} \leq \lim_{k \to +\infty} M_\infty^{1/k} \lambda_\infty < 1 \tag{18}$$

$\forall\, \alpha \geq \alpha_0$ and hence $r_\infty \leq \lambda_\infty < 1$. Further, $\forall\, \lambda \in (\lambda_\infty, 1)$ and any $|z| \geq \lambda$

$$\|(zI - L_\alpha)^{-1}\| \le \frac{1}{\lambda} \sum_{j=0}^{\infty} \left(\frac{\lambda_\infty}{\lambda}\right)^j M_\infty$$

$$= \frac{M_\infty}{\lambda(1 - \frac{\lambda_\infty}{\lambda})}, \quad \forall \alpha \ge \alpha_0 \tag{19}$$

For a proof of sufficiency, suppose that $r_\infty < 1$ and consider the contour in the complex plane defined by $M = \{z : |z| = \lambda\}$. Then use standard results from operational calculus in Banach spaces to write $L_\alpha{}^k$ as the contour integral

$$L_\alpha^k = \frac{1}{2\pi i} \int_M z^k (zI - L_\alpha)^{-1} dz \tag{20}$$

where $z = e^{i\theta}$ and i denotes 'the square root of minus one'. Taking norms now yields

$$\|L_\alpha^k\| \le \frac{1}{2\pi} \| \int_0^{2\pi} \lambda^k e^{ik\theta} (\lambda e^{i\theta} - L_\alpha)^{-1} \lambda i e^{i\theta} d\theta \|$$

$$\le \frac{1}{2\pi} \int_0^{2\pi} \lambda^{k+1} \|(\lambda e^{i\theta} - L_\alpha)^{-1}\| d\theta$$

$$\le (M_0 \lambda) \lambda^k \tag{21}$$

This verifies (17) with $M_\infty = M_0 \lambda$ and $\lambda = \lambda_\infty$ and the proof is complete.

Note: (16) is equivalent to asymptotic stability $\forall \alpha \ge \alpha_0$. Hence the reason for retaining the seperate identities of (a) and (b) in theorem 3, despite the fact that (b) does imply (a).

Stability of Differential Processes

Return now to the state-space model of the 'point' process of (4). Then to interpret theorem 1 for this case requires the calculation of the spectral values of L_α of (7). One approach to this is to consider the equation

$$(zI - L_\alpha) Y = \eta \tag{22}$$

and construct necessary and sufficient conditions on the complex scalar z to ensure that a solution exists $\forall \eta \in E_\alpha$ and that this solution is bounded in the sense that $\| Y \| \leq \gamma \| \eta \|$ for some real scalar $\gamma > 0$ and $\forall\ \eta \in E_\alpha$. This yields the following result which is again proved in [1].

Theorem 4: The process of (4) is asymptotically stable if, and only if,

$$r(D_1) < 1 \tag{23}$$

The weakness of asymptotic stability is now immediate since it is independent of the matrices A, B, B_0 and C and, in particular, independent of the eigenvalues of A. This is due entirely to the fact that α is finite and would permit 'exponential growth' terms in the dynamics along any pass. Such a situation is clearly unacceptable in practical applications and will be removed by the requirement of stability along the pass.

Suppose that theorem 4 holds. Then an immediate corollary is that the limit profile Y_∞ is a standard linear system with state space model parameterised by the triple $\{A + B_0 (I_m - D_1)^{-1} C, B, (I_m - D_1)^{-1} C\}$. Hence if the process is asymptotically stable it's repetitive dynamics can, after 'sufficiently large' number of passes, be replaced by those of a standard linear system. The implications of this fact in terms of the formulation and solution of practically feasible control policies is discussed in detail in [3].

To illustrate the weakness of asymptotic stability consider the case when $A = -1$, $B = 1, B_0 = 1 + \beta, \beta > 0, C = 1, D_1 = 0$. This process is asymptotically stable with limit profile dynamics

$$\dot{Y}_\infty(t) = \beta Y_\infty(t) + U_\infty(t), 0 \leq t \leq \alpha \tag{24}$$

Setting $U_{k+1}(t) \equiv 1, k \geq 0$ and $Y_0(t) \equiv 0$ yields

$$Y_1(t) = 1 - e^{-t}, 0 \leq t \leq \alpha \tag{25}$$

$$Y_\infty(t) = \beta^{-1}(e^{\beta t} - 1), 0 \leq t \leq \alpha \tag{26}$$

Hence, despite the fact that $Y_1(t)$ is a quite acceptable 'classical' response to a unit step demand, $Y_\infty(t)$ is clearly 'unstable along the pass' in an obvious intuitive sense.

The 'boundedness' condition (b) of theorem 3 is equivalent to the existence of a $\lambda \in (r_\infty, 1)$ such that (22) has a uniformly bounded, with respect to α, solution $Y \in E_\alpha$ $\forall$ choices of $\eta \in E_\alpha$ satisfying $\sup_\alpha \| \eta \| < +\infty$ and $\forall\ |z| \geq \lambda$. Further, it is clear that, in general, this condition could prove very difficult to interpret. In the case of (4), however, the following result is obtained (see again [1] for a proof and alternatives).

Theorem 5: Suppose that the pair $\{A, B_0\}$ is controllable and the pair $\{C, A\}$ is observable and let $G(s)$ denote the transfer-function matrix of the state-space quadruple (A, B_0, C, D_1). Then $S(E_\alpha, W_\alpha, L_\alpha)_{\alpha \geq \alpha_0}$ generated by (4) with $\alpha \geq \alpha_0$ is stable along the pass if, and only if,

(*a*) $$r(D_1) < 1 \tag{27}$$

(*b*) $$|sI_n - A| \neq 0, Re(s) \geq 0 \tag{28}$$

and

(*c*) $$|zI_m - G(i\omega)| \neq 0, \forall |z| = 1, \omega \geq 0 \tag{29}$$

These conditions can be tested using standard software. Further, it is clear that they should be tested in the order of (a) followed by (b) followed by (c) with termination if the one just tested does not hold, i.e. computationally the most intensive last and only if required. The use of these tests in the controller design environment is currently under investigation with preliminary results in [3].

One major defficiency with the 'Nyquist like' tests of theorem 5 is that they do not provide applicable measures of expected performance (compare with the standard linear systems case). Further, the interpretation of (b) of theorem 3, in particular, is a non-trivial task for the interpass smoothing model. The remainder of this section develops an alternative test which extends naturally to the interpass smoothing model and provides computable performance measures. This requires the following preliminary results which are proved in [4].

Lemma 1: Suppose that f is a scalar continuous function defined on $[0, +\infty]$ and of bounded variation on any finite interval $[0, T]$. Then the norm of f on $[0, T]$, denoted $N_T(f)$, is given by

$$N_T(f) = |f(0+)| + \sum_{j=1}^{k} |f(t_j) - f(t_{j-1})|$$

$$+ |f(T) - f(t_k)| \tag{30}$$

where $0 = t_0 < t_1 < t_2 < \ldots$ are the local maxima and minima of f on $[0, +\infty]$ and $t_k \leq T < t_{k+1}$. In the case of $T = +\infty$,

$$N_\infty(f) = \sup_{T \geq 0} N_T(f) = \operatorname{limit}_{k \to +\infty} N_T(f) \tag{31}$$

whenever the limit exists.

Lemma 2: Suppose that L is a bounded linear operator mapping $L_\infty(0, +\infty)$ into itself of the form

$$(LY)(t) = \int_0^\alpha K(t, \tau)\, Y(\tau)\, d\tau \tag{32}$$

Then

$$\|L\| \leq \sup_{t \geq 0} \int_0^\infty |K(t, \tau)|\, d\tau \tag{33}$$

and equality holds in (33) if $K \geq 0, \forall\, t, \tau$.

Return now to the 'point' case of (4) and suppose that the matrix A is stable - a necessary condition for stability along the pass. Suppose also that the step response matrix of the system parameterised by the state space quadruple (A, B_0, C, D_1) is available from appropriate simulation studies i.e.

$$W(t) = C \int_0^t e^{At^1} B_0\, dt^1 + D_1, t \geq 0 \tag{34}$$

with elements denoted by $W_{ij}(t)$. Further, apply the formula (31) of lemma 1 to each element in turn of $W(t)$ and denote the resulting matrix by $\| L \|_p$. Then the following result gives a sufficient condition for stability along the pass of (4).

Theorem 6: $S(E_\alpha, W_\alpha, L_\alpha)_{\alpha \geq \alpha_0}$ generated by (4) with $\alpha \geq \alpha_0$ is stable along the pass if

$$r(\|L\|_p) < 1 \tag{35}$$

Tests of this nature are 'robust' in a well defined numerical sense and are easily [3] incorporated into a computer aided analysis environment. Further, they produce, at negligible extra computational cost, unique computable information concerning the expected performance of (4). Complete details can be found in [1] and [3].

Return now to the interpass smoothing model of (9) and re-write it as

$$X_{k+1}(t) = AX_{k+1}(t) + BU_{k+1}(t) + B_0 V_k(t)$$

$$Y_{k+1}(t) = CX_{k+1}(t) \tag{36}$$

where

$$V_k(t) = \int_0^{\alpha} K(t,\tau)\, Y_k(\tau)\, d\tau \tag{37}$$

The Laplace transform of the contribution of the previous pass dynamics acting alone is

$$G_1(s) = G(s)\, V(s) \tag{38}$$

where $V(s)$ is the Laplace transform of

$$V(t) = \int_0^{\infty} K(t,\tau)\, Y(\tau)\, d\tau := (K\, Y)(t) \tag{39}$$

and K is the integral operator with kernel $K(t, \tau)$.

It now follows that the L_∞ induced norm of GK is bounded above by

$$\| L_s \|_p = \| L \|_p \| K \|_p \tag{40}$$

where

$$K = \sup_{t \geq 0} \int_0^{\infty} K(t,\tau)\, d\tau \tag{41}$$

Here the supremum is interpreted with respect to the partial ordering on $n_1 \times n_2$ matrices defined by $A \leq B$ if, and only if, (with the subscript denoting the elements) $A_{ij} \leq B_{ij}$, $\forall\ i, j$. The following result now provides a computable sufficient condition for stability along the pass of (9). (For others, again see [1] and [3]).

Theorem 7: $S(E_\alpha, W_\alpha, L_\alpha)_{\alpha \geq \alpha_0}$ generated by (9) with $\alpha \geq \alpha_0$ is stable along the pass if

$$r(\| L_s \|_p) < 1 \tag{42}$$

Proof: Omitted for brevity and detailed in [4].

Given this basic stability result, it is now possible to extend the performance bounds associated with the 'point' special case of theorem 3 to processes described by (9). This general area requires much further development work, which is on-going and will be reported in due course.

Conclusions

A rigorous stability theory for the class of 2D linear systems known as linear repetitive processes has been detailed in an abstract Banach space setting. Application of this theory to the sub-class of differential linear processes has produced conditions which can be tested using standard software, with consequent benefits in terms of controller design. Some basic results on the extension of these tests to interpass smoothing effects, often of major significance in terms of the systems dynamic behaviour, have also been developed. Output from the indepth development and/or extension of the results presented here will be reported in due course.

References

[1] Rogers E and Owens D H. 'Stability Analysis for Linear Repetitive Processes', Springer Verlag Lecture Notes in Control and Information Sciences Series, Vol 175, 1992.

[2] Edwards, J B et al. 'Modelling the Dynamics of Long-wall Coal Cutting', Research Report, Dept of Control Engineering, University of Sheffield, U.K, 1974.

[3] Smyth, K J. 'Computer Aided Analysis for Linear Repetitive Processes', PhD Thesis, Dept of Mechanical Engineering, University of Strathclyde, UK. 1992.

[4] Rogers, E and Owens D H. 'Stability Theory for Repetitive Processes with Interpass Smoothing Effects'. Research Report, Dept of Aeronautics and Astronautics, University of Southampton, UK. 1993.

4th Int. Coll. on Differential Equations, pp. 265-275
D. Bainov and V. Covachev (Eds)

Analyticity of the Solutions of the Heat Equation on the Half Space $\mathbb{R}^n_+$

SABUROU SAITOH

Department of Mathemaics, Gunma University, Kiryu 376, Japan

Abstract. Analyticity of the solutions of the heat equation on the half space R^n_+ is examined.

Keywords: Heat equation, analyticity, half space, reproducing kernel, analytic extension

1. INTRODUCTION

In [12, 14], we examined several isometrical identities in the one-dimensional heat equation. The results turned to analytic extension formulas of new type from several points of view in [1, 2, 3, 8, 9, 15]. Furthermore, these analytic extension formulas are applied to the investigations of analyticity of the solutions of nonlinear partial differential equations in [6, 7, 8, 9].

In this connection, we shall examine a higher dimensional case. As we see later, the results will be quite different from the one-dimensional case. In particular, we shall deal with the typical case as follows :

For $\mathbb{R}^n_+ = \{\mathbf{x} = (x_1, \mathbf{x}') = (x_1, x_2, \cdots, x_n) | x_1 > 0\}$, we examine the solutions $u(\mathbf{x}, t)$ of the heat equation

$$u_t(\mathbf{x}, t) = \Delta u(\mathbf{x}, t) \quad \text{on} \quad \mathbb{R}^+ \times \mathbb{R}^n_+ \tag{1.1}$$

subject to

$$u(0, \mathbf{x}', t) = F(\mathbf{x}', t) \quad \text{on} \quad \mathbb{R}^+ \times \mathbb{R}^{n-1}, \tag{1.2}$$

$$u(\mathbf{x}, 0) = 0 \quad \text{on} \quad \mathbb{R}^n_+, \tag{1.3}$$

$$u(\mathbf{x}, t) \to 0 \quad \text{as} \quad \mathbf{x} \to \infty \tag{1.4}$$

for functions F satisfying

$$\int_0^t \int_{\mathbb{R}^{n-1}} |F(\mathbf{x}', T)|^2 d\mathbf{x}' dT < \infty. \tag{1.5}$$

Of course, we need a suitable interpretation for the boundary values (1.2) and we also need a smoothness condition for F for the classical solutions $u(\mathbf{x}, t)$. However we shall examine the integral transform itself (see (2.1)) in the representation of $u(\mathbf{x}, t)$ in terms of F satisfying (1.5). In particular, by applying a general method for integral transforms using the theory of reproducing kernels, we shall give a proper identification of the images $u(\mathbf{x},t)$ and establish a natural isometrical identity between $u(\mathbf{x},t)$ and F. Furthermore, we shall establish related analytic extension formulas as in the one-dimensional case. However, the formulas obtained are quite different from the case of one-dimensional case. In particular, for any fixed $t>0$ the natural existence domain of $u(\mathbf{x},t)$ will be

$$D_H = \{(Rez_1)^2 > (Imz_1)^2 + \cdots + (Imz_n)^2\} \tag{1.6}$$

on $\mathbb{C}^n$, and the images $u(\mathbf{x},t)$ are analytic on D_H. In Section 2 we shall examine representations of $u(\mathbf{x},t)$ and their analyticity, and in Section 3 we shall establish analytic extension formulas to D_H.

2. REPRESENTATION OF $u(\mathbf{x},t)$ AND ANALYTICITY

We first recall the integral representation of $u(\mathbf{x},t)$ in terms of F, for $\mathbb{X} = (X_1, X_2, \cdots, X_n) = (X_1, \mathbb{X}')$

$$u(\mathbf{x},t) = -\int_0^t \int_{\mathbb{R}^{n-1}} -\frac{\partial}{\partial X_1} G(\mathbf{x},(X_1,\mathbb{X}');t-T)|_{X_1=0} F(\mathbb{X}',T)d\mathbb{X}'dT. \tag{2.1}$$

Here, $G(\mathbf{x},\mathbb{X};t)$ is the Green's function of (1.1)-(1.4), and for the fundamental solution $G^*(\mathbf{x},\mathbb{X};t)$ of the heat equation on $\mathbb{R}^+ \times \mathbb{R}^n$

$$\begin{aligned} G^*(\mathbf{x},\mathbb{X};t) &= \frac{1}{(2\pi)^n}\int_{\mathbb{R}^n} e^{-p^2 t} e^{i\mathbf{p}\cdot(\mathbf{x}-\mathbb{X})} d\mathbf{p} \\ &= \left(\frac{1}{4\pi t}\right)^{\frac{n}{2}} \exp\left\{-\frac{|\mathbf{x}-\mathbb{X}|^2}{4t}\right\} \qquad (p^2 = \sum_{j=1}^n p_j^2), \end{aligned}$$

$$G(\mathbf{x},\mathbb{X};t) = G^*(\mathbf{x},\mathbb{X};t) - G^*(\mathbf{x},(-X_1,\mathbb{X}');t)$$

(cf. [10, pp. 248-251]). We thus have the representation

$$\begin{aligned} u(\mathbf{x},t) = \int_0^t \int_{\mathbb{R}^{n-1}} \{ \frac{2}{(2\pi)^n i} \int_{\mathbb{R}^n} e^{-p^2(t-T)} \\ \cdot \exp[i\sum_{j=2}^n p_j(x_j - X_j)] e^{ip_1 x_1} p_1 d\mathbf{p}\} F(\mathbb{X}',T) d\mathbb{X}'dT. \end{aligned} \tag{2.2}$$

Following the general method ([11, 13]) for integral transforms, we examine the complex kernel form

$$(2.3)\qquad K_n(\mathbf{z},\overline{\mathbf{w}};t) = \int_0^t \int_{\mathbb{R}^{n-1}} [\frac{2}{(2\pi)^n i}\int_{\mathbb{R}^n} e^{-p^2(t-T)} \cdot \exp\{i\sum_{j=2}^n p_j(z_j - X_j)\} e^{ip_1 z_1} p_1 d\mathbf{p}] \cdot \overline{[\frac{2}{(2\pi)^n i}\int_{\mathbb{R}^n} e^{-\hat{p}^2(t-T)} \cdot \exp\{i\sum_{j=2}^n \hat{p}_j(w_j - X_j)\} e^{i\hat{p}_1 w_1}\hat{p}_1 d\hat{\mathbf{p}}]} d\mathbb{X}' dT$$

$$= \frac{16\overline{w}_1 z_1}{\pi^{\frac{n+1}{2}}} \int_{\frac{1}{8t}}^{\infty} \xi^{\frac{n+1}{2}} \cdot \exp[-\xi\{2z_1^2 + 2\overline{w}_1^2 + \sum_{j=2}^n z_j^2 + \sum_{j=2}^n \overline{w}_j^2 - 2\sum_{j=2}^n \overline{w}_j z_j\}]d\xi.$$

This kernel $K_n(\mathbf{z},\overline{\mathbf{w}};t)$ exists and is analytic on $D_H \times \overline{D_H}$ for the complex domain D_H on $\mathbb{C}^n$ defined by (1.6). The images $u(\mathbf{x},t)$ of (2.1) for functions F satisfying (1.5) are characterized as the members of the Hilbert space H_{K_n} admitting the reproducing kernel $K_n(\mathbf{z},\overline{\mathbf{w}};t)$. In particular, the images $u(\mathbf{x},t)$ can be extended analytically to the domain D_H precisely and for the analytic extension we obtain the isometrical identity

$$(2.4)\qquad \int_0^t \int_{\mathbb{R}^{n-1}} |F(\mathbb{X}',T)|^2 d\mathbb{X}' dT = \|u(\mathbf{z},t)\|^2_{H_{K_n}}$$

([11, 13]).

We shall consider a natural representation of the norm $\|u\|_{H_{K_n}}$ in terms of $u(\mathbf{z},t)$. From the final expression of $K_n(\mathbf{z},\overline{\mathbf{w}};t)$ in (2.3) we see that $u(\mathbf{z},t)$ is expressible in the form

$$(2.5)\qquad u(\mathbf{z},t) = \frac{16 z_1}{\pi^{\frac{n+1}{2}}} \int_{\frac{1}{8t}}^{\infty} \xi^{\frac{n+1}{2}} \cdot \exp[-\xi\{2z_1^2 + \sum_{j=2}^n z_j^2\}] h(\mathbf{z}';\xi)d\xi, \quad \mathbf{z}' = (z_2,\cdots,z_n)$$

where for almost all ξ $h(\mathbf{z}';\xi)$ are entire functions in $\mathbf{z}'$ and belong to the Fischer space $\mathcal{F}(\xi)$ admitting the reproducing kernel

$$\exp\left(2\xi\sum_{j=2}^n \overline{w_j} z_j\right)$$

and its norm is given by

$$(2.6)\qquad \|h\|^2_{\mathcal{F}(\xi)} = \frac{(2\xi)^{n-1}}{\pi^{n-1}} \int_{\mathbf{C}^{n-1}} |h(\mathbf{z}';\xi)|^2 \exp(-2\xi|\mathbf{z}'|^2) dm(\mathbf{z}').$$

Here, of course, $dm(\mathbf{z}')$ is the Lebesgue measure on $\mathbb{C}^{n-1}$. Furthermore, we have the identity

$$(2.7)\qquad \begin{aligned}\|u(\mathbf{z},t)\|^2_{H_{K_n}} &= \frac{16}{\pi^{\frac{n+1}{2}}}\int_{\frac{1}{8t}}^{\infty}\xi^{\frac{n+1}{2}}\|h\|^2_{\mathcal{F}(\xi)}d\xi \\ &= \frac{2^{n+3}}{\pi^{\frac{3n-1}{2}}}\int_{\frac{1}{8t}}^{\infty}\xi^{\frac{3n-1}{2}}\{\int_{\mathbf{C}^{n-1}}|h(\mathbf{z}';\xi)|^2 \\ &\cdot\exp(-2\xi|\mathbf{z}'|^2)dm(\mathbf{z}')\}d\xi.\end{aligned}$$

In (2.5) we shall represent $h(\mathbf{z}';\xi)$ in terms of $u(\mathbf{z},t)$; that is, we shall get the inverse transform of (2.5) as the integral transform of parameter z_1.

In order to apply Fourier's transform, we set

$$2z_1^2 = z = x + iy.$$

Then, from (1.6)

$$(2.8)\qquad x > 2(y_2^2+\cdots+y_n^2)\qquad (z_j = z_j + iy_j)$$

and so we can take a single-valued branch of $z^{\frac{1}{2}}$ with $1^{\frac{1}{2}} = 1$. Hence, we have

$$(2.9)\qquad \frac{u(\sqrt{\frac{z}{2}},\mathbf{z}',t)}{\sqrt{\frac{z}{2}}} = \frac{16}{\pi^{\frac{n+1}{2}}}\int_{\frac{1}{8t}}^{\infty}\xi^{\frac{n+1}{2}}e^{-\xi(x+iy)}\exp(-\xi\sum_{j=2}^{n}z_j^2)h(\mathbf{z}';\xi)d\xi.$$

Under the condition (2.8), there exists a positive constant M such that for $\mathbf{y}' = (y_2,\cdots,y_n)$

$$\xi^{n+1}e^{-2\xi x}\exp(2\xi|\mathbf{y}'|^2) \le M\xi^{\frac{3n-1}{2}}\exp(-2\xi|\mathbf{y}'|^2)\qquad \text{on}\quad (\frac{1}{8t},\infty).$$

Hence we have in (2.9)

$$(2.10)\qquad \begin{aligned}&\int_{\frac{1}{8t}}^{\infty}|\xi^{\frac{n+1}{2}}e^{-\xi x}\exp\{-\xi\sum_{j=2}^{n}z_j^2\}h(\mathbf{z}';\xi)|^2d\xi \\ &\qquad\le M\int_{\frac{1}{8t}}^{\infty}\xi^{\frac{3n-1}{2}}\exp(-2\xi|\mathbf{z}'|^2)|h(\mathbf{z}';\xi)|^2d\xi.\end{aligned}$$

By Fubini's theorem in (2.7), the last integrals in (2.10) are finite for almost all points $\mathbf{z}'$ satisfying (2.8). Hence, for almost all points $\mathbf{z}'$ satisfying (2.8)

$$u(\frac{\sqrt{z}}{2},\mathbf{z}',t)/\sqrt{\frac{z}{2}}$$

is $L_2(\mathbb{R},dy)$ integrable and, by Fourier's integral transform in the framework of the L_2 space, we obtain

$$(2.11)\qquad h(\mathbf{z}';\xi) = \frac{\pi^{\frac{n-1}{2}}}{32}\xi^{-\frac{n+1}{2}}e^{\xi x}\exp(\xi\sum_{j=2}^{n}z_j^2)\int_{-\infty}^{\infty}\frac{u(\sqrt{\frac{x+iy}{2}},\mathbf{z}',t)}{\sqrt{\frac{x+iy}{2}}}e^{iy\xi}dy.$$

In (2.7) we need almost all values on $\mathbb{C}^{n-1}$ of $h(\mathbf{z}';\xi)$ but (2.11) gives only the values on the subset satisfying (2.8) in terms of $u(\mathbf{z},t)$. In this connection, recall the representation of the norm in the Fischer space in terms of the trace $h(\mathbf{x}';\xi)$ to $\mathbb{R}^{n-1}$

(2.12)

$$\|h\|^2_{\mathcal{F}(\xi)} = \frac{(2\xi)^{n-1}}{\pi^{n-1}} \int_{\mathbf{C}^{n-1}} \left| h(\mathbf{z}';\xi)\exp(-\xi\sum_{j=2}^{n} z_j^2)\right|^2 \exp(-4\xi|\mathbf{y}'|^2)dm(\mathbf{z}')$$

$$= (\frac{\xi}{\pi})^{\frac{n-1}{2}} \sum_{j_2,\cdots,j_n=0}^{\infty} \frac{(4\xi)^{-(j_2+\cdots+j_n)}}{j_2!\cdots j_n!}$$

$$\cdot \int_{\mathbf{R}^{n-1}} \left|\partial_{x_2}^{j_2}\cdots\partial_{x_n}^{j_n}(h(\mathbf{x}';\xi)\exp(-\xi|\mathbf{x}'|^2))\right|^2 d\mathbf{x}'$$

([8]). This representation is also obtained directly by using Fourier's transform and Plancherel's theorem. We thus obtain, from (2.7), (2.11) and (2.12)

THEOREM 2.1. For any fixed $t > 0$ and for the images $u(\mathbf{z},t)$ of the integral transform (2.1) for functions F satisfying (1.5), $u(\mathbf{x},t)$ are extensible analytically to the domain D_H on $\mathbb{C}^n$ and for any fixed x satisfying (2.8), the functions

$$\frac{u(\sqrt{\frac{x+iy}{2}},\mathbf{z}',t)}{\sqrt{\frac{x+iy}{2}}}$$

are $L_2(\mathbb{R},dy)$ integrable and their Fourier's integrals in the framework of the L_2 space are extensible analytically onto the entire space $\mathbb{C}^{n-1}$ in $\mathbf{z}'$ (except for a null Lebesgue set) in the sense of (2.11) with (2.6). Furthermore, we have the isometrical identity

(2.13)

$$\int_0^t \int_{\mathbf{R}^{n-1}} |F(\mathbf{X}',T)|^2 d\mathbf{X}'dT = \frac{1}{2^6\pi}\int_{\frac{1}{8t}}^{\infty} \xi^{-1}e^{2\xi x} \sum_{j_2,\cdots,j_n=0}^{\infty} \frac{(4\xi)^{-(j_2+\cdots+j_n)}}{j_2!\cdots j_n!}$$

$$\cdot \int_{\mathbf{R}^{n-1}} \left|\partial_{x_2}^{j_2}\cdots\partial_{x_n}^{j_n}\left\{\int_{-\infty}^{\infty} \frac{u(\sqrt{\frac{x+iy}{2}},\mathbf{x}',t)}{\sqrt{\frac{x+iy}{2}}} e^{iy\xi}dy\right\}\right|^2 d\mathbf{x}'d\xi,$$

which is independent of x satisfying (2.8).

3. ANALYTIC EXTENSION THEOREM TO D_H

We shall consider another inverse transform of (2.5) as the function in z_1 in terms of the real variable functions $u(\mathbf{x},t)$. For this purpose we first examine the functions $u(z_1,\mathbf{x}',t)$ as the functions in z_1.

For any fixed $\mathbf{x}'$ and $t > 0$, we examine the characteristic property of the images $u(z_1,\mathbf{x}',t)$ in (2.2). We compute the complex kernel form in the integral transform (2.2)

(3.1)
$$k_n(z_1,\overline{w_1};t) = K_n((z_1,\mathbf{x}'),(\overline{w_1},\mathbf{x}');t)$$
$$= \frac{16 z_1\overline{w_1}}{\pi^{\frac{n+1}{2}}}\int_{\frac{1}{8t}}^{\infty} \xi^{\frac{n+1}{2}}\exp[-2\xi(z_1^2+\overline{w_1}^2)]d\xi.$$

The images $u(z_1, \mathbf{x}', t)$ belong intuitively to the Hilbert space H_{k_n} admitting the reproducing kernel $k_n(z_1, \overline{w_1}; t)$. We see that the construction of the Hilbert spaces H_{k_n} of even n-dimensional spaces is much more complicated than that of odd n-dimensional spaces. For $n = 1$ and 3, we have

$$k_1(z_1, \overline{w_1}; t) = z_1 \exp(-\frac{z_1^2}{4t}) \overline{w_1 \exp(-\frac{w_1^2}{4t})} \left[\frac{4}{\pi(z_1^2 + \overline{w_1}^2)^2} + \frac{1}{t\pi(z_1^2 + \overline{w_1}^2)}\right]$$

and

$$k_3(z_1, \overline{w_1}; t) = \frac{6}{\pi^2} z_1 \exp(-\frac{z_1^2}{4t}) \overline{w_1 \exp(-\frac{w_1^2}{4t})}$$
$$\left[\frac{1}{(z_1^2 + \overline{w_1}^2)^4} + \frac{1}{2^2 t (z_1^2 + \overline{w_1}^2)^3} + \frac{1}{2^5 t^2 (z_1^2 + \overline{w_1}^2)^2} + \frac{1}{3 \cdot 2^7 t^3 (z_1^2 + \overline{w_1}^2)}\right]$$

([5, p. 310, 3.351.2]). By applying the theory ([4]) of Aronszajn, we can investigate the Hilbert spaces H_{k_n} as in [14], but the construction of H_{k_n} is involved for a higher dimensional space.

However, we are now interested in the integral transform (2.5) for fixed points $\mathbf{x}'$ and $t > 0$ for functions h satisfying (2.6). From (2.7) and (2.12) for $j_2 = j_3 = \cdots = j_n = 0$, we have

$$\int_{\frac{1}{8t}}^{\infty} \xi^n \{\int_{\mathbb{R}^{n-1}} |h(\mathbf{x}'; \xi)|^2 \exp(-2\xi|\mathbf{x}'|^2) d\mathbf{x}'\} d\xi < \infty.$$

Hence, we see that from Fubini's theorem

$$\int_{\frac{1}{8t}}^{\infty} |h(\mathbf{x}'; \xi)|^2 \exp(-2\xi|\mathbf{x}'|^2) \xi^n d\xi < \infty \tag{3.3}$$

for almost all points $\mathbf{x}' \in \mathbb{R}^{n-1}$. For any fixed point $\mathbf{x}'$ satisfying (3.3), we shall examine a characteristic property of $u(z_1, \mathbf{x}', t)$ in z_1. So, for functions $H(\xi) = h(\mathbf{x}'; \xi)$ satisfying (3.3), we examine the integral transform (2.5) in the form

$$u(z_1, \mathbf{x}', t) = \frac{16}{\pi^{\frac{n+1}{2}}} \int_{\frac{1}{8t}}^{\infty} H(\xi) \cdot z_1 \exp(-2\xi z_1^2) \cdot \xi^{\frac{-n+1}{2}} \exp(\xi|\mathbf{x}'|^2) \cdot \xi^n \exp(-2\xi|\mathbf{x}'|^2) d\xi. \tag{3.4}$$

We compute the kernel form

$$\begin{aligned} k_n^*(z_1, \overline{w_1}; t) &= \frac{16}{\pi^{\frac{n+1}{2}}} \int_{\frac{1}{8t}}^{\infty} z_1 \exp(-2\xi z_1^2) \xi^{\frac{-n+1}{2}} \exp(\xi|\mathbf{x}'|^2) \\ &\quad \cdot \overline{w_1 \exp(-2\xi w_1^2) \xi^{\frac{-n+1}{2}}} \exp(\xi|\mathbf{x}'|^2) \cdot \xi^n \exp(-2\xi|\mathbf{x}'|^2) d\xi \\ &= \frac{1}{\pi^{\frac{n-1}{2}}} k_1(z_1, \overline{w_1}; t). \end{aligned} \tag{3.5}$$

We thus see that the images $u(z_1, \mathbf{x}', t)$ of (3.4) for functions $H(\xi)$ satisfying (3.3) belong to the Hilbert space H_{k_1} which is independent of the dimensions n and the points $\mathbf{x}'$.

The expression (3.2) of $k_1(z_1, \overline{w_1}; t)$ shows that the images $u(z_1, \mathbf{x}', t)$ are separated into the two components as in [14]. Hence, in the sequel we shall split our problem to the two components.

In (2.1) we split $u(\mathbf{x}, t)$ into the two components

$$u(\mathbf{x}, t) = \varphi(\mathbf{x}, t) + \phi(\mathbf{x}, t)$$

where

$$\varphi(\mathbf{x}, t) = -\int_0^t \int_{\mathbb{R}^{n-1}} -\frac{\partial}{\partial X_1} G(\mathbf{x}, (X_1, \mathbb{X}'); t - T)\big|_{X_1=0} \cdot F(\mathbb{X}', T) d\mathbb{X}' \frac{TdT}{t} \tag{3.6}$$

and

$$\phi(\mathbf{x}, t) = -\int_0^t \int_{\mathbb{R}^{n-1}} -\frac{\partial}{\partial X_1} G(\mathbf{x}, (X_1; \mathbb{X}'), t - T)\big|_{X_1=0} \cdot F(\mathbb{X}', T) d\mathbb{X}' (1 - \frac{T}{t}) dT. \tag{3.7}$$

We can apply the arguments in Section 2 for these integral transforms by substituting

$$u(\mathbf{x}, t), \quad dT, \quad d\xi$$

as follows :

$$\varphi(\mathbf{x}, t), \frac{TdT}{t}, (1 - \frac{1}{8t\xi}) d\xi,$$

and

$$\phi(\mathbf{x}, t), (1 - \frac{T}{t}) dT, \frac{1}{8t\xi} d\xi,$$

respectively. We can also obtain Theorem 2.1 for these modifications. Furthermore, (3.5) is also separated into the two components

$$\begin{aligned} k_n^*(z_1, \overline{w_1}; t) &= \frac{16 z_1 \overline{w_1}}{\pi^{\frac{n+1}{2}}} \int_{\frac{1}{8t}}^{\infty} \xi e^{-2\xi(z_1^2 + \overline{w_1}^2)} \{(1 - \frac{1}{8t\xi}) + \frac{1}{8t\xi}\} d\xi \\ &= K_1(z_1, \overline{w_1}) + K_2(z_1, \overline{w_1}) \end{aligned}$$

for

$$K_1(z_1, \overline{w_1}) = \frac{1}{\pi^{\frac{n-1}{2}}} z_1 \exp(-\frac{z_1^2}{4t}) \overline{w_1 \exp(-\frac{w_1^2}{4t})} \frac{4}{\pi(z_1^2 + \overline{w_1}^2)^2}$$

and

$$K_2(z_1, \overline{w_1}) = \frac{1}{\pi^{\frac{n-1}{2}}} z_1 \exp(-\frac{z_1^2}{4t}) \overline{w_1 \exp(-\frac{w_1^2}{4t})} \frac{1}{t\pi(z_1^2 + \overline{w_1}^2)}.$$

These $K_1(z_1, \overline{w_1})$ and $K_2(z_1, \overline{w_1})$ are the reproducing kernels determing the Hilbert spaces H_{K_1} and H_{K_2} which are formed by the images $\varphi(z_1, \mathbf{x}', t)$ and $\phi(z_1, \mathbf{x}', t)$, respectively. Furthermore, we have the isometrical identities

$$\|\varphi(z_1, \mathbf{x}', t)\|_{H_{K_1}}^2 = \int_0^t \int_{\mathbb{R}^{n-1}} |F(\mathbb{X}', T)|^2 d\mathbb{X}' \frac{TdT}{t} \tag{3.8}$$

and

$$\|\phi(z_1,\mathbf{x}',t)\|^2_{H_{K_2}} = \int_0^t \int_{\mathbb{R}^{n-1}} |F(\mathbb{X}',T)|^2 d\mathbb{X}'(1-\frac{T}{t})dT. \tag{3.9}$$

Hence, in particular, we have the identity

$$\|u(z_1,\mathbf{x}',t)\|^2_{H_{k_n^*}} = \|\varphi(z_1,\mathbf{x}',t)\|^2_{H_{K_1}} + \|\phi(z_1,\mathbf{x}',t)\|^2_{H_{K_2}}. \tag{3.10}$$

Note that $u(\mathbf{x},t) = t\varphi(\mathbf{x},t)$ are related to the solutions of the heat equation

$$\partial_t u(\mathbf{x},t) = \Delta u(\mathbf{x},t) \quad \text{on} \quad \mathbb{R}^+ \times \mathbb{R}_n^+$$

subject to

$$u(0,\mathbf{x}',t) = tF(\mathbf{x}',t) \quad \text{on} \quad \mathbb{R}^+ \times \mathbb{R}^{n-1},$$
$$u(\mathbf{x},0) = 0 \quad \text{on} \quad \mathbb{R}^n_+,$$

and

$$u(\mathbf{x},t) \to 0 \quad \text{as} \quad \mathbf{x} \to \infty.$$

On the other hand, $u(\mathbf{x},t) = -\frac{2t}{x_1}\phi(\mathbf{x},t)$ are related to the solutions of the heat equation

$$\partial_t u(\mathbf{x},t) = \Delta u(\mathbf{x},t) \quad \text{on} \quad \mathbb{R}^+ \times \mathbb{R}^n_+$$

subject to

$$u_{x_1}(0,\mathbf{x}',t) = F(\mathbf{x}',t) \quad \text{on} \quad \mathbb{R}^+ \times \mathbb{R}^{n-1},$$
$$u(\mathbf{x},0) = 0 \quad \text{on} \quad \mathbb{R}^n_+,$$

and

$$u(\mathbf{x},t) \to 0 \quad \text{as} \quad \mathbf{x} \to \infty.$$

Let Δ be the sector

$$\Delta = \{-\frac{\pi}{4} < \arg z_1 < \frac{\pi}{4}\}.$$

Then, we have the representations

$$\|\varphi\|^2_{H_{K_1}} = \pi^{\frac{n-1}{2}} \iint_\Delta |\varphi(z_1,\mathbf{x}',t)\exp(\frac{z_1^2}{4t})|^2 dm(z_1) \tag{3.11}$$

and

$$\|\phi\|^2_{H_{K_2}} = t\pi^{\frac{n-1}{2}} \int_{\partial\Delta} |\frac{\phi(z_1,\mathbf{x}',t)}{\sqrt{z_1}} \exp(\frac{z_1^2}{4t})|^2 |dz_1| \tag{3.12}$$

([14, Theorems 2.1 and 2.2]). From the representations of the Bergman and Szegö norms on Δ in terms of the trace to the positive real line ([3]), we furthermore obtain

$$\|\varphi\|^2_{H_{K_1}} = \pi^{\frac{n-1}{2}} \sum_{j=0}^\infty \frac{2^j}{(2j+1)!} \int_0^\infty |\partial^j_{x_1}(\varphi(x_1,\mathbf{x}',t)\exp(\frac{x_1^2}{4t}))|^2 x_1^{2j+1} dx_1 \tag{3.13}$$

and

$$(3.14)\qquad \|\phi\|^2_{H_{K_2}} = \sqrt{2}t\pi^{\frac{n-1}{2}}\sum_{j=0}^{\infty}\frac{2^j}{(2j)!}\int_0^\infty |\partial^j_{x_1}(\frac{\phi(x_1,\mathbf{x}',t)}{\sqrt{x_1}}\exp(\frac{x_1^2}{4t}))|^2 x_1^{2j}\,dx_1.$$

By using the representations (3.13) and (3.14), we shall give the inverse transforms of the integral transforms corresponding to (2.5)

$$(3.15)\quad \varphi(z_1,\mathbf{x}',t) = \frac{16}{\pi^{\frac{n+1}{2}}}\int_{\frac{1}{8t}}^\infty h_\varphi(\mathbf{x}';\xi)$$
$$\cdot z_1\exp(-2\xi z_1^2)\xi^{\frac{-n+1}{2}}\exp(\xi|\mathbf{x}'|^2)\cdot\xi^n\exp(-2\xi|\mathbf{x}'|^2)(1-\frac{1}{8t\xi})d\xi$$

for functions h_φ satisfying

$$\frac{16}{\pi^{\frac{n+1}{2}}}\int_{\frac{1}{8t}}^\infty |h_\varphi(\mathbf{x}';\xi)|^2\xi^n\exp(-2\xi|\mathbf{x}'|^2)(1-\frac{1}{8t\xi})d\xi < \infty$$

and

$$(3.16)\quad \phi(z_1,\mathbf{x}',t) = \frac{16}{\pi^{\frac{n+1}{2}}}\int_{\frac{1}{8t}}^\infty h_\phi(\mathbf{x}';\xi)$$
$$\cdot z_1\exp(-2\xi z_1^2)\xi^{\frac{-n+1}{2}}\exp(\xi|\mathbf{x}'|^2)\cdot\xi^n\exp(-2\xi|\mathbf{x}'|^2)\frac{1}{8t\xi}d\xi$$

for functions h_ϕ satisfying

$$\frac{16}{\pi^{\frac{n+1}{2}}}\int_{\frac{1}{8t}}^\infty |h_\phi(\mathbf{x}';\xi)|^2\xi^n\exp(-2\xi|\mathbf{x}'|^2)\frac{1}{8t\xi}d\xi < \infty.$$

By Theorem 4.1 in [13], we obtain

$$(3.17)\qquad h_\varphi(\mathbf{x}';\xi) = \pi^{\frac{n-1}{2}}\xi^{\frac{-n+1}{2}}\exp(\xi|\mathbf{x}'|^2)$$
$$\cdot s-\lim_{N\to\infty}\sum_{j=0}^{N}\frac{2^j}{(2j+1)!}\int_0^\infty \partial^j_{x_1}[\varphi(x_1,\mathbf{x}',t)\exp(\frac{x_1^2}{4t})]$$
$$\cdot\partial^j_{x_1}[x_1\exp\{(\frac{1}{4t}-2\xi)x_1^2\}]x_1^{2j+1}dx_1$$

and

$$(3.18)\qquad h_\phi(\mathbf{x}';\xi) = \sqrt{2}t\pi^{\frac{n-1}{2}}\xi^{\frac{-n+1}{2}}\exp(\xi|\mathbf{x}'|^2)$$
$$\cdot s-\lim_{N\to\infty}\sum_{j=0}^{N}\frac{2^j}{(2j)!}\int_0^\infty \partial^j_{x_1}[\frac{\varphi(x_1,\mathbf{x}',t)}{\sqrt{x_1}}\exp(\frac{x_1^2}{4t})]$$
$$\cdot\partial^j_{x_1}[\sqrt{x_1}\exp\{(\frac{1}{4t}-2\xi)x_1^2\}]x_1^{2j}dx_1$$

in the sense of the strong convergence in the spaces

$$L_2\left((\frac{1}{8t},\infty),\xi^n\exp(-2\xi|\mathbf{x}'|^2)(1-\frac{1}{8t\xi})d\xi\right)$$

and

$$L_2\left((\frac{1}{8t},\infty),\xi^n\exp(-2\xi|\mathbf{x}'|^2)\frac{1}{8t\xi}d\xi\right),$$

respectively. In particular, in the proof of Theorem 4.1 in [8], we now need to replace the representation

$$\|f\|^2_{H_K}=\lim_{N\to\infty}\int_{E_N}|f(q)|^2d\mu(q)$$

, for example, by the representation

$$\|\varphi\|^2_{H_{K_1}}=\pi^{\frac{n-1}{2}}\lim_{N\to\infty}\sum_{j=0}^{N}\frac{2^j}{(2j+1)!}\int_0^\infty\left|\partial^j_{x_1}\{\varphi(x_1,\mathbf{x}',t)\exp(\frac{x_1^2}{4t})\}\right|^2x_1^{2j+1}dx_1.$$

We thus obtain by using the modified Theorem 2.1 for $\varphi(\mathbf{x},t)$ and $\phi(\mathbf{x},t)$, respectively

THEOREM 3.1. In the integral transforms (3.6) and (3.7) we have the isometrical identities

(3.19)

$$\frac{1}{t}\int_0^t\int_{\mathbb{R}^{n-1}}|F(\mathbb{X}',T)|^2d\mathbb{X}'TdT$$
$$=\frac{16}{\pi}\int_{\frac{1}{8t}}^\infty\{\xi\sum_{j_2,\cdots,j_n=0}^{\infty}\frac{(4\xi)^{-(j_2+\cdots+j_n)}}{j_2!\cdots j_n!}$$
$$\cdot\int_{\mathbb{R}^{n-1}}|\partial^{j_2}_{x_2}\cdots\partial^{j_n}_{x_n}(\exp(\xi|\mathbf{x}'|^2)s-\lim_{N\to\infty}\sum_{j=0}^{N}\frac{2^j}{(2j+1)!}$$
$$\cdot\int_0^\infty\partial^j_{x_1}[\varphi(\mathbf{x},t)\exp(\frac{x_1^2}{4t})]$$
$$\cdot\partial^j_{x_1}[x_1\exp\{(\frac{1}{4t}-2\xi)x_1^2\}]x_1^{2j+1}dx_1)|^2d\mathbf{x}'\}$$
$$\cdot(1-\frac{1}{8t\xi})d\xi,$$

and

$$\int_0^t\int_{\mathbb{R}^{n-1}}|F(\mathbb{X}',T)|^2d\mathbb{X}'(1-\frac{T}{t})dT \tag{3.20}$$
$$=\frac{4t}{\pi}\int_{\frac{1}{8t}}^\infty\{\xi\sum_{j_2,\cdots,j_n=0}^{\infty}\frac{(4\xi)^{-(j_2+\cdots+j_n)}}{j_2!\cdots j_n!}$$
$$\cdot\int_{\mathbb{R}^{n-1}}|\partial^{j_2}_{x_2}\cdots\partial^{j_n}_{x_n}(\exp(\xi|\mathbf{x}'|^2)s-\lim_{N\to\infty}\sum_{j=0}^{N}\frac{2^j}{(2j)!}$$
$$\cdot\int_0^\infty\partial^j_{x_1}[\frac{\varphi(\mathbf{x},t)}{\sqrt{x_1}}\exp(\frac{x_1^2}{4t})]$$
$$\cdot\partial^j_{x_1}[\sqrt{x_1}\exp\{(\frac{1}{4t}-2\xi)x_1^2\}]x_1^{2j}dx_1)|^2d\mathbf{x}'\}d\xi,$$

respectively.

The isometrical identities (3.19) and (3.20) give the representations of the norms $\|\varphi\|_{H_{K_1}}$ and $\|\phi\|_{H_{K_2}}$ in terms of the traces $\varphi(\mathbf{x},t)$ and $\phi(\mathbf{x},t)$ to $\mathbb{R}^n_+$, respectively. Therefore, in particular, we have

COROLLARY 3.1. For smooth functions $\varphi(\mathbf{x})$ and $\phi(\mathbf{x})$ on $\mathbb{R}^n_+$, $\varphi(\mathbf{x})$ and $\phi(\mathbf{x})$ are extensible analytically to the domain D_H on $\mathbb{C}^n$, and the analytic extensions $\varphi(\mathbf{z})$ and $\phi(\mathbf{z})$ belong to the reproducing kernel Hilbert spaces H_{K_1} and H_{K_2} if and only if they have finite values in the right hand sides in (3.19) and (3.20), respectively.

In Corollary 3.1 we can also give the formulas such that for any point $\mathbf{z} \in D_H$ we represent $\varphi(\mathbf{z})$ and $\phi(\mathbf{z})$ in terms of the traces $\varphi(\mathbf{x})$ and $\phi(\mathbf{x})$ to $\mathbb{R}^n_+$ by using the representations (3.19) and (3.20) and the reproducing properties in the spaces H_{K_1} and H_{K_2}, respectively.

ACKNOWLEDMENT

The travel and living expenses of the author for attending to the Fourth International Colloquium on Differential Equations, Plovdiv, August 18-23, 1993 were supported by the Japanese Ministry of Education, Science and Culture.

REFERENCES

1. H. Aikawa, Complex Variables, 18, 253-266(1992).
2. H. Aikawa, N. Hayashi and S. Saitoh, Complex Variables, 15, 27-36(1990).
3. H. Aikawa, N. Hayashi and S. Saitoh, J. Math. Soc. Japan, 43, 195-201(1991).
4. N. Aronszajn, Trans. Amer. Math. Soc., 68, 337-404(1950).
5. I. S. Gradshteyn and I. M. Ryzhik, Table of Integrals, Series and Products, Academic Press, (1980).
6. N. Hayashi, Duke Math. J., 60, 717-727(1990).
7. N. Hayashi, Duke Math. J., 62, 575-591(1991).
8. N. Hayashi and S. Saitoh, Ann. Inst. Henri Poincaré, Physique théorique, 52, 163-173(1990).
9. N. Hayashi and S. Saitoh, Commun. Math. Phys., 129, 27-41(1990).
10. G. F. Roarch, Green's Functions, Cambridge University Press, (1982).
11. S. Saitoh, Proc. Amer. Math. Soc., 89, 74-78 (1983).
12. S. Saitoh, Applicable Analysis, 16, 1-6(1983).
13. S. Saitoh, Theory of reproducing kernels and its applications, Pitman Res. Notes in Math. Series. 189, Longman Scientific & Technical, England, (1988).
14. S. Saitoh, Applicable Analysis, 40, 139-149(1991).
15. S. Saitoh, Complex Variables, 19, 231-241(1992).
16. L. Schwartz, J. Analyse Math., 13, 115-256(1964).

4th Int. Coll. on Differential Equations, pp. 277-285
D. Bainov and V. Covachev (Eds)

Recent Advances in the Method of Quasilinearization

A. S. Vatsala
Department of Mathematics, P.O. Box 41010
University of Southwestern Louisiana,
Lafayette, LA 70504
USA

Abstract. The method of quasilinearization provides lower or upper bounds for the solution of nonlinear differential equations. In this paper we survey some recent results of several authors concerning the extension and generalization of the method of quasilinearization under weaker assumptions. The extension of quasilinearization method also provides lower and upper bounds simultaneously as well as, converge quadratically to the unique solution.

1. Introduction.

In this paper, we present some recent developments concerning the method of quasilinearization and its extensions and generalization. We indicate the proof of one generalization theorem. The method of quasilinearization is an effective tool to obtain lower or upper bounds for the solutions of nonlinear differential equations. For this purpose consider the IVP

$$u' = f(u), u(0) = u_0 \text{ on } J = [0,T]. \tag{1.1}$$

If f is convex, then one can find a $g(u,v)$ which is linear in u such that $f(u) = \max_v g(u,v)$. Using $g(u,v)$, one can construct a monotone sequence, provided a suitable initial approximation is known, and that the sequence converges quadratically to the unique solution of (1.1). Moreover, the sequence provides good lower bounds for the solution. On the other hand, if f is concave, a dual result holds that offers monotone approximations having similar properties. See [1, 2, 3, 4, 8, 12] for details. Also it is known that Newton-Raphson developed the procedure of quadratic convergence to the solution of polynomial equations. An extensive study of this procedure was first carried out by

Kantorovich for functional equations, hence the method in this context is usually called the Newton-Raphson-Kantorovich method. In connection with Newton-Raphson method for polynomials, the method of quasilinearization yields quadratic convergence for the solution of differential equations. This is very important in many applications where the time of computation is a factor.

A natural question that would arise are the following relative to the forcing function $F(t,u)$. If

(a) $F(t,u)$ is not convex or concave,

(b) $F(t,u) = f(t,u) + g(t,u)$, where $f(t,u)$ is convex and $g(t,u)$ is concave,

(c) $F(t,u) = f_1(t,u) + f_2(t,u) + f_3(t,u)$ where f_1 is convex, f_2 is Lipschitzian and f_3 is continuous,

then, can we generalize and extend the method of quasilinearization to yield quadratic convergence? The answers to these questions are positive [5, 7, 10, 11, 13, 14, 15] except in (c) where $f_3(t,u)$ is included. In this case we obtain weak quadratic convergence instead of quadratic convergence. We recall these results in the next three sections. The situation where $F(t,u)$ is split as the sum of a convex and a concave function, the extension of quasilinearization method [7], yields lower and upper bounds simultaneously that converge quadratically to the unique solution of the nonlinear differential equation.

2. Generalized quasilinearization.

Consider the IVP

$$u' = f(t,u), t \in J = [0,T] \tag{2.1}$$
$$u(0) = u_0$$

where $f \in C[J \times R, R]$.

A function $v \in C^1[J,R]$ is said to be an upper solution of (2.1) if $v' \geq f(t,v)$ and a lower solution of (2.1) if the reverse inequalities hold. The following known comparison result [6] is used in the main result of this section.

Theorem 2.1. Let α, $\beta \in C^1[J,R]$ be lower and upper solutions of (2.1) on J. Suppose that

$$f(t,u) - f(t,v) \leq L(u-v), L \geq 0$$

whenever $v \leq u$. Then $\alpha(0) \leq \beta(0)$ implies $\alpha(t) \leq \beta(t)$ on J.

Next we recall a known existence result [6] which proves the existence of a solution $u(t)$ of (2.1) in the closed set

$$\Omega = [(t,u) : \alpha(t) \leq u \leq \beta(t), t \in J].$$

Theorem 2.2. Let $f \in C[\Omega, R]$ and $\alpha,\ \beta \in C^1[J,R]$ be lower and upper solutions of (2.1) such that $\alpha(t) \leq \beta(t)$ on J. Then there exists a solution of (2.1) satisfying $\alpha(t) \leq u(t) \leq \beta(t)$ on J.

We now have the main result of [10].

Theorem 2.3. Assume that

(A_1) $\alpha,\ \beta \in C^1[J,R]$ are lower and upper solutions of (2.1) such that $\alpha(t) \leq \beta(t)$ on J;

(A_2) $f_u(t,u)$, $f_{uu}(t,u)$ exist and are continuous on Ω such that $f_{uu}(t,u) + 2M > 0$, $M > 0$.

Then there exists a monotone sequence $\{\alpha_n(t)\}$ which converges uniformly to the solution $u(t)$ of (2.1) and the convergence is quadratic.

Proof. Define, for any $\eta \in C[J,R]$ such that $\alpha \leq \eta \leq \beta$ on J,

$$g(t,u,\eta) = f(t,\eta) + [f_u(t,\eta) + 2M\eta](u-\eta) - M(u^2 - \eta^2). \tag{2.2}$$

In view of (A_2) one can easily see

$$f(t,u) \geq g(t,u,\eta) \tag{2.3}$$

for any $\eta \in \Omega$, and therefore it follows that

$$f(t,u) = \max_{\eta} g(t,u,\eta).$$

Set $\alpha(t) = \alpha_0(t)$ on J. Then because of (A_1), (2.2) and (2.3) we get

$$\alpha_0' \leq f(t,\alpha_0) \equiv g(t,\alpha_0,\alpha_0),\ \alpha_0(0) \leq u_0$$
$$\beta' \geq f(t,\beta) \geq g(t,\beta,\alpha_0),\ \beta(0) \geq u_0.$$

This proves α_0, β are lower and upper solutions of

$$\alpha_1' = g(t,\alpha_1,\alpha_0),\ \alpha_1(0) = u_0. \tag{2.4}$$

Further g satisfies Lipschitz condition on Ω. Now using Theorem 2.2 and (2.1), it follows the solution $\alpha_1(t)$ of (2.4) exists and is unique such that $\alpha_0(t) \leq \alpha_1(t) \leq \beta(t)$ on J.

Suppose now that for $n > 1$, $\alpha_0(t) \leq \alpha_{n-1}(t) \leq \alpha_n(t) \leq \beta(t)$ where $\alpha_n(t)$ is the solution of

$$\alpha_n' = g(t, \alpha_n, \alpha_{n-1}), \ \ \alpha_n(0) = u_0.$$

Then using meanvalue theorem and (A_2) we can prove

$$\alpha_n' \leq f(t, \alpha_n) \equiv g(t, \alpha_n, \alpha_n), \ \ \alpha_n(0) = u_0. \tag{2.5}$$

Since

$$\alpha_{n+1}' = g(t, \alpha_{n+1}, \alpha_n), \ \ \alpha_n(0) = u_0$$
$$\beta' \geq g(t, \beta, \alpha_n),$$

it follows using Theorem 2.1 and Theorem 2.2 that, there exists a unique $\alpha_{n+1}(t)$ such that

$$\alpha_n(t) \leq \alpha_{n+1} \leq \beta(t).$$

Now by induction principle, we obtain

$$\alpha(t) = \alpha_0(t) \leq \alpha_1(t) \leq \cdots \leq \alpha_n(t) \leq \beta(t) \ on \ J.$$

Also from (2.4) and Theorem 2.2 it follows that $\alpha(t) \leq u(t)$ where $u(t)$ is the solution of (2.1). Consequently, the monotone sequence $\alpha_n(t)$ satisfies

$$\alpha(t) = \alpha_0(t) \leq \cdots \leq \alpha_n(t) \leq u(t) \leq \beta(t) \ on \ J,$$

and as a result, employing standard arguements (See[6]) it is easy to show that the sequence $\alpha_n(t)$ converges uniformly and monotonically to the unique solution $u(t)$ of (2.1) on J.

Next we show that the convergence of $\alpha_n(t)$ to $u(t)$ is quadratic. For this purpose, consider $m(t) = u(t) - \alpha_n(t) \geq 0$ for each n and note that $m(0) = 0$. Setting $F(t,u) = f(t,u) + Mu^2$, one can show that

$$m' \leq Km + H(u - \alpha_{n-1})^2,$$

where $|F_{uu}(t,u)| \leq H$, $|F_{uu}(t, \alpha_{n-1}) - M(u + \alpha_n)| \leq K$ on Ω. As a result treating $H(u - \alpha_{n-1})^2$ as the forcing term and applying Gronwall's inequality, it follows that

$$0 \leq u(t) - \alpha_n(t) \leq H \int_0^t |u(s) - \alpha_{n-1}(s)|^2 \, e^{K(t-s)} ds \ on \ J.$$

This yields the desired result

$$\max_J |u(t) - \alpha_n(t)| \leq HK^{-1}e^{KT} \max_J |u(t) - \alpha_{n-1}(t)|^2 .$$

Remark 2.1. The above result is considered as generalized quasilinearization since the method considered works for nonconvex functions that have the property $f_{uu}(t,u) \geq -2M$, $M > 0$. In fact the theorem is valid if

$$f_{uu}(t,u) \geq -\varphi_{uu}(t,u)$$

where $\varphi(t,u)$ is any convex function.

The above result has been extended to (i) Stochastic initial value problem in [13] , (ii) first order systems with initial conditions in [5] and (iii) second order boundary value problem in [14]. Here we recall the result of first order systems with initial conditions. The method of quasilinearization for systems as in [1] is the following result. For that purpose consider the system

$$x' = g(t,x), x(0) = x_0. \tag{2.6}$$

Theorem 2.3. Let $g(t,x)$ be a strictly convex function of x for $0 \leq t \leq t_0$. Further if $\dfrac{\partial g_i}{\partial x_j} \geq 0$, $i \neq j$. Then there exists a sequence $\alpha_n(t)$ which converges quadratically to the solution of (2.6).

3. An extension of the method of quasilinearization

Let α_0, $\beta_0 \in C^1[J,R]$ such that $\alpha_0(t) \leq \beta_0(t)$ on $J = [0,T]$. Define the closed set

$$\Omega = [(t,u) : \alpha_0(t) \leq u(t) \leq \beta_0(t), t \in J].$$

We consider the IVP

$$u' = F(t,u), \ u(0) = u_0 \tag{3.1}$$

where $F \in C[\Omega, R_+]$.

Definition. Suppose that F admits a decomposition

$$F(t,u) = f(t,u) + g(t,u) \tag{3.2}$$

and that α_0, $\beta_0 \in C^1[J,R]$ such that $\alpha_0(t) \le \beta_0(t)$ on J. Then we say that α_0, β_0 are coupled lower and upper solutions of (3.1), if

$$\begin{aligned} \alpha_0' &\le f(t,\alpha_0) + g(t,\beta_0), \alpha_0(0) \le u_0 \\ \beta_0' &\ge f(t,\beta_0) + g(t,\alpha_0), \beta_0(0) \ge u_0. \end{aligned} \tag{3.3}$$

We recall the following known comparison result which is a special case of Theorem 1.3.1 in [6].

Theorem 3.1. Assume that

$$m' \le Am + \sigma$$

where $m \in C^1[J,R^n]$, $A = (a_{ij})$ is a n×n matrix such that $a_{ij} \ge 0$, $i \ne j$ and $\sigma \in C[J,R^n]$. Then

$$m(t) \le m(0)e^{At} + \int_0^t e^{A(t-s)}\sigma(s)ds, t \in J,$$

where the inequalities are to be understood componentwise.

The result which yields two sided bounds for the solution of (3.1) which is an extension of the method of quasilinearization is given in the following Theorem.

Theorem 3.2. Assume that

(A_1) $F = f + g$, f, g are uniformly convex and concave in u for $t \in J$ respectively and $g_u(t,u) \le (\beta_0 - \alpha_0)g_{uu}(t,u)$;

(A_2) α_0, $\beta_0 \in C^1[J,R]$ such that $\alpha_0(t) \le \beta_0(t)$ on J are coupled lower and upper solutions of (3.1) on J.

Then there exist monotone sequences $\{\alpha_n(t)\}$, $\{\beta_n(t)\}$ that converge uniformly to the unique solution of (3.1) on J. Moreover the convergence of these sequences is quadratic.

Theorem 3.1 is used to prove the above result. See [7] for details of the proof.

The results of Theorem 3.2 has been extended to the scalar diffusion equation of the form

$$\begin{aligned} \mathcal{L}u = \frac{\partial u}{\partial t} - Lu &= F(t,x,u) on\ Q_T \\ u(0,x) &= u_0(x) on\ \bar{\Omega} \\ Bu(t,u) &= \varphi(x) on\ \Gamma_T \end{aligned}$$

where Ω is a bounded domain in R^m with boundary $\partial\Omega \in C^{2+\alpha}$ and closure $\bar{\Omega}$, $Q_T = (0,T] \times \Omega$, $\Gamma_T = (0,T) \times \partial\Omega$, $\bar{Q}_T = [0,T] \times \bar{\Omega}$, $\bar{\Gamma}_T = [0,T] \times \partial\Omega$, $T > 0$. Here $\mathcal{L}$ is a second order differential operator defined by

$$\mathcal{L} = \frac{\partial}{\partial t} - L$$

where $L = \sum_{i,j=1}^{m} a_{ij}(t,x)\frac{\partial^2}{\partial x_i \partial x_j} + \sum_{i=1}^{m} b_i(t,x)\frac{\partial}{\partial x_i}$, and B is a boundary operator given by $Bu = p(t,x)u + q(t,x)\frac{du}{d\nu}$ where $\frac{du}{d\nu}$ denotes the normal derivative of u and $\nu(t,x)$ is the unit outward normal vector on $\partial\Omega$ for $t \in [0,T]$. This is proved by using the comparison theorems of [6, 9] and using existence results for the corresponding systems where the lower and upper solutions are of Müler's type [6] for parabolic equations. See [14] for details.

4. Lower bounds for Minimal solution.

In this section the following two situations are considered. That is (a) when the forcing function is the sum of convex function and Lipschitzian and (b) the forcing function is the sum of continuous function and a convex function. See [11] for details.

For this purpose consider

$$u' = F(t,u), u(0) = u_0, t \in J, J = [0,T]$$

where $F \in C[\Omega, R]$. Let α, $\beta \in C^1[J,R]$ such that $\alpha \le \beta$ on J. Define the set

$$\Omega = [(t,u) : \alpha(t) \le u(t) \le \beta(t), t \in J].$$

Then if (a) holds, we have the following result [11].

Theorem 4.1. Assume that

(i) α, $\beta \in C^1[J,R]$, such that $\alpha \le \beta$ on J and

$$\alpha' \le F(t,\alpha)$$
$$\beta' \ge F(t,\beta), \quad \textit{on } J;$$

(ii) $F \in C[\Omega, R]$, F admits a decomposition $F = f + g$ where $f(t,u)$ is Lipschitzian in u, $g(t,u)$ is uniformly convex in u for $t \in J$.

Then there exists a monotone sequence $\{\alpha_n(t)\}$ which converges to the unique solution $u(t)$ of (4.1) and the convergence is quadratic.

In the next result we consider the situation when (b) holds. In this case the convergence of the sequence is weakly quadratic and the sequence converges to the minimal solution of (4.1).

Definition 4.1. Suppose $\{\alpha_n(t)\}$ is a sequence that converges to α, if positive constants λ_1, λ_2 and λ_3 exist with

$$\max_{t\in J} |\alpha(t) - \alpha_n(t)| \leq \lambda_1[\lambda_2 + \lambda_3 \max_{t\in J} |\alpha(t) - \alpha_{n-1}(t)|^2],$$

then as $n \to \infty$, $\alpha_n \to \alpha$ weakly quadratically.

Theorem 4.2. In addition to the assumption (i) of Theorem 4.1, suppose that $F \in C[\Omega, R]$, F admits decomposition $F = f + g$ where $g(t,u)$ is uniformly convex in u for $t \in J$, and $f(t,u)$ is continuous in u on Ω.

Then there exists a monotone sequence $\{\alpha_n(t)\}$, which converges to the minimal solution α^* of (4.1) and the convergence is weakly quadratic.

The third question raised in the introduction can now be answered by combing the results of Theorem 4.1 and Theorem 4.2. However in this case the convergence is weakly quadratic and it converges to the minimal solution of (4.1).

REFERENCES

[1] R. Bellman, Methods of Nonlinear Analysis, Vol. II, Academic Press, New York (1973).

[2] R. Bellman and R. Kalaba, Quasilinearization and Nonlinear Boundary Value Problems, American Elsevier, New York (1985).

[3] C. Y. Chan, Positive Solutions for Nonlinear Parabolic Second Initial Boundary Value Problems, Quarterly of Applied Mathematics, Jan (1974).

[4] C. Y. Chan, Existence, uniqueness, upper and lower bounds of solutions of nonlinear space-time Nuclear reactor kinetics, Siam J. Appl. Math., Vol. 27, No 1, July (1974).

[5] S. Köksal and A. S. Vatsala, Generalized Quasilinearization for first order systems with initial conditions (submitted).

[6] G. S. Ladde, V. Lakshmikantham A. S. Vatsala, Monotone Iterative Techniques for Nonlinear Differential Equations, Pitman, Boston (1985).

[7] V. Lakshmikantham, An Extension of the Method of Quasilinearization, to appear in Journal of Optimization Theory and Applications.

[8] V. Lakshmikantham and S. Leela, Nonlinear Differential Equations in Abstract spaces, Pergmon Press, Oxford (1981).

[9] V. Lakshmikantham and S. Leela, Differential and Integral Inequalities, Vol. I, II, Academic Press, New York (1968).

[10] V. Lakshmikantham and S. Malek, Generalized Quasilinearization, to appear in Nonlinear World Vol. 1, (1994).

[11] V. Lakshmikantham and S. Köksal, Obtaining a lower solutions for the Nonlinear differential equations by the method of quasilinearization, to appear in the proceedings of the conference 'Dynamic Systems and Applications' (1993).

[12] V. Lakshmikantham, A. R. Mitchell and D. Sety, Method of Quasilinearization and positivity of solutions in abstract cones, Journal of Optimization Theory and Applications 22, 353-372(1977).

[13] F. A. Mcrae, Generalized Quasilinearization of Stochastic first order initial value problems, to appear in Stochastic Analysis and Applications.

[14] A. S. Vatsala, An extension of the Method of Quasilinearization for reaction diffusion equations, to appear in symposium proceedings on 'Comparison Methods and Stability Theory'.

[15] A. S. Vatsala, Method of generalized quasilinearization for second order boundary value problems (submitted).

4th Int. Coll. on Differential Equations, pp. 287-294
D. Bainov and V. Covachev (Eds)

On a radially symmetric minimal surface with non-empty free boundary

YOSHIHIKO YAMAURA
Department of Mathematics, Faculty of Sience and Technology
Keio University, Hiyoshi Yokohama 223 Japan

Abstract We treat the variational problem concerning a free boundary problem for minimal surface equations in the radially symmetric case. We finally show the existence of minimizers with non-empty free boundaries in the 2-dimensional case.

Keywords: variational problem, minimal surface, free boundary.

0. Introduction

In this paper we deal with a problem of minimal surface with free boundary, the formulation of which has been stimulated by the work [1] due to H.W. Alt and L.A. Caffarelli. We begin by the description of our problem: Let $B_R = B_R(0) := \{x \in \mathbf{R}^n : |x| < R\}$. For a non-negative constant h, we consider the variational problem

$$(P) \qquad \begin{cases} \text{Minimize the functional} \\ J(u) \;=\; \displaystyle\int_{B_R} \sqrt{1+|Du|^2} \;+\; \int_{B_R} Q^2 \chi_{u>0}\, d\mathcal{L}^n + \int_{\partial B_R} |u-h|\, d\mathcal{H}^{n-1} \\ \text{in the class } BV(B_R), \end{cases}$$

where Q is a constant, and $\chi_{u>0}$ is the characteristic function of the set $B_R(u>0) := \{x \in B_R : u(x) > 0\}$. The admissible function space $BV(B_R)$ is the space of functions of bounded variation in B_R, and the first term of $J(u)$ is defined as the total variation of the vector valued measure $(1, D_1 u, \cdots, D_n u)$ (refer to the book of E.Giusti [3]).

The existence of a minimizer of (P) is known in the paper [6, Theorem 1.1] and the non-negativity of minimizer is also proved in [6, Theorem 1.2].

Let u be a minimizer and the free boundary $\partial B_R(u>0) := B_R \cap \partial[B_R(u>0)]$ be empty, then u is the usual area minimizing function known to be smooth. We are thus mainly interested in the regularity of the minimizers with non-empty free boundaries. In fact, we obtain the following regularity result:

Main Theorem. *Let u be a minimizer with non-empty free boundary. Then we have the following:*

(i) *In case* $Q^2 < 1$, *u is Lipschitz continuous in* B_R;

(ii) *In case* $Q^2 = 1$, *u is continuous, but not Lipschitz continuous in* B_R;

(iii) *In case* $Q^2 > 1$, *u has the jump-discontinuity on the surface of an n-dimensional ball concentric with* B_R, *the radius being determined by* Q^2, R *and* h.

In the first half of this paper, the outline of the proof of this Main Theorem is stated. In the second half, we investigate, only in the case of 2 dimension, a sufficient condition for minimizers to have the non-empty free boundaries.

1. The outline of the proof of Main Theorem

The proof of Main Theorem is based on the two lemmata. The first one is as follows (see [7, Main lemma] for the proof):

Lemma A. *Let u be a minimizer of* (P). *Let* $U := \{(x,t) \in B_R \times \mathbf{R} : t < u(x)\}$. *Then U minimizes the Radon measure, denoted by* $(|D\chi_U| + Q^2|\partial_t \chi_U| \lfloor B_R \times \mathbf{R}_+)$ *and defined for any open set* $G \subset B_R \times \mathbf{R}$ *in the following way:*

$$(|D\chi_U| + Q^2|\partial_t \chi_U| \lfloor B_R \times \mathbf{R}_+)(G)$$
$$:= \sup_{\substack{\zeta \in C_0^1(G,\mathbf{R}^{n+1}) \\ |\zeta| \le 1}} \int_G \chi_U \mathrm{div}\, \zeta \, d\mathcal{L}^{n+1} + \sup_{\substack{\zeta \in C_0^1(G) \\ |\zeta| \le 1}} \int_{G \cap B_R \times R_+} \chi_U \partial_t \zeta \, d\mathcal{L}^{n+1}, \qquad (1)$$

where $\mathbf{R}_+$ *is the set of all positive numbers.*

The introduction of the first term of (1) is known in the paper [5] due to M. Mirand, and the regularity of the set minimizing such measure is known in the paper [2] due to E. DeGiorgi.

Now by applying to this measure the symmetrization of Radon measures ([4]), we obtain that any minimizer is radially symmetric. Thus, by 1-dimensional argument based on the minimality we can obtain the following properties of minimizers (see [8, section 2 and 3] for the precise argument and the proof):

Proposition. *Let u be a minimizer of* (P). *Then there exists* $\rho \in (0, R)$ *such that* $u = 0$ *in* B_ρ *and* $u > 0$ *in* $B_R \backslash \overline{B_\rho}$, *where* B_ρ *is the concentric ball of radius* ρ, *and*

(i) *The graph has the non-negative jump, the height of it being denoted by* h_0 *in the follwing, on* $\partial B_R(u > 0)$;

(ii) *u takes the value h on* ∂B_R;

(iii) *u classically satisfies the minimal surface equation in* $B_R(u > 0)$;

(iv) *If* $h_0 > 0$, *then the gradient of u on* $\partial B_R(u > 0)$ *is* $+\infty$.

Especially, by making use of (iii), we can calculate the first variation formula, which is the second lemma being used in the proof of Theorem:

Lemma B. *Let u be a minimizer with non-empty free boundary. Then we have that the equality*

$$\lim_{\delta \downarrow h_0} \int_{\partial B_R(u>\delta)} \left(Q^2 - 1 + \frac{1}{\sqrt{1+|Du|^2}}\right)\langle \eta, \nu_\delta \rangle d\mathcal{H}^{n-1} = (n-1)\omega_n h_0 \rho^{n-2}$$

holds uniformly on $\{\eta \in C_0^\infty(B_R, \mathbf{R}^n) : |\eta| + |D\eta| \leq M \text{ in } B_R\}$, where M is a positive number and ν_δ is the unit outer normal to the set $B_R(u > \delta) := \{x \in B_R : u(x) > \delta\}$.

This equality is calculated by the comparison function used as in the proof of [1, Theorem 2.5] (see [6, Theorem 2.1] for the precise calculation).

Now since minimizers are radially symmetric as stated above, we can substitute ν_δ to η, and finally by considering (iv) in Proposition, we infer from Lemma B the following: *If $Q > 1$, then $h_0 > 0$. If $Q \leq 1$, then $h_0 = 0$. Moreover, if $Q < 1$, then $|Du| = \sqrt{1-(1-Q^2)^2}/(1-Q^2)$ on $\partial B_R(U > 0)$, and if $Q = 1$, then $|Du| = \infty$ on $\partial B_R(U > 0)$.* We thus complete the proof of Main Theorem.

2. On the existence of minimizers with non-empty free boundaries

Let u be a minimizer of (P) with non-empty free boundary. Then there exists a $\rho \in (0, R)$ such that $B_R(u > 0) = B_R \setminus \overline{B}_\rho$. As stated in (iii) of Proposition, u is the solution of the minimal surface equation in $B_R \setminus \overline{B}_\rho$. By noting that u is radially symmetric, it follows that u satisfies

$$u''(r) + \frac{n-1}{r} u'(r) + \frac{n-1}{r}(u'(r))^3 = 0 \qquad \text{in } (\rho, R),$$

where $r = |x|$. Owing to Main Theorem and Lemma B we have the following representation of the minimizer by integration:

Lemma C. (i) *Let $Q < 1$. Then there exists a number ρ, $\rho \in (0, R)$, satisfying the equation*

$$\int_\rho^R \frac{C_Q\, \rho^{n-1}}{\sqrt{r^{2(n-1)} - C_Q^2\, \rho^{2(n-1)}}}\, dr = h$$

such that

$$u(x) = \begin{cases} 0 & \text{for } x \in B_\rho, \\ \displaystyle\int_\rho^{|x|} \frac{C_Q\, \rho^{n-1}}{\sqrt{r^{2(n-1)} - C_Q^2\, \rho^{2(n-1)}}}\, dr & \text{for } x \in B_R \setminus \overline{B}_\rho, \end{cases}$$

where $C_Q = \sqrt{1-(1-Q^2)^2}$.

(ii) *Let* $Q \geq 1$. *Then there exists a number* ρ, $\rho \in (0,R)$, *satisfying the equation*

$$\int_\rho^R \frac{\rho^{n-1}}{\sqrt{r^{2(n-1)}-\rho^{2(n-1)}}}\,dr + \left(\frac{Q^2-1}{n-1}\right)\rho = h$$

such that

$$u(x) = \begin{cases} 0 & \text{for } x \in B_\rho, \\ \displaystyle\int_\rho^{|x|} \frac{\rho^{n-1}}{\sqrt{r^{2(n-1)}-\rho^{2(n-1)}}}\,dr + \left(\frac{Q^2-1}{n-1}\right)\rho & \text{for } x \in B_R\backslash\overline{B}_\rho. \end{cases}$$

We now restrict ourselves to the case $n = 2$. Then the integrations in Lemma C are represented by the logarithmic function. In fact, for arbitrary number c smaller than ρ, we derive

$$\int_\rho^{|x|} \frac{c}{\sqrt{r^2-c^2}}\,dr = \log\left(\frac{|x|+\sqrt{|x|^2-c^2}}{\rho+\sqrt{\rho^2-c^2}}\right),$$

which satisfies the minimal surface equation in $B_R \setminus \overline{B}_\rho$.

Thus we have only to find ρ as stated in Lemma C. For this purpose we shall define for each ρ, $\rho \in (0,R)$, the function u_ρ belonging to $BV(B_R)$. More precisely, we give the definition to u_ρ in a different way according to the sign of $(h - H_{\rho,R})$, where

$$H_{\rho,R} := \rho\log\left(\frac{R+\sqrt{R^2-\rho^2}}{\rho}\right).$$

We fix ρ in $(0,R)$. Then

I. In case $h \leq H_{\rho,R}$, we define

$$u_\rho(x) = \begin{cases} 0 & \text{if } |x| \leq \rho, \\ c\log\left(\dfrac{|x|+\sqrt{|x|^2-c^2}}{\rho+\sqrt{\rho^2-c^2}}\right) & \text{if } \rho < |x| < R, \end{cases}$$

where c is the constant determined by the equation

$$c\log\left(\frac{R+\sqrt{R^2-c^2}}{\rho+\sqrt{\rho^2-c^2}}\right) = h. \tag{2}$$

II. In case $h > H_{\rho,R}$, we define

$$u_\rho(x) = \begin{cases} 0 & \text{if } |x| \leq \rho, \\ c\log\left(\dfrac{|x|+\sqrt{|x|^2-\rho^2}}{\rho}\right) + (h - H_{\rho,R}) & \text{if } \rho < |x| < R. \end{cases}$$

We remark that if $h > H_0$, $h > H_{\rho,R}$ holds for all $\rho \in (0, R)$, and if $h < H_0$, $H_0 = \max_{0<\rho<R} H_{\rho,R}$, then there exist ρ_H and ρ^H, belonging to $(0, R)$, with $\rho_H < \rho^H$ such that $h \leq H_{\rho,R}$ holds for $\rho \in (\rho_H, \rho^H)$ and $h > H_{\rho,R}$ otherwise.

By considering the properties stated in Proposition, it follows that the minimizer of (P) is one of the function belonging to

$$\{u_\rho\}_{0<\rho<R} \cup \{0\} \cup \{h\},$$

where u_ρ is a function defined as above.

$J(u_\rho)$ can be concretely calculated as follows: $J(u_\rho) =$

$$\begin{cases} \pi\Big(R\sqrt{R^2-c^2} - \rho\sqrt{\rho^2-c^2} + ch + (1-Q^2)\rho^2 + Q^2R^2\Big) \\ \hfill \text{if } h \leq H_{\rho,R}; \\ \pi\left(R\sqrt{R^2-c^2} + 2\rho h - \rho^2 \log\left(\dfrac{R+\sqrt{R^2-\rho^2}}{\rho}\right) + (1-Q^2)\rho^2 + Q^2R^2\right) \\ \hfill \text{if } h > H_{\rho,R}, \end{cases}$$

where c is a constant in (2). Since, as noted above, $h > H_{\rho,R}$ holds in neighbourhood of 0 and R and moreover

$$J(u_\rho)\Big|_{\rho=0} = J(h) \qquad \text{and} \qquad J(u_\rho)\Big|_{\rho=R} = J(0),$$

in order to obtain the minimizer of (P), it is sufficient to derive ρ in $[0, R]$ which gives the minimum of the function $J(\rho) := J(u_\rho)$, in $[0, R]$. The elementary, but rather complicated, calculations for investigating the behaviour of the value of $J(\rho)$ enables us to obtain the following concrete description of minimizers of (P):

Theorem. (i) *In case $Q^2 \leq 1$, a minimizer of (P) is of the form*

$$\begin{cases} u & \text{when } h \leq h_{Q^2}R, \\ h & \text{when } h_{Q^2}R \leq h, \end{cases}$$

where u is defined as follows:

$$u(x) = \begin{cases} 0 & \text{for } x \in B_\rho, \\ C_Q\rho \log\left(\dfrac{|x| + \sqrt{|x|^2 - C_Q^2\rho^2}}{\rho + \sqrt{\rho^2 - C_Q^2\rho^2}}\right) & \text{for } x \in B_R \setminus \overline{B}_\rho \end{cases} \tag{3}$$

(see Lemma C for C_Q), and h_{Q^2} is the solution of

$$1+h^2+2\log\left\{\left(\frac{2-Q^2}{C_Q}\right)h\right\} = 0 \tag{4}$$

and ρ is the larger solution of

$$C_Q\rho \log\left(\frac{R+\sqrt{R^2-C_Q^2\rho^2}}{\rho+\sqrt{\rho^2-C_Q^2\rho^2}}\right) = h. \tag{5}$$

Specially in case $h = h_{Q^2}R$, both functions u and h are minimizers of (P).

(ii) *In case $1 < Q^2 < 2$, a minimizer of (P) is of the form*

$$\begin{cases} 0 & \text{when } h \leq (Q^2-1)R, \\ u & \text{when } (Q^2-1)R < h \leq h_{Q^2}R, \\ h & \text{when } h_{Q^2}R \leq h, \end{cases}$$

where u is defined as follows:

$$u(x) = \begin{cases} 0 & \text{for } x \in B_\rho, \\ \rho\log\left(\frac{|x|+\sqrt{|x|^2-\rho^2}}{\rho}\right) + (Q^2-1)\rho & \text{for } x \in B_R\backslash\overline{B}_\rho \end{cases}$$

and h_{Q^2} is the solution of

$$(3-2Q^2+h^2)+2\log h = 0$$

and ρ is the larger solution of

$$\rho\log\left(\frac{R+\sqrt{R^2-\rho^2}}{\rho}\right)+(Q^2-1)\rho = h.$$

Specially in case $h = h_{Q^2}R$, both functions u and h are minimizers of (P).

(iii) *In case $2 \leq Q^2$, a minimizer of (P) is of the form*

$$\begin{cases} 0 & \text{when } h \leq Q^2 R/2, \\ h & \text{when } h \geq Q^2 R/2. \end{cases}$$

Specially in case $h = Q^2 R/2$, both functions 0 and h are minimizers of (P).

References

[1] H.W.ALT – L.A.CAFFARELLI, *Existence and regularity for a minimum problem with free boundary*, J.Reine Angew. Math. **325** (1981) 105-144.

[2] E.DEGIORGI, *Frontiere orientate di misura minima*, Sem. Mat. Scuola Norm. Sup. Pisa, 1960-61 Editrice Tecnico scinetifica, Pisa, 1961.

[3] E.GIUSTI, Minimal Surfaces and functions of bounded variation, Birkhäuser, 1984.

[4] E.H.A.GONZALEZ, *Sul Problema della goccia appoggiata*, Rend. Sem. Mat. Univ. Padova, **58** (1977).

[5] M.MIRANDA, *Superfici cartesiane generalizatte ed insieme de perimetro finito sui prodotti cartesiani*, Ann. Sc. Norm. Sup. Pisa **18** (1964) 515-542.

[6] Y.YAMAURA, *A free boundary problem for minimal surface equation*, to appear in Bollettino Della Unione Matematica Italiana.

[7] ——————, *A uniform estimate of the perimeter for minimizers of a free boundary problem*, to appear in Rendiconti di Matematica e Delle Sue Applicazioni, Univ. Roma.

[8] ——————, *The regularity of minimizers of a radially symmetric free boundary problem*, to appear in Annali dell'Universita di Ferrara.

Appendix. The computer graphics of the graph of minimizers

In the following, we set $R = 1$. Let $Q^2 = 1/2$. Then we obtain an approximate value of h_{Q^2} by solving numerically equation (4) with $Q^2 = 1/2$. In fact, we have $h_{Q^2} \approx 0.331$. Substituting 0.153 to h, we obtain the approximate value of ρ by solving numeically equation (5), so that we have $\rho \approx 0.692$. Figure 1 in the next page is the three dimensional computer graphics of the graph of the minimizer given by (3) with $h = 0.330$ and $\rho = 0.692$.

Figures 2 and 3 are the graphics of the graphs of the minimizers in the case of $Q^2 = 1$ and $12/10$, respectively, where the boundary conditions are taken such that the minimizers have the non-empty free boundaries in the same way as in the case of Figure 1 The value of h and approximate values of h_{Q^2} and ρ are stated under each figure.

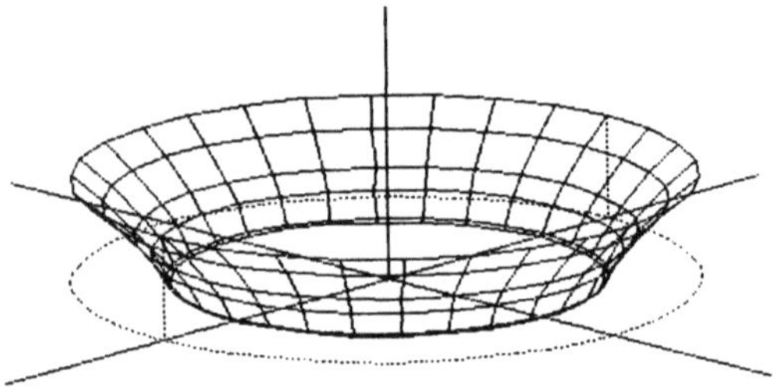

Fig. 1

The graph of the unique minimizer in case $Q^2=1/2$ and $h=0.330$, where h_{Q^2}, as in Theorem, approximately equals to 0.331 and the radius of the free boundary approximately equals to 0.692.

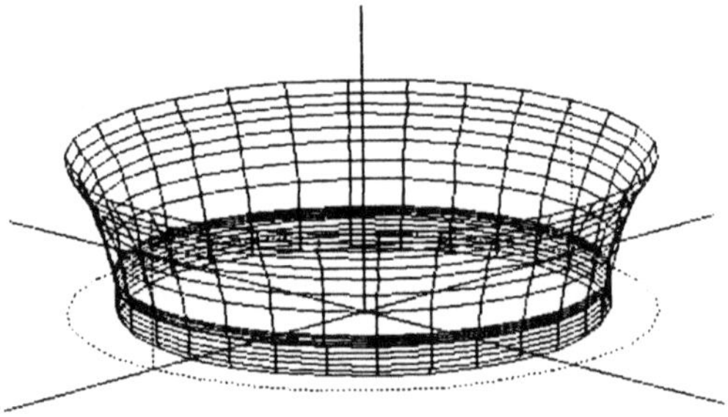

Fig. 2

The graph of the unique minimizer in case $Q^2=1$ and $h=0.526$, where h_{Q^2}, as in Theorem, approximately equals to 0.528 and the radius of the free boundary approximately equals to 0.827.

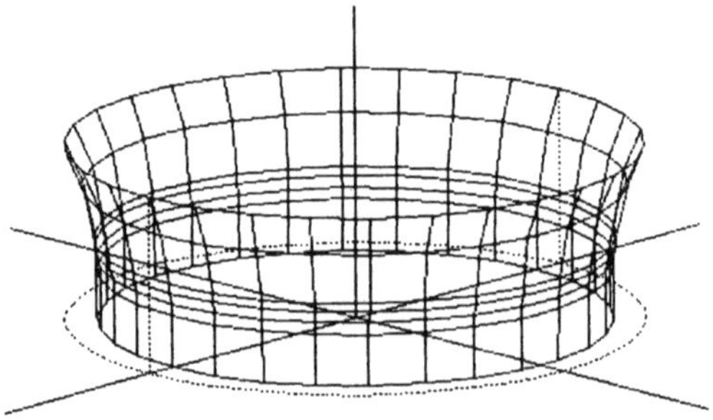

Fig. 3

The graph of the unique minimizer in case $Q^2=1.2$ and $h=0.612$, where h_{Q^2}, as in Theorem, approximately equals to 0.614 and the radius of the free boundary approximately equals to 0.893. Furthermore, the height of the jump of the graph is equal to 0.179.

4th Int. Coll. on Differential Equations, pp. 295-305
D. Bainov and V. Covachev (Eds)

Ahmed I. Zayed

Department of Mathematics
University of Central Florida
Orlando, FL 32816, USA

ABSTRACT

Sampling theorems have been used successfully in communication engineering for over 40 years. Their relationship with boundary-value problems was first reported in the early fifties; however, in the last five years new connections have been discovered. The aim of this paper is to report on some of these new discoveries and to discuss various sampling theorems associated with boundary-value problems.

1. Introduction

One of the most important mathematical techniques used in communication engineering is *Sampling Theory*. The fundamental result in Sampling Theory is the celebrated Whittaker-Shannon-Kotel'nikov (WSK) sampling theorem, which in Shannon's original wording reads as follows [1]:

> "If a function of time is limited to the band from 0 to W cycles per second, it is completely determined by giving its ordinates at a series of discrete points spaced $1/2W$ seconds apart in the manner indicated by the following result: If $f(t)$ has no frequencies over W cycles per second, then
>
> $$f(t) = \sum_{n=-\infty}^{\infty} f\left(\frac{n}{2W}\right)\frac{\sin\pi(2Wt-n)}{\pi(2Wt-n)}.\text{"}$$

Although this result is known in communication and electrical engineering literature as Shannon's sampling theorem, we shall call it from now on, the WSK sampling theorem. Shannon's result can be rephrased as follows.

THEOREM 1 (Whittaker-Shannon-Kotel'nikov)

If $f(t)$ is a signal (function) band-limited to $[-\sigma,\sigma]$, i.e.,

$$f(t) = \frac{1}{\sqrt{2\pi}}\int_{-\sigma}^{\sigma} F(\omega)e^{it\omega}d\omega, \tag{1.1}$$

for some $F \in L^2(-\sigma,\sigma)$, then it can be reconstructed from its sampled values at the points $t_k = k\pi/\sigma$, $k = 0, \pm 1, \pm 2, \ldots$, via the formula

$$f(t) = \sum_{k=-\infty}^{\infty} f(t_k)\frac{\sin\sigma(t-t_k)}{\sigma(t-t_k)}, t \in \Re \tag{1.2}$$

with the series being absolutely and uniformly convergent on compact sets.

We shall call the points $\{t_k\}_{k=-\infty}^{\infty}$ the sampling points and the functions

$$S_k(t) := \frac{\sin \sigma(t - t_k)}{\sigma(t - t_k)}, \tag{1.3}$$

the sampling functions. The sampling frequency σ/π in (1.2) is known as the Nyquist rate, which is the minimum rate at which the signal needs to be sampled in order to reconstruct it completely.

Since band-limited signals are represented by Fourier integrals of functions with compact support, we can say that the WSK sampling theorem provides a sampling series expansion for the reconstruction of certain Fourier transforms taken over finite intervals. If a signal is represented by another type of integral transforms, the WSK sampling theorem is no longer applicable; a more general sampling theorem for the reconstruction of integral transforms of types other than the Fourier one is needed.

The idea of reconstructing integral transforms from a discrete set of data stems from the work of J. M. Whittaker [2] who had suggested a sampling series expansion for an integral transform taken over a finite interval with the Bessel function, instead of the exponential function, as its kernel. Nevertheless, the primary result in this direction is attributed to H. Kramer and is known as Kramer's sampling theorem.

THEOREM 2 (Kramer [3])

Let there exist a function $K(x,t)$ continuous in t such that $K(x,t) \in L^2(I)$ for every real number t. Assume that there exists a sequence of real numbers $\{t_k\}_{k=-\infty}^{\infty}$ such that $\{K(x,t_k)\}_{k=-\infty}^{\infty}$ is a complete orthogonal family of functions in $L^2(I)$. Then for any $f(t)$ of the form

$$f(t) = \int_a^b F(x) K(x,t)\, dx, \tag{1.4}$$

where $F(x) \in L^2(I)$ and I is a finite closed interval $[a,b]$, we have the sampling series representation

$$f(t) = \sum_{k=-\infty}^{\infty} f(t_k) S_k^*(t), \tag{1.5}$$

with

$$S_k^*(t) = \frac{\int_a^b K(x,t)\overline{K(x,t_k)}\, dx}{\int_a^b |K(x,t_k)|^2\, dx}. \tag{1.6}$$

2. Sampling and Boundary-Value Problems

To see how Theorem 2 generalizes the WSK sampling theorem, just take

$$I = [-\sigma,\sigma], \quad K(x,t) = e^{ixt} \quad \text{and} \quad t_k = \frac{k\pi}{\sigma},$$

then it easily follows that $S_k^*(t) = \frac{\sin\sigma(t-t_k)}{\sigma(t-t_k)}$. Hence, if

$$f(t) = \int_{-\sigma}^{\sigma} F(x)e^{ixt}dx\,, \quad \text{with} \quad F(x) \in L^2(-\pi,\pi)\,,$$

then

$$f(t) = \sum_{k=-\infty}^{\infty} f(t_k)\frac{\sin\sigma(t-t_k)}{\sigma(t-t_k)}\,,$$

which is the WSK sampling theorem.

Yet, it is not really clear how the function $K(x,t)$ and the sequence $E = \{t_k\}_{-\infty}^{\infty}$ in Theorem 2 can be found. Kramer has indicated that if the regular self-adjoint boundary-value problem

$$Ly = ty\,, \quad x \in I \tag{2.1}$$

$$U_j(y) = 0\,; \quad j = 1,2,\dots,n\,, \tag{2.2}$$

possesses a function $\phi(x,t)$ that generates the eigenfunctions of the problem $\{\phi_k(x)\}$ when the parameter t is replaced by the eigenvalues $\{t_k\}$, i.e., $\phi_k(x) = \phi(x,t_k)$, where $I = [a,b]$ is some finite closed interval,

$$L = p_0(x)\frac{d^n}{dx^n} + p_1(x)\frac{d^{n-1}}{dx^{n-1}} + \dots + p_n(x)\,, \tag{2.3}$$

$p_k(x)$ being complex-valued function with $n-k$ continuous derivatives, $k = 0,1,\dots,n, p_0(x) \neq 0$ for any $x \in I$ and $U_j(y) = 0,\ j = 1,\dots,n$, are linearly independent homogeneous self-adjoint boundary conditions independent of t, then we can choose $\phi(x,t)$ as $K(x,t)$ and the eigenvalues $\{t_k\}$ as the sampling points E.

To illustrate the idea, let us consider the following regular Sturm-Liouville boundary-value problem:

$$-y'' = ty\,, \quad x \in [0,\pi] \tag{2.4}$$

$$y'(0) = 0 = y'(\pi)\,. \tag{2.5}$$

The function $\phi(x,t)$ that generates the eigenfunctions $\{\cos kx\}$ is $\cos\sqrt{t}\,x$, where $t_k = k^2$ are the eigenvalues.

It is easily seen that

$$S_k^*(t) = \frac{2\sqrt{t}\sin\pi(\sqrt{t}-k)}{\pi(t-k^2)} \quad \text{if} \quad k = 1,2,\dots, \quad \text{and} \quad S_0^*(t) = \frac{\sin\pi\sqrt{t}}{\pi\sqrt{t}}\,.$$

Therefore, Kramer's theorem now takes the form:

If

$$f(t) = \int_0^{\pi} F(x)\cos\sqrt{t}\,x\,dx\,, \tag{2.6}$$

for some $F \in L^2(0,\pi)$, then

$$f(t) = f(0)\frac{\sin \pi\sqrt{t}}{\pi\sqrt{t}} + \sum_{k=1}^{\infty} f(k^2)\frac{2\sqrt{t}\sin\pi(\sqrt{t}-k)}{\pi(t-k^2)}. \tag{2.7}$$

If we replace t by t^2 and set $\tilde{f}(t) = f(t^2)$, we obtain

$$\tilde{f}(t) = \tilde{f}(0)\frac{\sin \pi t}{\pi t} + 2\sum_{k=1}^{\infty} \tilde{f}(k)\frac{t\sin\pi(t-k)}{\pi(t^2-k^2)},$$

which, in turn, can be put in a more familiar form

$$\tilde{f}(0) = \tilde{f}(0)\frac{\sin \pi t}{\pi t} + \sum_{k=-\infty}^{\infty}{}' \tilde{f}(k)\frac{\sin\pi(t-k)}{\pi(t-k)}.$$

The prime on the summation sign indicates that the term corresponding to $k = 0$ is omitted.

As it turns out, Kramer's assumption that the eigenfunctions are all generated by one single function $\phi(x,t)$ is very restrictive, especially when a general boundary-value problem like (2.1) and (2.2) is considered. In fact, this assumption does not hold if the order n of the differential operator (2.3) is greater than 2, except in some very special cases.

Worse yet, even if $n = 2$ this assumption is not always valid under general boundary-conditions, as seen from the following simple example:

$$-y'' = ty, \quad x \in [0,\pi] \tag{2.8}$$

$$y(0) = y(\pi) \quad \text{and} \quad y'(0) = y'(\pi). \tag{2.9}$$

The eigenfunctions of this problem are $\sin 2nx$ and $\cos 2nx$; $n = 0,1,2,\ldots$, which are not generated by one single real-valued function.

Finally, Kramer concluded his paper by considering the boundary-value problem:

$$y'' - \frac{(\nu^2 - 1/4)}{x^2}y = ty, \quad 0 \le x \le 1, \tag{2.10}$$

with

$$y(0) = 0 \quad \text{and} \quad y(1) = 0, \tag{2.11}$$

from which he derived the sampling series expansion

$$f(t) = \sum_{k=1}^{\infty} f(t_k)\frac{2\sqrt{t_k}J_\nu(\sqrt{t})}{(t_k - t)J_\nu'(\sqrt{t_k})}, \tag{2.12}$$

for

$$f(t) = \int_0^1 F(x)\sqrt{x}J_\nu(x\sqrt{t})dx, \quad \text{with} \quad \sqrt{x}F(x) \in L^2[0,1],$$

where $J_\nu(z)$ is the Bessel function of the 1st kind and order ν, and t_k is the kth non-null zero of $J_\nu(\sqrt{z})$.

A close inspection of this innocuous example reveals that the boundary-value problem (2.10) and (2.11) does not really satisfy Kramer's hypotheses because of the singularity at $x = 0$ in the differential equation. Nevertheless, Kramer was able to derive the sampling series expansion (2.12). Therefore, it is fair to say that Kramer knew—although he never mentioned it explicitly—that his theorem could be extended to some singular boundary-value problems.

This raises the question of whether Kramer's theorem could be extended to singular boundary-value problems. Since in sampling series expansions associated with boundary-value problems the samples are taken at the eigenvalues of the problem, and since singular problems, generally speaking, have continuous spectra, there is no hope to have an affirmative answer.

Well, is it then safe to say that Kramer's sampling theorem holds for singular boundary-value problems with discrete spectra? Unfortunately, the answer is still negative. A counter example is provided by the singular Sturm-Liouville problem

$$y'' - x^2 y = -ty, \quad -\infty < x < \infty$$

$$\lim_{|x| \to \infty} |y(x)| < \infty.$$

The spectrum is discrete; in fact, the eigenvalues and eigenfunctions are $t_k = 2k+1$ and $\phi_k(x) = \exp(-x^2/2)H_k(x)$, $k = 0, 1, \ldots$ respectively, where $H_k(x)$ is the Hermite polynomial of degree k. Kramer's assumption that the eigenfunctions are all generated by one single function seems to hold in this case as well. In fact, the parabolic cylindrical function $\phi(x,t)$ is such a function. This appears to be in contradiction with Kramer's theorem; yet a careful analysis of the problem reveals that there is no contradiction because Kramer's proof does not necessarily hold for singular problems. For more details, see [4].

The questions of whether Theorem 2 gives essentially new results and whether it is an improvement over the WSK sampling theorem or not, were partially answered by Campbell [5], who showed that if the kernel $K(x,t)$ arises from either a regular Sturm-Liouville boundary-value problem or from the first order boundary-value problem

$$-iy' + q(x)y = ty, \quad x \in [-\sigma, \sigma]$$

$$y(-\sigma) = y(\sigma)e^{i\phi}, \quad -\pi < \phi \leq \pi,$$

then any function that can be represented by Kramer's sampling series (1.5), can also be represented by the WSK sampling series (1.2).

3. Sampling Theorems Associated With Sturm-Liouville Boundary-Value Problems

In this section we discuss sampling theorems associated with Sturm-Liouville boundary-value problems. First note that the series (1.2) can be written in the form

$$f(t) = \sum_{k=-\infty}^{\infty} f(t_k) \frac{G(t)}{(t - t_k)G'(t_k)},$$

where $G(t) = \sin \sigma t$. That is the series (1.2) is nothing more than a Lagrange-type interpolation series. In a recent paper [6], we proved the following.

THEOREM 3 (Zayed, [6])

Consider the following regular Sturm-Liouville problem:

$$y'' - q(x)y = -\lambda y = -t^2 y, \quad -\infty < a \leq x \leq b < \infty, \tag{3.1}$$

$$y(a)\cos\alpha + y'(a)\sin\alpha = 0, \tag{3.2}$$

$$y(b)\cos\beta + y'(b)\sin\beta = 0, \tag{3.3}$$

where $q(x)$ is continuous on (a,b) and tends to finite limits as $x \to a^+$ and $x \to b^-$. Let $\phi(x,\lambda)$ and $\chi(x,\lambda)$ be the solutions of (3.1) such that

$$\phi(a,\lambda) = \sin\alpha, \quad \phi'(a,\lambda) = -\cos\alpha, \tag{3.4}$$

$$\chi(b,\lambda) = \sin\beta, \quad \chi'(b,\lambda) = -\cos\beta. \tag{3.5}$$

Let $F(x) \in L^2(a,b)$,

$$f(\lambda) = \int_a^b F(x)\phi(x,\lambda)\,dx \tag{3.6}$$

and

$$f^*(\lambda) = \int_a^b F(x)\chi(x,\lambda)\,dx. \tag{3.7}$$

Then, $f(\lambda)$ and $f^*(\lambda)$ are entire functions of order 1/2 and type η with $0 \leq \eta \leq b - a$ that admit the following sampling series representations:

$$f(\lambda) = \sum_{n=0}^{\infty} f(\lambda_n)\,\frac{G(\lambda)}{(\lambda - \lambda_n)G'(\lambda_n)}, \tag{3.8}$$

$$f^*(\lambda) = \sum_{n=0}^{\infty} f^*(\lambda_n)\,\frac{G(\lambda)}{(\lambda - \lambda_n)G'(\lambda_n)}, \tag{3.9}$$

where $\{\lambda_n\}_{n=0}^{\infty}$ are the eigenvalues of the problem (3.1) to (3.3) and $G(\lambda)$ is the Wronskian $W(\phi,\chi) = \phi(x,\lambda)\chi'(x,\lambda) - \phi'(x,\lambda)\chi(x,\lambda)$ of the two functions ϕ and χ which, without loss of generality, may be written in the form

$$G(\lambda) = \begin{cases} \prod_{n=0}^{\infty}\left(1 - \dfrac{\lambda}{\lambda_n}\right) & \text{if none of the eigenvalues is zero}, \\ \lambda \prod_{n=1}^{\infty}\left(1 - \dfrac{\lambda}{\lambda_n}\right) & \text{if one of the eigenvalues say } \lambda_0 = 0. \end{cases} \tag{3.10}$$

The two series (3.8) and (3.9) converge uniformly on any compact subset of the complex plane and they are related via the relation $f^*(\lambda_n) = k_n f(\lambda_n)$ for all n, where $k_n = \phi(x,\lambda_n)/\chi(x,\lambda_n)$ which is neither zero nor infinity.

For more on sampling theorems associated with regular and singular Sturm-Liouville problems, see [4], [6] and [7].

To obtain a sampling theorem associated with boundary-value problems such as (2.8) and (2.9), we shall use a different approach based on the Green's function of the problem.

4. Sampling With The Green's Function

Consider the differential operator

$$L = p_0(x)\frac{d^n}{dx^n} + p_1(x)\frac{d^{n-1}}{dx^{n-1}} + \dots + p_n(x), \tag{4.1}$$

where $p_0, p_1, \dots, p_n$ are continuous, complex-valued functions defined on a finite real interval $I = [a,b]$, with $p_0(x) \neq 0$ for any $x \in I$. The adjoint differential operator L^* of L is defined by

$$L^* = (-1)^n\left(\frac{d^n}{dx^n}\right)(\overline{p}_0(x)\,\cdot) + (-1)^{n-1}\left(\frac{d^{n-1}}{dx^{n-1}}\right)(\overline{p}_1(x)\,\cdot) \cdots + \overline{p}_n(x). \tag{4.2}$$

This is to be interpreted as follows: if g is any function on I such that $\overline{p}_k g (k = 0, 1, \dots, n)$ has $n-k$ derivatives on I, then

$$L^* g = (-1)^n \frac{d}{dx^n}(\overline{p}_0 g) + (-1)^{n-1}\frac{d^{n-1}}{dx^{n-1}}(\overline{p}_1 g) + \dots + \overline{p}_n g .$$

The equation $L^* y = 0$ is called the adjoint equation of $Ly = 0$.

From now on we shall assume that $p_k \in C^{n-k}(I)$, $k = 0, 1, \dots, n$. It can be shown that ([8], p. 9) if $u, v \in C^n(I)$, then

$$\int_{x_1}^{x_2}\left(\overline{v}Lu - u\overline{L}^* v\right) dx = [uv]_{x=x_2} - [uv]_{x=x_1}, \tag{4.3}$$

for any $x_1, x_2 \in I$, where

$$[uv] = \sum_{m=1}^{n} \sum_{\substack{j+k=m-1 \\ j \geq 0, k \geq 0}} (-1)^j u^{(k)} (p_{n-m}\overline{v})^{(j)} .$$

Let

$$U_j(y) = \sum_{k=1}^{n} \alpha_{j,k} y^{(k-1)}(a) + \beta_{j,k} y^{(k-1)}(b), \quad j = 1, 2, \dots, m (m \leq 2n)$$

be linear forms in $y^{(k-1)}(a)$, $y^{(k-1)}(b)$, $k = 1, 2, \dots, n$, and assume that they are linearly independent.

It is known [8] that given any boundary forms $U_1, \dots, U_m$ and their complementary forms $U_{m+1}, \dots, U_{2n}$, we can find a unique system of linearly independent boundary forms $V_1, \dots, V_{2n}$, called the adjoint boundary forms, such that

$$\int_a^b (Ly)\overline{z}\, dx = U_1 V_{2n} + \dots + U_m V_{2n-m+1} + U_{m+1} V_{2n-m} + \dots + U_{2n} V_1 + \int_a^b y(\overline{L}^* z) dx . \tag{4.4}$$

Now consider the boundary-value problems:

$$\Pi_m: \begin{cases} Ly = \lambda y\,, & x \in I \qquad (4.5) \\ U_j(y) = 0\,, & j = 1,2,\dots,m\,, \qquad (4.6) \end{cases}$$

and

$$\Pi^*_{2n-m}: \begin{cases} L^*y = \lambda y\,, & x \in I \qquad (4.7) \\ V_j(y) = 0\,, & j = 1,2,\dots,2n-m\,, \qquad (4.8) \end{cases}$$

where L and L^* are given by (4.1) and (4.2) respectively.

From now on we shall assume that $m = n$, and denote the boundary-value problems Π_n, Π^*_n by Π and Π^*, respectively, but we shall not assume that problem Π is self-adjoint. Since both problems Π and Π^* always have the trivial solution $y \equiv 0$, we shall be concerned only with the nontrivial solutions.

Let $y_1(x,\lambda),\dots,y_n(x,\lambda)$ denote the fundamental system of solutions of equation (4.5) that satisfy the initial conditions

$$y_j^{(\nu-1)}(a,\lambda) = \begin{cases} 0 & \text{for} \quad j \neq \nu \\ 1 & \text{for} \quad j = \nu \end{cases}, \quad j\,,\nu = 1,2,\dots,n\,. \tag{4.9}$$

Consider the boundary-value problem Π, and let $\{y_i(x,\lambda)\}_{i=1}^n$ be the fundamental system of solutions of (4.5) that satisfy (4.9). Let W be their Wronskian

$$W(x) = \begin{vmatrix} y_1^{(n-1)}(x) & y_2^{(n-1)}(x) & \dots & y_n^{(n-1)}(x) \\ y_1^{(n-2)}(x) & y_2^{(n-2)}(x) & \dots & y_n^{(n-2)}(x) \\ \dots & \dots & \dots & \\ y_1(x) & y_2(x) & \dots & y_n(x) \end{vmatrix}. \tag{4.10}$$

Set

$$g(x,\xi) = \pm\frac{1}{W(\xi)}\begin{vmatrix} y_1(x) & y_2(x) & \dots & y_n(x) \\ y_1^{(n-2)}(\xi) & y_2^{(n-2)}(\xi) & \dots & y_n^{(n-2)}(\xi) \\ \dots & \dots & \dots & \\ y_1(\xi) & y_2(\xi) & \dots & y_n(\xi) \end{vmatrix}, \tag{4.11}$$

where the positive sign is taken if $x > \xi$, and the negative sign if $x < \xi$; $x,\xi \in I$.

It is shown in [8, p. 37] that the Green's function $G(x,\xi,\lambda)$ of problem Π can be given by

$$G(x,\xi,\lambda) = \frac{(-1)^n}{\Delta(\lambda)}H(x,\xi,\lambda)\,, \tag{4.12}$$

where

$$\Delta(\lambda) = \begin{vmatrix} U_1(y_1) & U_1(y_2) & \dots & U_1(y_n) \\ U_2(y_1) & U_2(y_2) & \dots & U_2(y_n) \\ \dots & \dots & \dots & \\ U_n(y_1) & U_n(y_2) & \dots & U_n(y_n) \end{vmatrix}, \tag{4.13}$$

and

$$H(x,\xi,\lambda) = \begin{vmatrix} y_1(x) & y_2(x) & \dots & y_n(x) & g(x,\xi) \\ U_1(y_1) & U_1(y_2) & \dots & U_1(y_n) & U_1(g) \\ U_2(y_1) & U_2(y_2) & \dots & U_2(y_n) & U_2(g) \\ \dots & \dots & \dots & \dots & \dots \\ U_n(y_1) & U_n(y_2) & \dots & U_n(y_n) & U_n(g) \end{vmatrix} . \tag{4.14}$$

Let us denote the poles of G, or equivalently the eigenvalues of Π, by $\{\lambda_k\}_{k=0}^{\infty}$. Although all are assumed to be simple poles of G, some or all of the λ_k's may not be simple zeros of $\Delta(\lambda)$; hence, they may not be simple eigenvalues of Π, i.e., their multiplicities as eigenvalues may be greater than one.

We shall now assume that the asymptotic behavior of the eigenvalues $\{\lambda_k\}_{k=0}^{\infty}$ of problem Π is given by $\lambda_k^{1/n} = 0(k\pi/(b-a))$ as $|k| \to \infty$ where n is the order of the differential operator (4.1). Such an asymptotic behavior is guaranteed if the boundary conditions are regular. For the definition of regular boundary conditions, we refer the reader to [8, p. 56]. As a consequence of this, for $n > 1$, the canonical product

$$P(\lambda) = \begin{cases} \prod_{k=1}^{\infty}\left(1-\dfrac{\lambda}{\lambda_k}\right) , & \text{if none of the eigenvalues is zero} , \\ \\ \lambda\prod_{k=2}^{\infty}\left(1-\dfrac{\lambda}{\lambda_k}\right), & \text{if one of the eigenvalues, say } \lambda_1 ,\ \text{is zero} , \end{cases} \tag{4.15}$$

is well defined since $\sum_{k=2}^{\infty} 1/\lambda_k < \infty$. For $n = 1$, if the product in (4.15) does not converge, we set

$$P(\lambda) = \begin{cases} \prod_{k=1}^{\infty}\left(1-\dfrac{\lambda}{\lambda_k}\right)\exp(\lambda/\lambda_k) , & \text{if none of the eigenvalues is zero} , \\ \\ \lambda\prod_{k=2}^{\infty}\left(1-\dfrac{\lambda}{\lambda_k}\right)\exp(\lambda/\lambda_k), & \text{if one of the eigenvalues, say } \lambda_1,\ \text{is zero} . \end{cases} \tag{4.16}$$

To simplify the proofs, we shall assume, without loss of generality, that none of the eigenvalues is zero.

Now fix $\xi_0 \in [a,b]$ and define

$$\Phi(x,\lambda) = P(\lambda)G(x,\xi_0,\lambda) , \tag{4.17}$$

where G is given by (4.12) and P is given by (4.15) or (4.16) depending on whether n is larger than or equal to 1. Since G has only a simple pole at $\lambda = \lambda_k$, $k = 1,2,\dots$, $\Phi(x,\lambda)$ is an entire function in λ.

THEOREM 4 (Zayed, [9])

Let

$$f(\lambda) = \int_a^b \overline{F}(x)\Phi(x,\lambda)\,dx , \tag{4.18}$$

for some $F \in L^2(a,b)$. Then f is an entire function of order not exceeding $1/n$ (n is the order of the differential operator L), which admits the following sampling series representation

$$f(\lambda) = \sum_{k=1}^{\infty} f(\lambda_k) \frac{P(\lambda)}{(\lambda - \lambda_k) P'(\lambda_k)}. \tag{4.19}$$

Moreover, if problem Π is self-adjoint, or if F satisfies the adjoint boundary conditions (4.8), then the series converges uniformly on compact subsets of the complex λ-plane.

5. Sampling Theorems Associated With Other Types of Boundary-Value Problems

Kramer's sampling theorem has been generalized to higher dimensions in [10], where it has also been shown that the N-dimensional sampling series associated with Kramer's theorem is an N-dimensional Lagrange-type series, when the underlaying boundary-value problem is separable on a hypercube. This gives a generalization of Theorem 3 to N-dimensions.

Another extension of Theorem 3 was reported in [11], where the kernel $K(x,t)$ is assumed to arise from a second order eigenvalue problem with the eigenvalue parameter λ appearing in the boundary conditions. More precisely, the kernel $K(x,t)$ is associated with the boundary-value problem

$$(p(x)y')' + q(x)y = -\lambda y, \quad -\infty < a \le x \le b < \infty,$$

$$\cos \alpha y(a) + \sin \alpha y'(a) = 0, \quad \alpha \in [0,\pi)$$

$$-[\beta_1 y(b) - \beta_2 y'(b)] = \lambda[\beta_1' y(b) - \beta_2' y'(b)],$$

with

$$\beta_1, \beta_2, \beta_1', \beta_2' \in \Re \quad \text{and} \quad \beta_1'\beta_2 - \beta_2'\beta_1 > 0,$$

where $p(x) \neq 0$, $1/p(x)$ and $q(x)$ are assumed to be real-valued and integrable on $[a,b]$.

Finally, this short survey article would not be complete without mentioning a very recent and interesting paper by W. Everitt, G. Schöttler and P. Butzer [12], in which they studied sampling theorems associated with more general types of boundary-value problems. For example, they derived sampling theorems associated with self-adjoint differential operators (in the sense of Naimark) determined by separated symmetric boundary conditions.

References

1. C. E. Shannon, Proc. IRE, 137, pp. 10-21 (1949).
2. J. M. Whittaker, Interpolatory Function Theory, Cambridge University Press, Cambridge, England (1935), p. 71.
3. H. P. Kramer, J. Math. Phys., 63, pp. 68-72 (1957).
4. A. I. Zayed, Advances in Shannon's Sampling Theory, CRC Press, Boca Raton, Florida, USA (1993).
5. L. L. Campbell, SIAM J. Appl. Math., 16, pp. 626-636 (1968).
6. A. I. Zayed, SIAM J. Appl. Math., 51, pp. 575-604 (1991).
7. A. Zayed, G. Hinsen and P. Butzer, SIAM J. Appl. Math., 50, pp. 893-909 (1990).
8. M. Naimark, Linear Differential Operators, Vol. 1, Elementary Theory of Linear Differential Operators, George Harrap & Co., London, England (1967).
9. A. I. Zayed, J. Math. Anal. Appls., 175, pp. 222-238 (1993).
10. A. I. Zayed, J. Multidimensional Systems and Signal Processing, 3, pp. 323-339 (1992).
11. M. H. Annaby, Kramer's Expansion and Lagrange Interpolation Theorem Associated with General Self-adjoint Differential Operators, M.S. Thesis, A. I. Zayed and M. A. El-Sayed supervisors, Cairo University, Egypt (1991).
12. W. N. Everitt, G. Schöttler, and P. L. Butzer, Sturm-Liouville Boundary-Value Problems and Lagrange Interpolation Series, manuscript.

Bei Fragen zur Produktsicherheit wenden Sie sich bitte an:
If you have any questions regarding product safety,
please contact:

Walter de Gruyter GmbH
Genthiner Straße 13
10785 Berlin
productsafety@degruyterbrill.com